OXFORD PAPERBACK REFERENCE

A Guide to Countries of the World

Peter Stalker is a writer based in Oxford, who works as a consultant to many UN agencies. He is a former editor of the *Human Development Report*, which is produced by the United Nations Development Programme, and has written a series of books, and a website, on international migration. His website is at www.pstalker.com.

The most authoritative and up-to-date reference
books for both students and the general reader.

Oxford Paperback Reference

A Guide to Countries of the World

REVISED SECOND EDITION

PETER STALKER

OXFORD
UNIVERSITY PRESS

OXFORD
UNIVERSITY PRESS

Great Clarendon Street, Oxford OX2 6DP

Oxford University Press is a department of the University of Oxford.
It furthers the University's objective of excellence in research, scholarship,
and education by publishing worldwide in

Oxford New York

Auckland Cape Town Dar es Salaam Hong Kong Karachi
Kuala Lumpur Madrid Melbourne Mexico City Nairobi
New Delhi Shanghai Taipei Toronto

With offices in

Argentina Austria Brazil Chile Czech Republic France Greece
Guatemala Hungary Italy Japan Poland Portugal Singapore
South Korea Switzerland Thailand Turkey Ukraine Vietnam

Oxford is a registered trade mark of Oxford University Press
in the UK and in certain other countries

Published in the United States
by Oxford University Press Inc., New York

© Oxford University Press 2000, 2004, 2007

The moral rights of the author have been asserted
Database right Oxford University Press (maker)

First published 2000 as *Handbook of the World*
Second edition 2004
Revised second edition 2007

British Library Cataloguing in Publication Data

Data available

Library of Congress Cataloguing in Publication Data

Data available

Printed in Great Britain by
Clays Ltd. Bungay, Suffolk

ISBN 978-0-19-920271-3
3

Key for world maps inside the front and back covers

A.	Albania	Lie.	Liechtenstein
Afg.	Afghanistan	Lu.	Luxembourg
An.	Andorra	M.	Macedonia
Ar.	Armenia	Mo.	Montenegro
Au.	Austria	Mol.	Moldova
Az.	Azerbaijan	N.	Netherlands
B.	Belgium	Nor.	Norway
Ba.	Bahrain	Nth. Korea	North Korea
Ban.	Bangladesh	Pol.	Poland
Be.	Benin	Q.	Qatar
BH.	Bosnia and Hervegovina	R.	Romania
Bots.	Botswana	S.	Slovakia
Bu.	Burkina Faso	Se.	Senegal
Bul.	Bulgaria	Ser.	Serbia
C.	Croatia	Sl.	Slovenia
CAR.	Central African Republic	S.M.	San Marino
		St.K. & N.	St. Kitts & Nevis
CZ	Czech Republic	St.L.	St. Lucia
Den.	Denmark	St.V.	St. Vincent & the Grenadines
El Salv.	El Salvador		
Eq. Guinea	Equatorial Guinea	Sth. Korea	South Korea
Fin.	Finland	Sw.	Swaziland
G.	Germany	Sw.	Switzerland
G.-B.	Guinea-Bissau	Swe.	Sweden
Gab.	Gabon	Taj.	Tajikistan
Gam.	The Gambia	T.L.	Timor-Leste
Gh.	Ghana	Tu.	Turkmenistan
Gre.	Grenada	U.A.E.	United Arab Emirates
H.	Hungary	U.K.	United Kingdom
Is.	Israel	Ug.	Uganda
Kazak.	Kazakhstan	Ukr.	Ukraine
Kir.	Kyrgyzstan	Uzb.	Uzbekistan
L.	Lebanon	Zam.	Zambia
Li.	Lithuania	Zimb.	Zimbabwe

Preface

In an era of globalization the nation-state seems anachronistic. Huge quantities of data, funds, and consumer products are rushing around the world contemptuous of national borders. People too are travelling more rapidly and casually than ever before. So the principle of organizing the world according to a specific group of people occupying a designated part of the earth's surface seems increasingly old-fashioned. Do we really need all these countries?

It seems we do. Indeed, the nation-state appears to be in remarkably vigorous health. Not only do we have far more states, they have been reproducing rapidly. In the first half of the twentieth century, new states materialized at the rate of around one per year. Between 1950 and 1990 the rate was over two. In the 1990s, new states appeared on average at a rate of almost three per year. Since then the rate of reproduction has declined but new states continue to emerge.

In large part, this has been a process of fragmentation. Many groups and people now believe that the only way to protect their interests and establish a more cohesive national identity is to secede and to create a state of their own, however minuscule. Ultimately, they are chasing a myth. No country of whatever size or type conforms to the ideal model of the nation-state—a well-defined territory that forms the homeland of a single people united by a common history, language, and culture. All states are home to more than one nation; all nations are dispersed across more than one state.

In one sense, globalization actually makes it easier to form new states. Cross-national groupings such as the EU, for example, or agencies of the UN, now offer many more facilities. So smaller states can now contract out difficult functions—like defence, or pensions management, or disaster relief. And transnational corporations will be only too happy to print the banknotes, manage the customs service, or run the prisons, for a suitable fee. Rather than disappearing, therefore, the nation-state is merely taking a more modern form.

Certainly, the UN is becoming an increasingly popular club. In 1945, the UN was founded with 45 members. In 2006, with the arrival of Montenegro, the total membership reached 192.

This makes the world a more interesting place, but also more confusing. The average newspaper reader might once have been satisfied to identify the vast land mass of the Soviet Union, say, on a globe, without being too curious about its component parts. Now the headlines can raise more awkward questions. Where is Tajikistan, or Turkmenistan? And where, for that matter, did Tuvalu come from, or Kiribati? Citizens of these countries are in no doubt. People who live further away may be less sure.

This *Guide to Countries of the World* offers some help—a compact digest of the economic and political situation in 231 countries. The treatment is inevitably sketchy, and destined to be left further and further behind by the rapid march of events.

But a tenuous and fleeting grasp may be better than none.

Inevitably, too, the perspective is a personal one. The data and events included here accord with some globally accepted priorities and concerns. But the selection and the treatment reflect my own interests and inclinations.

I am grateful to Yusuf Bangura, John Black, and John Connell, who reviewed some of the material for the first edition of this book, the *Handbook of the World*. All errors and misconceptions that remain, however, are entirely my own work.

For the names of countries I have used the conventional short form in the title rather than the official name.

Thus the Libyan Arab Jamarhiya is referred to as Libya, and the Republic of Korea as South Korea. Most countries are covered in two pages. If each profile were weighted for population size, then a page for Nauru, say, would imply 120,000 pages for China. Nevertheless, as a gesture to proportionality, the largest countries, with populations of significantly more than 120 million, are accorded a generous three pages, and the smallest, which are collected together at the end of the book, only half a page. This arrangement may seem strange and arbitrary. To that extent at least, the book does match the world it attempts to describe.

Introduction

Each country has an accompanying box summarizing some of the more significant data. Such information, here and elsewhere, should be treated with caution. Many countries do not collect data in a consistent and comprehensive fashion. The size and distribution of population, for example, can be considered politically sensitive and thus subject to manipulation.

Less reliable still are the estimates on such issues as language, ethnic group structure, and religion. These take little account of the fact, for example, that many people speak a number of languages. People can also profess multiple and overlapping ethnic identities, and while for some people such matters are central to their lives, for others they may be of scant significance. Nevertheless these issues can have political consequences so some indication, however approximate, can be useful. In each case the data are for the latest available year. Where a figure is quoted in dollars, this refers to US dollars.

Sources of information

Economic and social data have been taken from various editions of the UNDP *Human Development Report,* the ILO *World Employment Report*, and the *World Bank World Development Report*. Also useful have been the national *Human Development Reports* prepared locally for individual countries. For consistency, most of the estimates for ethnic structures and language come from the *CIA World Factbook*.

More recent political developments have been traced through a number of printed publications. These include principally: the *Economist,* the *Economist Intelligence Unit* country reports, the *Far Eastern Economic Review*, the *Financial Times*, the *Guardian,* and the *New Internationalist*.

Of the sources available on the internet, among the most useful ones have been: *OneWorld Online* (oneworld.net) and the *Pacific Islands Report* (http://pidp.eastwestcenter. org/pireport/).

Terms used in the text

This book is intended for the general reader so tries to avoid most jargon. It may be useful, however, to indicate the usage of some of the terms that crop up regularly.

Gross Domestic Product (GDP). This is the total value of all the goods and services produced in a country. This can then be allocated to different forms of activity—agriculture, industry, and services, say—to give an indication of the structure of the economy.

Gross National Product (GNP). This is similar to GDP but adds to it the net income from abroad—from overseas investments, say. The GNP is thus equivalent to a country's total income. Dividing this by the population, to give the GNP per capita, gives an indication of how rich or poor a country is.

Human development. This is an all-encompassing view of human well-being, that takes into account not just income but many other issues such as health and standards of education.

Human development index (HDI). This is an attempt to measure human development. It combines data on income, educational attainment, and life expectancy into a single composite figure. This can then be used to rank countries in order of success. Not all countries can provide sufficient data to do this. The 2005 *Human Development Report* from the United Nations Development Programme was able to rank 177 countries—Norway was at the top; Niger at the bottom.

Poverty rate. This is the proportion of people living on less than a certain sum. For international comparisons of the situation in developing countries, this is sometimes set at $1 per day. This *Guide*, however, uses national poverty rates. In developing countries these are based on the proportion of people who can afford to consume a certain number of calories daily along with essential non-food items. National poverty rates, being based on national standards, better reflect local circumstances and aspirations, though may not be internationally comparable. In richer countries poverty rates are set differently—typically the proportion of people who earn less than a certain proportion, often half, of the average income. These thus measure 'relative poverty'.

Purchasing power parity (PPP). If you compare the per capita GDP or GNP between different countries using the standard exchange rate to convert the local currency into US dollars you can get a misleading result. This is because the cost of living is lower in some countries than others. Purchasing power parity takes into account the real purchasing power of the local currency. So comparisons of GNP as $PPP give a better indication of relative national incomes.

Contents

Smaller countries

Indicator tables

Afghanistan

Afghanistan's nascent democracy is again threatened by the Taliban

Land area: 652,000 sq. km.	
Population: 27 million—urban 23%	
Capital city: Kabul, 3.0 million	
People: Pashtun 42%, Tajik 27%, Hazara 9%, Uzbek 9%, other 13%	
Language: Afghan Persian (Dari) 50%, Pashtu 35%, Turkic languages 11%, other 4%	
Religion: Muslim: Sunni 84%, Shia 16%	
Government: Islamic republic	
Life expectancy: 46 years	
GDP per capita: $PPP 800	
Currency: Afghani	
Major exports: Opium, fruits, carpets	

Afghanistan is a largely mountainous country: more than half the territory is above 2,000 metres. Its central highlands are the western end of the Himalayan chain and include the Hindu Kush mountain range. The richest and most intensively cultivated agricultural land is in the lowland plains of the north. The land to the south and south-west is mostly desert or semi-desert. In recent years the environment has been devastated by conflict and drought.

The main ethnic group in Afghanistan are the Pashtun. They live throughout the country, but particularly in the south and east. Though they are the dominant group, the Pashtun are by no means homogenous and there have been frequent conflicts between different tribes and sub-tribes. The second largest group are Tajiks. Not organized on tribal lines, they are found mainly in the north-east and the west, particularly in the capital, Kabul, where they have made up most of the educated élite. The third main group, and among the poorest, are the Hazara, who live in the central highlands or as labourers or servants in Kabul.

Other groups include Uzbeks and Turkmen, descendants of refugees from the former Soviet Union who live in the northern plain. Almost all Afghans are Muslims, mostly Sunni.

In 2003 the population was estimated at 23 million. Since 1979, at least one million people have been killed. Millions more become refugees; around 4.5 million have returned, but in mid-2006 there were still 2.6 million in Pakistan. Levels of human development in Afghanistan were never high and are now among the worst in the world. Rates of infant and maternal mortality are high and most parts of the country lack access to clean water, adequate sanitation, and sufficient food.

Around one-third of health facilities have been severely damaged and most doctors have left the country. There have been some successful interventions by aid agencies, including a child immunization campaign. But creating a functioning health system will take many years and billions of dollars in aid.

Education too is in a grim state. Most qualified teachers left the country during the Taliban period and girls' schools were closed. Schools have now been reopened but they are struggling to cope with the new influx of pupils. Universities have also reopened. But illiteracy remains the highest in the world.

Women during Taliban rule could not be seen in public unless almost completely covered. But now girls can go to school and a number of women hold national positions.

Most people have traditionally relied on subsistence agriculture, raising livestock or growing crops on irrigated land—chiefly wheat, corn, and rice. But incessant conflict, including the sowing of landmines, has undermined food production and also destroyed most industry and infrastructure. Around half the government's budget has to come from international aid.

The only thriving activity seems to be the cultivation of opium. The

Opium dominates the economy

Taliban had banned production, but since then prices have risen and by 2003 total production was around 4,000 tons—worth around half of GDP. Local warlords make a good income by taxing the trade.

Political power has never been highly centralized in Afghanistan. Allegiances have been based more on region or tribe or on occupational group. The most recent conflicts date from 1978, when Soviet-supported communists seized power

and launched an attack on Islam, provoking widespread tribal revolts. The Soviet Union, fearful of an Islamic republic on its doorstep, from 1979 sent in more than 100,000 troops to fight the Mujahideen (holy warriors). Localized rebellions then exploded into a more organized jihad, a holy war, in which the Mujahideen were backed by the USA and Pakistan.

After the Soviet Union withdrew in 1989, warlords representing different militias fought ferocious battles among themselves. Into this chaotic environment came the Taliban, a fundamentalist Islamic group which in 1996 seized Kabul, and imposed their authoritarian rule.

The Taliban soon found themselves in conflict with the United States for harbouring terrorists, notably Osama bin Laden. After the terrorist attack on New York on September 11, 2001, the US launched a sustained bombing campaign that allowed dissidents to oust the Taliban in November 2001.

UN-chaired peace talks resulted in an interim administration headed by Pashtun leader Hamid Karzai. Following the adoption of a new Constitution, in October 2004 Karzai overwhelmingly won a fairly peaceful presidential election. In September 2005, elections were held for the lower house, the Wolesi Jirga, where most people serve as individuals.

The arrival of a formal democracy brought some stability, and rapid economic growth, in Kabul at least. But the government's influence does not extend far beyond the capital and is especially weak in the south where the Taliban have been regrouping and are confronting the NATO troops who took over from the US in 2006.

Albania

One of the most isolated of the former communist states, Albania remains poor and lawless

Land area: 29,000 sq. km.
Population: 3.0 million—urban 44%
Capital city: Tirana, 0.3 million
People: Albanian
Language: Albanian 95%, Greek 3%, other 2%
Religion: Muslim 70%, Albanian Orthodox 20%, Roman Catholic 10%
Government: Republic
Life expectancy: 74 years
GDP per capita: $PPP 4,584
Currency: Lek
Major exports: Metals, electricity

Though Albania is a coastal country, not much of its territory is at sea level. Mountains make up more than two-thirds of the land area: north, west, and south—highest in the north and centre; lower in the south where they merge with the mountains of northern Greece. Even the coastal plain in the west and centre of the country is scattered with hills.

Most Albanians live in the coastal plain and are one of Europe's more homogenous populations—a product partly of the country's physical isolation. Nevertheless there are some language differences, since there are two dialects of Albanian. Most people in the north speak Gheg while those in the south, along with Albanians in Kosovo, speak Tosk.

Albania previously had very rapid population growth. The communist regime promoted fertility and in the 1980s achieved population growth of more than 2% per year. Following the collapse of communism, the birth rate plummeted as many people left the country in search of work. Between 1990 and 1997 around 40% of the population aged 19–40 emigrated, chiefly to Italy, former Yugoslavia, and Greece.

More than one million Albanians are thought to be overseas—most of them living and working unauthorized. Remittances from overseas workers also provide most of the foreign exchange—over $600 million annually.

Because of the way Albania's borders were delineated in 1921 many ethnic Albanians also finished up in neighbouring countries: around two million live in the Kosovo region of Serbia and hundreds of thousands in Macedonia.

Living standards were low during the communist era but subsequent economic collapse and social chaos saw standards slip further. Health indicators have improved, however, and life expectancy has increased in recent years.

Around one-third of GDP comes from agriculture, though despite extensive terracing only around one-quarter of the country is suitable for arable farming. By 1994, the collective farms had been dismantled and almost all land had been returned to private hands. Privatization boosted the output of grains, maize, and vegetables. But irrigation is limited, technology primitive, and productivity low.

Albanian manufacturing industry collapsed when exposed to world markets. Nevertheless, Albania's relatively educated and low-waged workforce has proved attractive to investors for the production of low-tech products like garments and footwear, in companies working as subcontractors for foreign corporations.

Prospects for the mining industry are reasonable. Albania was at one point the world's third largest producer of chrome ore and has significant deposits of other minerals such as copper. The industry has now been fully privatized and foreign companies now exploit the chrome and copper mines.

For 40 years until his death in 1985, Albania was in the grip of the repressive, and steadily more isolated, regime of Enver Hoxha. When the communist system collapsed, two main parties emerged. The first was the reformed communist party, now called the Socialist Party of Albania (SP). The second was the right-wing Democratic Party of Albania (DP).

Elections for the People's Assembly in 1992 resulted in a convincing victory for the DP, which used its majority to elect Sali Berisha as president. Berisha pushed through a number of economic reforms but his government became steadily more corrupt and repressive, persecuting former communists, and muzzling the press and the courts. Berisha and the DP also won the 1996 elections—though the opposition protested that these were fraudulent.

Albania's economic reforms had made the country feel more prosperous and this, combined with remittances, had tempted unsophisticated investors to sink their money into fraudulent financial 'pyramid' schemes. When these schemes collapsed spectacularly at the end of 1996 around half the population lost their savings. This provoked widespread riots and an insurrection that required the intervention of an Italian-led peacekeeping force. It also led to the collapse of the DP government, which had been linked to the bogus schemes.

Chaos as pyramid schemes collapse

The ensuing 1997 election was won by the SP. The new government was initially led by Fatos Nano who over the next few years had several spells as prime minister. The SP administrations were consistently undermined by internal feuds.

In July 2005, the centre-right DP, still led by Sali Berisha, returned to power when the parliamentary elections gave them and their allies around 60% of the seats—though the SP once again disputed the outcome of the elections.

Albania has ambitions to join the EU and as part of the preparations Berisha has been taking the advice of US public relations consultants. He being less hostile to the media and also vowed to tackle widespread corruption.

Algeria

Algeria is now much more peaceful, but its democracy is fragile and still vulnerable to occasional terrorism

Land area: 2,382,000 sq. km.
Population: 32 million—urban 59%
Capital city: Algiers, 4.6 million
People: Arab-Berber
Language: Arabic, Tamazight (Berber)
Religion: Sunni Muslim
Government: Republic
Life expectancy: 71 years
GDP per capita: $PPP 6,107
Currency: Algerian dinar
Major exports: Oil, gas

Algeria has a narrow coastal plain which is regularly broken by parts of the Atlas range of mountains which run across the country from east to west and between which there are high plateaux. To the south, the mountains give way to the arid, sandy expanse of the Sahara that accounts for around 90% of the territory.

Most Algerians, 90% of whom live in the coastal region, are of mixed Arab and Berber descent and are Sunni Muslims. But around one-fifth, particularly in the Kabylia region, consider themselves to be Berber rather than Arab and have protested against official efforts at 'Arabization'. In response, the government declared in 2002 that Tamazight would become an official language.

Although standards of human development have improved in

recent years, and primary school enrolment by 2005 had reached 99%, only around one-third of adults are literate. One of the main problems is unemployment, officially around 18%, but much higher for young people.

Algeria's rapid development after independence in 1962 was based on huge reserves of hydrocarbons in the Sahara. Initially the emphasis was on oil extraction, and Algeria still produces around 1% of the world's oil and has five refineries. But in recent decades the balance has swung in favour of natural gas which now makes up 80% of hydrocarbon production. Hydrocarbons account for more than 96% of exports and 30% of GDP.

Algeria also has extensive reserves of iron ore as well as smaller deposits of other metals. In the past, mining has been relatively neglected, but the government is now keen to develop the mining industry by inviting foreign investment for extracting phosphorus, gold, and diamonds.

In the 1970s, Algeria used its oil revenues to finance industrialization in a number of other sectors, including steel, vehicles, and cement. Almost all of this was state-owned, and remains so, though there have been efforts to attract foreign partners in oil, gas, and other industries. As a result, most people in formal employment work for the state.

Only a small proportion of the land, mostly in the coastal plains and valleys, is suitable for agriculture, which employs around 25% of the workforce. The country does grow cereals, but for many foods it depends on imports.

Algeria waged a long and bitter war for independence from France. In the decades following independence in 1962, political life in Algeria was dominated by the only legal party, the Front de libération national (FLN). Development initially took the form of state socialism but this started to unravel in the mid-1980s with the fall in oil prices, and the government started to liberalize the economy.

From 1988, the younger generation became increasingly resentful of the FLN's grip on power and protested at its austerity measures. The president, Colonel Chadli Benjedid, responded in 1989 with a new constitution that legalized other parties. Radical Islamic groups took full advantage, and in 1990 the Front Islamique du Salut (FIS) gained control of local government in most major cities. Then in 1992 they looked like taking power nationally. After the first round of voting in the parliamentary elections, the FIS took 188 of the 231 seats.

Radical Islamic groups win an election

Faced with the prospect of an Islamic government, the military leadership stepped in, cancelled the election, suspended parliament, and forced Benjedid to resign. This provoked a ferocious backlash. The armed wing of the FIS, the Islamic Salvation Army, targeted not just the military but most other secular groups. The government, led by a series of army appointees, responded in kind and armed village militias unleashed carnage on a horrific scale. The AIS agreed a truce in 1997, but the fighting continued with some of the worst atrocities now committed by other loosely organized groups: the Groupe Islamique Armée (GIA) and the Groupe Salafi pour la predication et le combat (GSPC) as well as by the army. By 1999, the war had cost more than 80,000 lives.

Prospects for peace improved in 1999 with the election of Algeria's first civilian president, Abdelaziz Bouteflika, a former exile whom the military had asked to return. In July 1999, he launched a 'civil accord' initiative which was approved by a referendum and around 3,000 rebels returned to their homes.

Life in Algiers is now more normal though there are still attacks across the country by Islamic militants of the GSPC. In 2001 there was also an uprising in the Kabylia region following the death of a youth in police custody.

In April 2002 the FLN won almost half the seats in the parliament, though turnout was low because of a boycott by the Kabyles. But Bouteflika, with the support of the army, was comfortably re-elected president in April 2004, defeating the FLN candidate Ali Benflis.

Since then Bouteflika has been demonstrating a degree of independence from the military and in 2005 introduced a Charter on Peace and National Reconciliation which grants an amnesty to Islamic militants who are prepared to disarm.

Angola

Angola's civil war is over. Now a devastated country faces a long period of reconstruction —physical and political

Land area: 1,247,000 sq. km.
Population: 15 million—urban 36%
Capital city: Luanda, 2.2 million
People: Ovimbundu 37%, Kimbundu 25%, Bakongo 13%, other 25%
Language: Portuguese, Bantu, and others
Religion: Indigenous beliefs 47%, Roman Catholic 38%, Protestant 15%
Government: Republic
Life expectancy: 41 years
GDP per capita: $PPP 2,344
Currency: Kwanza
Major exports: Oil, diamonds, gas

Angola has a dry coastal plain which is at its broadest in the north where it extends up to 200 kilometres inland. From this plain the territory rises through a number of escarpments to highlands that reach 2,600 metres. Beyond the highlands is a vast plateau that covers around two-thirds of the country. The population is concentrated in and around the highlands.

The largest ethnic group are the Ovimbundu, who are to be found mostly in the central highlands as well as in the coastal towns. The second largest are the Mbundu, who live more along the coast and in the north and north-west; they tend to be the most urbanized and many speak only Portuguese. The third group, the Bakongo, are also to be found in the north, as well as in neighbouring countries.

Decades of civil war devastated a potentially rich country. At least half a million people were killed in the fighting, and the disruption to food supplies and health services killed many more. Around half the population are malnourished. In many parts of Angola the health system has ceased to function and the only services are provided by aid organizations. Added to this is the largely unmeasured devastation caused by HIV/AIDS.

The war also destroyed much of the education system, but many schools have now reopened, and attendance is rising: 60% of children enrol in primary school.

Most people struggle to survive from subsistence agriculture. The land in the fertile highlands allows for ample food crops of cassava, beans, maize, millet, and sorghum. The fighting had a devastating effect on agriculture. And although food production is rising the country remains heavily dependent on food aid. The north-west of Angola used also to be a major coffee producer. The output of this and other cash crops like sisal and cotton also declined steeply, though here too production has started to recover.

Though agriculture was devastated by the war, two other areas of economic activity were less affected: oil extraction, which funds the government; and diamond mining, which funded the rebel UNITA army.

Oil was first produced in 1955 and Angola has become Sub-Saharan Africa's largest exporter after Nigeria. Vast new offshore areas continue to be discovered. In 2003, total reserves were over 13 billion barrels. Output has been growing and is expected to increase by 60% between 2005 and 2009. Oil accounts for half the GDP and more than 90% of exports, and for 75% of locally generated government revenue.

The national oil company Sonangol has joint ventures with many companies, the largest being with Chevron. Sonangol's accounts, however, are very opaque, and it plays an important part in the country's system of 'parallel finance' which

Oil riches siphoned off into private pockets
in recent years has siphoned off around $4.2 billion into private pockets. The seven richest people, with assets of over $100 million, are all former or current government officials.

Angola is also the world's fourth largest diamond producer. These alluvial deposits are spread over large areas, much of which were controlled by UNITA. Since the war hampered geological surveys, there are probably many more deposits to be discovered. This trade too is largely controlled by the army and government officials.

Angola's civil war started even before independence in 1975. There were three main independence

movements. The Movimento Popular de Libertação de Angola (MPLA) was more urban-based and Marxist and had Soviet and Cuban support. The União Nacional para a Independencia Total de Angola (UNITA) was supported by the Ovimbundu in the rural areas and had the backing in particular of South Africa. A third group, the Frente Nacional de Libertação de Angola (FNLA), represented the Bakongo in the north.

The MPLA took control of Luanda when the Portuguese left and gradually achieved international recognition while the UNITA–FNLA alliance fought a bitter rearguard action in the rural areas. A ceasefire in 1989 was followed by a presidential election in 1992 which was won by the no-longer-Marxist MPLA led by José Eduardo dos Santos. However, the defeated UNITA leader, Jonas Savimbi, refused to accept the result and plunged the country back into civil war. A war that started with some ideological basis had long since degenerated into a struggle for power and money.

The 27-year conflict finally ended in April 2002 after Savimbi was killed in an ambush. UNITA quickly capitulated and has since transformed itself into a political party, now led by Isaias Samakuva, a moderate and respected figure, and has joined a government of national unity.

Both presidential and parliamentary elections had been planned for 2006 but these were postponed until 2007 at least. Registration of voters proved very difficult given massive refugee movements. President dos Santos argues that viable elections would require much better infrastructure.

Argentina

Painful economic reforms have stimulated rapid economic growth, but economic and social stability remain elusive

Land area: 2,767,000 sq. km.
Population: 38 million—urban 90%
Capital city: Buenos Aires, 10 million
People: White 97%; mestizo, Indian and others 3%
Language: Spanish
Religion: Roman Catholic
Government: Republic
Life expectancy: 75 years
GDP per capita: $PPP 12,106
Currency: Peso
Major exports: Oil, cereals, processed food, vegetable oil

To the west, Argentina is bounded by the southernmost section of the Andes, but most of the country is a vast plain that descends gradually eastwards to sea level. The north-eastern part of this plain includes semi-tropical forests and Argentina's section of the arid Gran Chaco region, while to the south lies the semi-desert tableland of Patagonia. But it is the rich agricultural area of the central plain, the pampas, that is home to two-thirds of the population.

Argentines are of European descent, predominantly Italian. The country was originally settled by the Spanish who optimistically named it 'land of silver'. In fact, the territory had few precious metals though good prospects for agriculture. The largest wave of immigration, from the 1880s to the early years of the 20th century, brought people to work on the farms and ranches of the pampas. Around half were Italian, and most of the rest Spanish, though there were also Welsh who settled in Patagonia, as well as English investors and managers who developed many of the roads and railways. There are now few descendants of the original Indian inhabitants.

During its peak years of immigration Argentina was among the ten richest countries in the world, and a land of immense promise. But long decades of political upheaval and social stagnation have stifled development. Early in 2005, almost 40% of the population lived below the poverty line. As a result of a series of economic crises in the 1990s, many people left the country: more than half a million Argentines now live abroad.

In many ways Argentina still has great potential. It has rich agricultural resources. Cattle ranching, the basis of its early wealth, still provides substantial export income and beef remains a mainstay of the diet—Argentines eat three times as much beef per person as Europeans. In recent years the fertile soils of the pampas have also permitted a rapid increase in cereal and oilseed exports.

By 1996, however, the major

export was oil and the country is still the third largest producer in Latin America though largely for local use. Argentina is also an important source of natural gas which it exports to neighbouring countries, notably Brazil and Chile.

Argentina is also a major manufacturer. The opening up of the economy in the 1980s and 1990s boosted productivity, and output has increased substantially. The food industry, for example, was previously the preserve of family firms but has gradually been penetrated by multinationals such as Nabisco or Cadbury-Schweppes.

The most radical free-market reforms started in 1989. Previously, the country had been plagued by bouts of hyperinflation. To fight this tendency, the peso was in future to be pegged to the dollar and made freely convertible. Other reforms included privatizing banks and the national airline. This encouraged foreign investors and delivered price stability and growth but also exposed Argentina to the fluctuations of the international capital markets.

Historically Argentina's politics has been dominated by 'caudillos', strongmen. One of the most significant was General Juan Peron. Peron was a populist, who, with the support of his charismatic wife, Evita,

The legacy of Juan Peron and Evita took over as president in 1946. He worked closely with the strong labour unions and nationalized many industries, rising to semi-mythical status in Argentina's history. The Peronist party is still a significant force.

Peron's overthrow in 1955 launched Argentina on three decades of economic and political instability punctuated by military coups. One of the most violent periods followed a coup in 1976 when the military government launched the 'dirty war' against leftist guerrillas. This resulted in widespread torture and murder, with up to 30,000 'disappeared'.

But the military made a fatal miscalculation in its 1982 invasion of the Malvinas Islands ruled by the British as the Falklands. The British responded by sending a task force which retook the islands, a crushing defeat that effectively put paid to the era of military rule. The new presidency of Raul Alfonsín helped to re-establish democracy but he was replaced in 1989 by Carlos Menem.

Menem was a Peronist but rapidly set about dismantling Peron's statist legacy. His free-market policies ushered in a period of rapid economic growth—though at immense social cost as older people struggled to come to terms with a 'dollarized' economy.

The Peronists lost the 1999 presidential election to Fernando de la Rua of the Radical Party. He resigned in 2001, following violent demonstrations provoked by the continuing recession, his term being completed by Eduardo Duhalde.

The presidential election of 2003 was won by a Peronist, the governor of San Cruz, Néstor Kirchner. He has presided over an economic recovery so remains very popular and seems certain to seek a second term in 2007. But he could face a strong challenge from his former economy minister, Roberto Lavagna, who could also claim credit for Argentina's economic revival.

Armenia

Armenia is a poor and landlocked country with a worldwide diaspora and a distinctive culture

Land area: *30,000 sq. km.*
Population: *3 million—urban 65%*
Capital city: *Yerevan, 1.1 million*
People: *Armenian 98%, Yezidi (Kurd) 1%, other 1%*
Language : *Armenian 96%, Russian 2%, other 2%*
Religion: *Armenian Orthodox*
Government: *Republic*
Life expectancy: *72 years*
GDP per capita: *$PPP 3,671*
Currency: *Dram*
Major exports: *Jewellery, manufactured goods*

Occupying the north-western part of the Armenian Highlands, Armenia is almost entirely mountainous: the average elevation is 1,800 metres. The landscape is spectacular, with rushing rivers, deep valleys, and many lakes—the largest of which is Lake Sevana. The land can also be violent. The country is dotted with extinct volcanoes, and an earthquake in 1988 killed more than 25,000 people. But the country has few resources in terms of energy or minerals, and much of the land and water is heavily polluted.

Armenia is one of the world's oldest civilizations and its territory once extended from the Mediterranean in the west to the Caspian Sea in the east. Centuries of conquest and occupation have, however, shrunk it to more modest boundaries.

Even within these the population is fairly concentrated. Two-thirds of

Armenians live in towns and cities, with the greatest numbers in the Ararat plain along the south-western border with Turkey. There is also an extensive Armenian diaspora. Around 1.5 million ethnic Armenians live in other republics of the former Soviet Union and 2.5 million more are scattered around the world. Recent wars generated an influx of around 250,000 ethnic Armenian refugees from Azerbaijan, while ethnic Azeris moved in the other direction.

Armenia's distinctive culture also survived through the Soviet era. Ethnically the country is very homogenous. Most people are Christians—members of the Armenian Orthodox Church, and the Armenian language has retained its own unique alphabet.

During the Soviet era, education and health standards were high, but have deteriorated following sharp cuts in public services. In 2002 healthcare spending per capita was just $7.

The Soviet era also transformed Armenia into an industrial economy, but since then heavy industries have suffered a steep decline. One of the more productive areas remains precious stones and gem cutting—

based on skilled and low-cost labour, though often using imported diamonds. Other healthier industries in the past few years include energy, telecoms, and chemicals. Gold mining has also benefited from new flows of foreign investment.

Armenia's steep terrain does not make for much productive agriculture and the country has to import a high proportion of its grains and dairy products. Nevertheless, agriculture still employs around two-fifths of the population, working largely on irrigated land in the Ararat plain where the main crops include potatoes, grapes, and tobacco. Agriculture was boosted by land reform in the 1990s, and most land is now in private hands, though typically in small farms.

The economy shrank dramatically in the early 1990s but despite faster growth in recent years is still only around 70% of its 1990 size. Probably around one-third of the workforce is unemployed or underemployed and more than 40% of people live below the poverty line. This has driven even more people into the informal sector which is thought to account for around half of GDP.

Following the collapse of the Soviet Union, Armenia became independent in 1991. The new

The dispute over Nagorno-Karabakh

president was the leader of the Pan-Armenian National Movement, Levon Ter-Petrosian, who was elected on a platform of modest reform. Since then political developments have been shaped by the dispute over the Nagorno-Karabakh region of Azerbaijan, which is inhabited by ethnic Armenians. When, in 1992, Nagorno-Karabakh declared its independence, Armenia supplied it with weapons and eventually invaded. Armenia's soldiers had previously been among the élite of the Soviet Army, and in 1993 easily overcame the Azerbaijanis, seizing around 20% of Azerbaijan—including Nagorno-Karabakh. A ceasefire was agreed in 1994 but Armenian troops remain.

Ter-Petrosian was re-elected in 1996, but in 1997 he made the fatal mistake of softening the line on Nagorno-Karabakh. This outraged nationalist sentiment and helped provoke mass defections to the opposition, and in early 1998 forced Ter-Petrosian to resign.

The ensuing presidential election in April 1998 was won by his ex-prime minister Robert Kocharian, also an ex-leader of Nagorno-Karabakh.

On the domestic front, popular discontent with the government took a violent turn in October 1999 when five armed nationalists stormed the parliament, killing the prime minister and seven others, but this seems to have been an isolated incident.

Since then the opposition has remained weak and no single leader has emerged, which left the field open for Kocharian, who does not have any party affiliation, to be re-elected in March 2003. He does, however, get support from the Republican Party, which in 2003 gained the most seats in the National Assembly. Andranik Markarian was appointed prime minister at the head of a coalition led by the Republican Party. Both ballots, however, were considered suspect, and the opposition has called for fresh elections.

Australia

Australia has been establishing stronger links with Asia—but has been unable to shake off the British monarchy

Land area: 7,713,000 sq. km.
Population: 20 million—urban 92%
Capital city: Canberra, 324,000
People: Caucasian 92%, Asian 7%, Aboriginal and other 1%
Language: English
Religion: Roman Catholic 26%, Anglican 21%, other Christian 21%, other 32%
Government: Constitutional monarchy
Life expectancy: 80 years
GDP per capita: $PPP 29,632
Currency: Australian dollar
Major exports: Minerals, agricultural products, manufactures

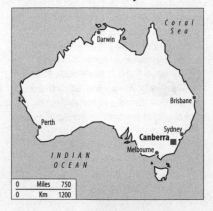

Australia's landmass—which can be viewed as the world's largest island—is dominated by a vast and largely empty interior of arid plains and plateaux, known as the outback. The most significant mountainous area is the Great Dividing Range, which runs down the east of the country, separating the outback from the coastal plain, re-emerging further south as the island of Tasmania. Most people are concentrated in the more fertile coastal areas, particularly in the east, south-east, and west.

Australia's population has been shaped by a long and continuing process of immigration: one-quarter of the current population were born overseas. Currently the immigration level is set at 120,000 per year. Until the mid-1960s the government had a 'white Australia' policy which ensured that immigration was largely from Europe and particularly from the British Isles. Since then, it has cast its net wider, seeking the people with the highest skills. This, combined with refugee flows, has tilted the balance in favour of Asia. Even so, the largest single source of immigrants is New Zealand.

Australia has an increasingly diverse population, but its most distinctive ethnic group are the original inhabitants—a quarter of a million Aborigines. They are the most marginalized part of Australian society—the majority living in desperate poverty, either in reserves in the outback or in urban slums. In recent decades they have become more assertive—and have had notable clashes with the mining companies who exploit their ancestral lands.

Australia's economy is dominated by service industries that account for more than 80% of GDP and employment. The largest employer is the retail sector, but many others work in catering—which also involves feeding more than five million tourists who arrive each year. Manufacturing has declined significantly. Until the 1990s, many Australian producers were protected by high tariffs, but most of these have now been cut, output has fallen, and in 2005 manufacturing employed only 11% of the workforce.

Australia remains a major producer and exporter of raw materials. Its extensive mineral wealth includes iron, aluminium, uranium, zinc, nickel, and lead. It also has gold and diamonds, as well as abundant energy sources in the form of coal, oil, and gas. For many minerals, Australia is the world's leading exporter. Mineral rights are generally vested in the Crown or in individual states which receive

Aboriginal communities demand their rights

royalties from mining companies. But land title is disputed; Aboriginal communities are demanding either a veto over mineral development or a share of the profits.

Much of the territory is arid, but Australia also has rich agricultural land. Agriculture has become a less significant part of the economy but it still generates around one-quarter of exports—chiefly from wool, cereals, meat, and sugar. A more recent success has been wine, of which exports are now worth more than $1.8 billion per year. Australia has some of the world's leading wine brands.

Australia is a federation of six states, each of which has its own government and governor. The head of state is still the British monarch, though there have been several attempts to despatch her. In a 1999 referendum, 55% of Australians rejected the proposal to become a republic, primarily because they disliked the form of republic on offer, or because they regarded the whole constitutional process as a waste of time and money.

Australia's political parties have traditionally pitched conservative forces against those of organized labour—though as elsewhere political parties of right and left have been moving to the centre ground. The conservative party is the Liberal Party, which draws its support from the urban middle class. The Labor Party has relied more on urban workers. The third major force is the National Party, which draws its support from the rural areas. Two minority parties are the Australian Democrats and the Greens. The racist and anti-immigration One Nation party, which emerged in the 1990s, has now largely disappeared.

Labor's longest period in government stretched from 1983 to 1996, initially headed by former trade union leader Bob Hawke. During this period, Labor continued the process of opening up the economy, floating the Australian dollar, dismantling tariffs, and privatizing utilities.

In the 1996 election, following an economic downturn, Labor was replaced by a Liberal–National coalition led by John Howard. He continued economic reforms and called a general election in 1998, achieving a narrow victory.

In the elections in November 2001, in the wake of September 11, Howard, to many people's surprise, was elected for a third term. He strongly supported the US invasion of Iraq and, despite popular opposition to the war, was returned for a fourth term in October 2004. A demoralized Labor Party turned to former leader, Kim Beazley, in an attempt to rebuild its support, but he has made little progress.

Austria

Decades of consensus have been shaken by a resurgence of the far right

Land area: *84,000 sq. km.*
Population: *8 million—urban 68%*
Capital city: *Vienna, 1.6 million*
People: *Austrian 91%, others 9%*
Language: *German*
Religion: *Roman Catholic 74%, Protestant 5%, Muslim and other 21%*
Government: *Republic*
Life expectancy: *79 years*
GDP per capita: *$PPP 30,094*
Currency: *Euro*
Major exports: *Metal goods, textiles, paper products, chemicals, electronics*

Austria is one of Western Europe's most mountainous countries. The western two-thirds of the country fall within the Alps, in ranges that run from east to west divided by broad river valleys. Another smaller and less dramatic highland region extends from the north of the country into the Czech Republic. Lowland Austria is largely in the east, including the Vienna basin through which flows the Danube on its way to Slovakia.

In ethnic terms, Austria is fairly homogenous, though there is some representation of neighbouring nationalities. The end of the cold war also saw a further surge of immigration from Eastern Europe, following the war in former Yugoslavia. As a result, between 1989 and 2001 the proportion of foreign-born rose from 4% to 9%—though these flows have eased now as border controls have been tightened.

Immigration did at least temporarily rejuvenate the population. Otherwise, with a fertility rate of only 1.4 children per woman of childbearing age, the population is ageing fast: by 2030 over-60s will be 32% of the population.

Austrians enjoy one of Europe's more generous systems of welfare. Parents are entitled to three years of maternity leave, for example. But many welfare benefits have recently been cut: retirement benefits currently absorb 9% of GDP, and a major reform of the state pension system was introduced in 2004 to reduce future liabilities.

Two-thirds of Austrians are now employed in service industries, from banking to transport. And following the opening up of Eastern Europe, many foreign companies use Vienna as a base. One of the more significant service industries is tourism. In 2004 Austria had 1.1 million 'tourist beds'. People from other European countries are drawn to Austria's rich cultural heritage and to the scenic and winter sports attractions of the Alps. Tourist density is high: 11 per inhabitant, a ratio two or three times greater than in other popular European destinations such as France or Spain.

Austria is also an important centre for manufacturing, which employs around 17% of the workforce and accounts for 21% of GDP. Traditional industries such as textiles and footwear have stagnated, but others have been expanding, including electronics, chemicals, and metals. Much of the output consists of medium-technology intermediate goods for other European countries, particularly Germany.

Agriculture now employs only 1% of the workforce, but Austria is still largely self-sufficient in food, much of it produced using 'organic' methods. Large herds of livestock, particularly in the mountainous west of the country, also allow for the export of some dairy products. And, with around two-fifths of the country forested, there are also exports of timber.

Austria has enjoyed steadier economic progress than many other **Austria has a system of social partnership** European countries. Much of this is due to its distinctive system of sozialpartnerschaft (social partnership). This involves regular discussions between employers, trade unions, academics, and agricultural representatives. Most major economic policies derive from this process of consultation which has contributed to stable labour relations, low inflation, and extensive welfare benefits—though it has also diminished the status of the parliament.

Until recently, consensus was also the major characteristic of Austrian politics. For most of the period since the Second World War, Austria has effectively been run by two parties

generally in a coalition. From 1970 onwards, the dominant partner has often been the centre-left Social Democratic Party, which draws its support from labour and trade unions. In 1997, Viktor Klima took over as party leader and became chancellor— head of the government. The second party was the centre-right People's Party, led by Wolfgang Schüssel, which is linked to the Roman Catholic Church and draws its support more from the middle class and business.

From the mid-1990s, however, the cosy arrangements between the two parties (which included 'Proporz'—sharing out government jobs between their party card-holders) were challenged from the far right by the Freedom Party and its aggressive former leader Jörg Haider who stood on a populist, nationalist, anti-EU, anti-immigrant platform.

The 1999 elections gave the two major parties a shock. The Social Democratic Party still had the largest share of the vote, but the Freedom Party came second with 29%—narrowly ahead of the People's Party. This resulted in a People's Party–Freedom Party coalition headed by Schüssel, though without Haider.

International opinion was outraged. The EU introduced diplomatic sanctions. These were subsequently lifted and since then the Freedom Party has declined in popularity.

In the 2002 election the People's Party gained the most votes and initially formed a coalition with the Freedom Party. In 2005 the Freedom Party split so the People's Party joined forces with a new party, the Alliance for the Future of Austria, made up of Freedom's more moderate elements.

Azerbaijan

Oil and gas-rich Azerbaijan could have a prosperous future if it made peace with its neighbours

Land area: 87,000 sq. km.
Population: 8 million—urban 50%
Capital city: Baku, 1.9 million
People: Azeri 91%, Dagestani 2%, Russian 2%, Armenian 2%, other 3%
Language: Azeri 89%, Russian 3%, Armenian 2%, other 6%
Religion: Muslim 93%, Russian Orthodox 3%, Armenian Orthodox 2%, other 2%
Government: Republic
Life expectancy: 67 years
GDP per capita: $PPP 3,617
Currency: Manat
Major exports: Oil, textiles, food

Around half of Azerbaijan comprises the central lowland area through which flow the two main rivers, the Aras and the Kura, which join before draining into the Caspian Sea. These plains are contained by spurs of two mountain ranges, the Greater Caucasus to the north and the Lesser Caucasus to the south-west. In addition, there is a small isolated section of its territory to the south-west, Nakhichevan, sandwiched between Armenia and Iran. Azerbaijan has a varied, and often strikingly beautiful, terrain.

Azerbaijan is ethnically fairly homogenous. Around 90% are Azeris who are Shia Muslims. But there have always been minority groups. The most prominent are the 100,000 or so Armenians, who are Christians, and who live in the Nagorno-Karabakh region in the south, which since 1994

has been occupied by neighbouring Armenia. There are also many ethnic Azeris elsewhere—indeed more live in Iran than in Azerbaijan itself.

In the past, Azerbaijan's population has achieved high standards of education and health but government services have been undermined by the conflict with Armenia, and by the emigration of many professionals. Although the healthcare system is officially free, in fact most people have to make informal payments to receive attention.

Around 11% of the workforce are unemployed and around 40% of the population live below the poverty line.

Over 40% of the workforce are employed in agriculture, largely in the fertile lowlands. Most of the former state and collective farms have now been privatized. But agriculture remains very inefficient, production has been falling, and the country has to import much of its food. In addition to growing grains Azerbaijan has also been a major cotton producer. However, the toxic chemicals used for this have polluted the land and the water. Poor drainage has also led to salination and a threat from the Caspian Sea which is rising by

25 centimetres per year.

The mainstay of the country's economy remains oil, three-quarters of which is extracted offshore in the Caspian Sea. This area has been an oil producer since the 1850s and by the early 1900s it supplied half the world's oil. Baku still has the ornate grand houses built by the first oil millionaires.

Azerbaijan's significance declined with the discovery of larger deposits elsewhere. Nevertheless, the international oil companies have still congregated here looking for new deposits. At the end of 2004 the country had 0.6% of world reserves of oil and 0.8% of reserves of gas.

In addition to extracting oil and gas, Azerbaijan was a major producer of equipment for the oil industry—and the city of Sumgait was also one of the main suppliers of chemicals to the Soviet Union; now much of the plant there is rusting in what is probably one of the most polluted places on earth.

One of the most polluted places on earth

Efforts have been made to switch to lighter industries, and the country has some textile factories that use local cotton. But production generally has been held back by the slow pace of economic reform and is weakened by extensive corruption. The economy started growing again after 1997, but with a much narrower base.

Given its ethnic fault lines, Azerbaijan's political development was always likely to be painful. Even when it was a Soviet republic, there had been a long-running dispute with Armenia over Nagorno-Karabakh and the difficulties in handling this crisis cost several leaders their jobs. Independence was declared in 1991 and there were initially some successes in consolidating control over the disputed enclave. But in 1993 the Armenians managed to seize this and neighbouring regions—around 20% of Azerbaijan.

The ensuing political turmoil led in October 1993 to the election as president of Heidar Aliev, a former communist leader. He survived a number of assassination attempts and consolidated his position with a new constitution in 1995 that granted him sweeping new powers. In 1996 Aliev's New Azerbaijan Party (NAP) won most of the seats in the parliamentary elections.

Fighting with Armenia continued until 1994 when a ceasefire was brokered by the Organization for Security and Co-operation in Europe—though Armenia has not withdrawn, and there have been subsequent outbreaks of fighting.

Meanwhile Aliev set about nurturing a personality cult and to no one's surprise was re-elected president in October 1998, after the opposition declared a boycott. The parliamentary elections of November 2000 resulted in a victory for the NAP, but this was largely the result of another blatantly fraudulent process, with widespread stuffing of ballot boxes.

In the presidential election of October 2003 the ailing Heidar Aliev was succeeded by his son Ilham, whom he had groomed as his successor (Heidar Aliev died in December 2003). Ilham has followed in his father's footsteps, rigging the 2005 parliamentary elections in favour of the NAP.

Bahamas

One of the richest countries in the Caribbean

Land area: 14,000 sq. km.
Population: 304,000—urban 89%
Capital city: Nassau, 212,000
People: Black 85%, white 12%, others 3%
Language: English, Creole
Religion: Baptist 35%, Anglican 15%, Roman Catholic 14%, Pentecostal 8%, Church of God 5%, Methodist 4%, other 19%
Government: Constitutional monarchy
Life expectancy: 70 years
GDP per capita: $PPP 17,159
Currency: Bahamas dollar
Major exports: Lobsters, rum

The Bahamas comprises more than 700 islands in the Caribbean, many with sandy beaches and clear water, although only around 30 are inhabited. The majority of people live on New Providence Island, particularly in the capital, Nassau.

Bahamians generally have good standards of education and, compared with many other Caribbean countries, incomes are high. Health standards too are good but around 3% of 15–49-year-olds are HIV-positive. The islands also attract unauthorized immigrants, particularly from Haiti, to work in hotels and restaurants, and are also a stopover for illegal Chinese migrants being smuggled to the USA.

Tourism accounts for around 27% of GDP and employs around half the workforce. Each year there are more than one million stopover visitors to the country's resorts, and about two million cruise passengers—the majority of visitors coming from the USA. Casino gambling now brings in more than $200 million. These numbers are set to rise following investment in resort hotels. The largest employer is the South African company Sun International.

But not everyone has gained from the tourist boom which still leaves many young people unemployed.

There are no rivers, and the soil is poor, so opportunities for agriculture

are limited. But there is a small fishing industry that catches lobsters for local consumption and export.

The Bahamas also has a significant offshore financial sector that contributes about 12% of GDP and employs around 4,000 people. The industry acquired a dubious reputation in the 1980s through association with money laundering and also the trade in cocaine and marijuana. Following further international criticism, the country took steps in 2001 to clean up the financial services industry.

The Bahamas is a parliamentary democracy whose head of state is the British monarch. In the decades after independence in 1973 the government was usually in the hands of Sir Lynden Pindling, leader of the Progressive Liberal Party (PLP). By the end of the 1980s, however, there was an economic downturn along with allegations of corruption. The elections in 1992 resulted in a victory for the Free National Movement (FNM), which won again in 1997.

The elections in 2002, however, saw the return of the PLP now led by Perry Christie. The PLP tends to favour state intervention in the economy but is now more welcoming to foreign investors.

Bahrain

Bahrain is now a constitutional monarchy, though the king still has most of the power

The state of Bahrain consists of around 30 islands in the Persian Gulf. Bahrain island itself, which has more than 85% of the territory, is mostly arid. Wells and springs in the north are used to irrigate fruit and vegetables, but the water table is sinking, so future agricultural prospects are poor.

The native population of Bahrain is largely Arab and divided between the majority Shia and minority Sunni sects. Bahrainis enjoy high standards of health and education but, unlike the citizens of most other Gulf oil states, many are not well off, and 16% were unemployed in 2002—with a higher rate among the Sunnis. Even so, Bahrainis employ immigrants, who are 60% of the workforce, primarily from the Indian sub-continent, to do most of the less popular work.

Bahrain's economy is based on oil, but its reserves are small and may not last beyond 2009. It processes this in its own refineries, along with oil from Saudi Arabia. Since it has less oil than other Gulf countries, Bahrain has had to develop a more diverse economy. Thus, it uses gas reserves to fuel one of the world's largest aluminium smelters and it also has a range of industries that include ship repairing, as well as light engineering and manufacturing.

Bahrain is also a financial, trading,

Land area: 1,000 sq. km.
Population: 0.7 million—urban 90%
Capital city: Manama, 153,000
People: Bahraini 63%, Asian 13%, other Arab 10%, Iranian 8%, other 6%
Language: Arabic, English, Farsi, Urdu
Religion: Muslim, 81%, other 19%
Government: Absolute monarchy
Life expectancy: 74 years
GDP per capita: $PPP 17,479
Currency: Bahraini dinar
Major exports: Oil products, aluminium, chemicals

and distribution hub for the other Gulf countries; it is linked by a 25-kilometre causeway to Saudi Arabia. Two-thirds of the workforce are employed in these and other services.

Until 2002, Bahrain was an absolute monarchy run by the al-Khalifa family in co-operation with the Sunni business élite. However, the emir had long faced opposition, both from within the country from the minority Shia, and from the London-based Bharaini Freedom Movement.

In 1999, Sheikh Hamad bin Isa al-Khalifa became emir and started a wave of reforms—transforming Bahrain into a constitutional monarchy with a parliament of two chambers: one elected, one appointed. On a day-to-day basis, however, the country is run by his uncle, the prime minister, Sheikh Khalifa bin Salman al-Khalifa, whom he has been trying to marginalize.

In 2002 Sheikh Hamad declared himself king but he also amended the constitution to give the appointed chamber powers equal to those of the elected chamber—effectively entrenching his own power. So when elections were held in October 2002 they were boycotted by opposition groups. The tension continues.

Bangladesh

Bangladesh's democracy is constantly undermined by bitter political feuding

> **Land area:** 144,000 sq. km.
> **Population:** 137 million—urban 24%
> **Capital city:** Dhaka, 10 million
> **People:** Bengali 98%, some tribal
> **Language:** Bangla
> **Religion:** Muslim 83%, Hindu 16%, other 1%
> **Government:** Republic
> **Life expectancy:** 63 years
> **GDP per capita:** $PPP 1,770
> **Currency:** Taka
> **Major exports:** Garments, jute products

Bangladesh consists of a vast alluvial plain. The waters of the Ganges and the Brahmaputra flow in via India to deposit billions of tons of silt each year across this delta before emerging sluggishly into the Bay of Bengal. The only higher areas are the Chittagong Hill Tracts on the eastern border with Burma.

The annual flooding is a welcome event. Even in a normal rainy season, 20% of the landscape is flooded. This recharges the underground aquifers, deposits the silt that makes for fertile and productive soil, and allows for the spawning and migration of 300 or more species of fish. But flooding can also be more destructive. In 1998 floods covered more than two-thirds of the country, killing 1,300 people and causing damage of around $900 million.

The other major natural hazard is cyclones. The funnel-shaped coastline regularly sucks in some of the world's most violent storms. In 1991, winds of up to 240 kilometres per hour combined with a tidal surge 25 feet high to kill around 140,000.

Survival in these climatic extremes has made Bangladesh's people remarkably resilient. As a group, they are also relatively homogenous, with a rich language and culture shared with the state of West Bengal in neighbouring India. Bangladesh means 'land of the Bengalis'. Most people are Muslim, though in practice they tend to have a fairly flexible attitude towards religion. The largest distinctive ethnic groups are the tribal peoples who live in the Chittagong Hill Tracts.

Bar some city-states, Bangladesh has the world's highest population density—834 per square kilometre. However, a massive, largely aid-financed, family planning programme slowed annual population growth from around 2.8% in the early 1970s to around 1.5% in 2006. The government's goal is for population growth to stop altogether by the year 2045, though by this time the population may well have doubled to around 250 million.

Bangladesh remains very poor and standards of health are low. A country where half of children are malnourished, and where half of the population do not use sanitary latrines, is unlikely to be healthy, and Bangladesh is not. Millions of children have their lives extinguished all too soon—infant mortality is 46 deaths per thousand live births. A further problem is contamination of groundwater by arsenic: around 200,000 people suffer from some kind of arsenic poisoning.

Bangladesh also has low standards of education, though by 2003 the adult literacy rate had increased to 41%. The government has increased investment in education and many more children are now enrolling in school. But most schools are in decrepit buildings, with overcrowded classrooms, and there are shortages of books and equipment. Not surprisingly, attendance is poor and around half of the children do not finish primary school.

Some of this is compensated for by non-formal schools run by non-governmental agencies (NGOs). Bangladesh has some of the world's largest and most enterprising NGOs, such as the Bangladesh Rural Advancement

Bangladesh has some of the world's largest NGOs

Committee (BRAC). They may offer little more than bamboo huts with mud floors, and teachers who themselves have only a basic education, but they have proved very effective at catching children who have missed out on primary school. Other major NGOs include Proshika, and the NGO-like Grameen Bank, which operates in more than half the country's villages with a system of micro-credit that is being emulated

all over the world—and now offers mobile phone business.

Most people live in the rural areas and around half the workforce make their living from agriculture. In many respects they have been successful. Over the past 25 years, rice production has more than doubled and the country is roughly self-sufficient. But harvests are still erratic and heavily dependent on the climate. Plots of land are very small—measured in tenths of a hectare. Some estimates suggest that over the next two decades output could increase by 50%. But yields are still lower than in neighbouring countries, and farmers living on a knife-edge of survival are reluctant to invest.

The country is now almost self-sufficient in food production—though many people still cannot afford to feed themselves properly.

Another crop that has traditionally been important is jute which employs around three million farm households. Bangladesh is the world's second largest producer. However, production of jute, along with that of another cash crop, tea, has been declining.

Those unable to find work in the rural areas are heading for the cities, or even overseas. There are probably around 2.8 million Bangladeshis working abroad, especially in the Gulf, sending home around $4 billion per year in remittances.

In terms of industry, Bangladesh's greatest progress has been in light manufacturing, particularly the garment sector, which employs around 1.5 million people, 85% of them women. Until recently Bangladesh enjoyed preferential access to Western markets but these preferences have now been withdrawn so Bangladesh will find it more difficult to compete.

One more optimistic development is the discovery of substantial natural gas reserves with the potential for exports to India.

At independence in 1971, the US Secretary of State famously dismissed the new country as an 'international basket-case'. Bangladeshis have defied this gloomy prognostication. Nevertheless, this remains one of the poorest developing countries. Currently about 40% of the population live below the poverty line. Economic growth was up to 6% in 2005, though needs to be higher to make a greater dent on poverty.

Bangladesh achieved independence in 1971 after a bloody war of secession from Pakistan. Elections in what was then East Pakistan had produced a victory for the Awami League (AL) led by Sheikh Mujibur Rahman. He demanded greater autonomy and when this was refused organized strikes and demonstrations. The Pakistani army attacked Dhaka and a full-scale civil war erupted in which the East Bengalis were supported by India. By the time the war ended, more than one million people had died.

A bloody war of secession

After independence, 'Sheikh Mujib' and the Awami League took power, but he became increasingly autocratic and replaced the parliament with a presidential system. The military grew dissatisfied and in 1975 a group of officers assassinated him.

Following this, the de facto leader of the government was the army chief, General Zia ur-Rahman, who took over the presidency in 1977 at the head of his new Bangladesh National Party. In 1981, he in turn was assassinated by rebel army officers led by General Hossain Mohammad Ershad, who took the presidency in 1983. Ershad's rule became increasingly unpopular. He was deposed in 1990—and subsequently served seven years in prison for corruption and murder.

A general election in 1991 led to a surprising but fair victory for the BNP, which was now led by Zia's widow, Begum Khaleda Zia. She defeated the Awami League, now led by Sheikh Mujib's daughter Sheikh Hasina Wajed. In the same year, a referendum altered the constitution to return executive power to the prime minister.

The Awami League embarked on a bitter campaign of destabilization, using 'hartals'—general strikes enforced by intimidation. The Awami League then boycotted the 1996 election demanding that further voting take place under a neutral administration. Eventually they got their way and the AL won the subsequent election. The feuding continued except that it was Khaleda Zia and the BNP who organized the hartals. Remarkably the government saw out its full term and went to the polls in 2001. This time the BNP, heading a four-party alliance, and following a characteristically violent campaign, won a crushing victory. True to form, Sheikh Hasina and the Awami League then claimed that the election was rigged and resumed their own hartals. This toxic political mix was further complicated in 2005 by a series of terrorist bombings, probably by the banned Islamic group Jamiat-ul-Mujahideen Bangladesh.

Barbados

A Caribbean success story with a growing economy and a stable system of governance

Barbados is the furthest east of the Caribbean islands and, except for low, central mountains, is largely flat. It has few natural resources apart from attractive scenery, a good climate, white beaches and coral reefs.

With close to 100% literacy and good health standards, levels of human development here rival those in many industrial countries. Education is compulsory until the age of sixteen and also free to university level. Nevertheless, unemployment remains high and many Barbadians emigrate to work abroad; their remittances are still an important source of income.

The country's original wealth was based on sugar cultivation, and sugar cane still provides much of the export income. But agriculture has long since given way to tourism as the main economic activity. Tourism, with around half a million visitors a year, provides half of foreign exchange.

Barbados has also made efforts to diversify into other services and manufacturing. It has, for example, built up a financial services industry and also attracted some of the 'back-office' data processing work for US companies. Like other Caribbean countries, Barbados has also become a transhipment point for drugs, though the government has taken a strong line

> **Land area:** 430 sq. km.
> **Population:** 0.3 million—urban 52%
> **Capital city:** Bridgetown, 108,000
> **People:** Black 90%, white 4%, other 6%
> **Language:** English
> **Religion:** Protestant 67%, other 33%
> **Government:** Constitutional monarchy
> **Life expectancy:** 75 years
> **GDP per capita:** $PPP 15,720
> **Currency:** Barbados dollar
> **Major exports:** Sugar, rum, other foods and beverages

against this and permits US officials to carry out hot-pursuit searches in Barbadian waters.

Barbados was a British colony. Though it had complete internal autonomy in 1961 it subsequently reverted to being a self-governing colony after the West Indies Federation dissolved in 1962. In 1966, Barbados achieved full independence and since then has enjoyed a stable parliamentary system. Two main parties have alternated in power. Between 1986 and 1994 the government was in the hands of the Democratic Labour Party.

The early 1990s, however, were difficult years. The sugar harvest was poor, unemployment was high, and increasing levels of violence were discouraging tourists. In 1994, power passed to the Barbados Labour Party led by Owen Arthur, who increased his majority in the 1999 elections and won again in 2003.

At present, the head of state is the British monarch in what has been one of the most British of the Caribbean islands. In 1998 a constitutional commission recommended switching to a republic with a ceremonial, elected president, and in 2005 the government restated its intention to hold a referendum on this issue.

Belarus

Belarus has changed little since the Soviet era and even cherishes ambitions to reunite with Russia

Land area: 208,000 sq. km.
Population: 10 million— urban 71%
Capital city: Minsk, 1.8 million
People: Belarusian 81%, Russian 11%, Polish Ukrainian and others 8%
Language: Belarusian, Russian
Religion: Eastern Orthodox 80%, other 20%
Government: Republic
Life expectancy: 68 years
GDP per capita: $PPP 6,052
Currency: Belarusian rubel
Major exports: Vehicles, oil, minerals

Belarus is largely flat—the highest point in this whole landlocked country is only around 350 metres above sea level. One-third of the territory is forested, including the primeval Belovezhskaya Forest on the western border with Poland. There are also extensive peat bogs and marshes, the largest of which are the Pripet Marshes on the southern border with Ukraine.

Most of the population are ethnic Belarusians, but during the Soviet era Belarus was a significant economic centre, and attracted Russian immigrants. Some have returned, but most remain, and there are also sizeable minorities of Poles and Ukrainians. Meanwhile more than two million Belarusians live in other former Soviet republics.

The Belarusians had achieved good standards of health and education but after 1990 infant mortality rose and life expectancy fell. Since 1993, death rates have exceeded birth rates and the population has been falling. Belarusians are also still suffering the effects of the 1986 Chernobyl disaster; though the explosion was just across the border in Ukraine, most of the fallout was in Belarus.

Belarus industrialized rapidly from the 1950s onwards, becoming an important centre for oil refining and petrochemical production. In addition, the Soviet Union used Belarus to manufacture military equipment and to supply trucks, tractors, and motor cycles. Unfortunately this involved importing oil, gas, and other raw materials from Russia and elsewhere, so when they had to pay market prices industries shrank rapidly.

By 1996, things had started to stabilize and nowadays industry accounts for around 30% of GDP. Nevertheless, the country has been accumulating an energy debt with Russia, which the Russian gas company Gazprom periodically has to write off.

Although the government professes to be following 'market socialism', in practice Belarus still has most of the elements of a command economy. Almost 40% of industrial output still comes from state-owned enterprises and most of the rest from enterprises that are effectively under state control. Russia takes around half of all exports, mostly machinery and foodstuffs, and this leaves the country very vulnerable to fluctuations in Russia's economy.

Agriculture, too, is still mostly a state activity. Small farms have proliferated since independence but most output still comes from state and collective farms, the majority of which are insolvent and need regular injections of state funds.

Prior to 1991, Belarus had never been an independent state. It did have separate UN membership but this, like the membership for Ukraine, was a legal fiction invented by Stalin to give the Soviet Union another UN vote.

Belarus had never been independent

Most of the impetus for Belarus's independence came from the Belarusian Popular Front (BFP) whose nominee, Stanislau Shushkevich, was appointed head of state. But it did not take long for the communists to reassert their authority. In 1994, a commission headed by a former state farm manager, Alexander Lukashenka, ousted Shushkevich on charges of corruption.

Lukashenka was able to lever the exposure this gave him into a victory in the 1994 presidential election, and since then he has proved an effective populist, and an astute political operator. In 1996 he successfully conducted a referendum on a new constitution that extended his term of office and created a new bicameral assembly which the president can dissolve if it 'systematically or seriously violates the constitution'. Most of his support comes from unreformed communists.

Opposition has been heavily repressed: in 1996, protests against his regime were crushed and the BFP leader Zyanon Paznyk sought political asylum in the USA. Opposition radio stations and newspapers were also closed down.

One of Lukashenka's main aims is to unite Belarus with Russia, preferably with him as leader. In December 1999, this appeared to move forward with the signing of yet another union treaty with Russia. But genuine union between Belarus and Russia is some way off. Some communists in Russia are attracted by the kudos that this expansion would bring them, but others have balked at the economic implications of fully absorbing its backward neighbour, especially with its eccentric leader. Lukashenka's prospects of leading a united country shrank even further when Vladimir Putin was elected president of Russia. He has shown little appetite for unification.

So far, both the autocratic president and his Russophile inclinations are supported by most people—though it is difficult to gauge opposition strength. For the 2000 parliamentary elections, the opposition were denied access to government-controlled media, so boycotted the polls.

There was a similar atmosphere of fear and intimidation in September 2001 when Lukashenka was re-elected president with 76% of the vote. The process and the result were similar in the 2006 presidential election, following which leading opposition figures were arrested.

Belgium

Belgium has conflicts between its language communities and has been shaken by political scandals

Belgium has three main geographical regions running from north-west to south-east. From the sand-dunes and dykes that fringe the North Sea coast the coastal plain extends inland for up to 50 kilometres—with land that is often marshy and intercut with shipping canals. Then the land rises to the rolling hills of the fertile central plateau, and finally to the dense forests of the Ardennes highlands in the south-east.

In addition to Belgium's political boundaries there is also an invisible cultural frontier running across the country from west to east, passing about 20 kilometres south of the capital, Brussels. To the north of this line is Flanders, home to just over half the population who are Flemish, and speak a dialect of Dutch. To the south is Wallonia, home to the one-third of the population who are Walloons

Land area: 31,000 sq. km.
Population: 10 million—urban 97%
Capital city: Brussels, 1.0 million
People: Fleming 58%, Walloon 31%, mixed or other 11%
Language: Dutch 60%, French 40%
Religion: Roman Catholic 75%, Protestant or other 25%
Government: Constitutional monarchy
Life expectancy: 79 years
GDP per capita: $PPP 28,335
Currency: Euro
Major exports: Transportation equipment, chemicals, manufactured goods

and speak French. There is also a third group, though much smaller, of German speakers on the eastern border.

Superimposed on these divisions is the one million population of Brussels, which, though within Flanders, is a largely French-speaking enclave. It is also home to many of Belgium's immigrants. Belgium has one of Europe's larger immigrant communities—around 9% of the total population. The inflows continue though the proportion of non-nationals has remained fairly stable because it has become easier to obtain Belgian nationality. Around 60% of immigrants are EU citizens attracted by the status of Brussels as unofficial capital of Europe. Brussels is home to the European Commission and to other international organizations such as NATO that generate around 10% of employment. Of the other immigrants, the largest groups are Moroccan and Turkish. The presence of all these people makes Belgium one of Europe's most densely populated, and most urbanized, countries.

Belgium was one of Europe's first countries to industrialize, taking advantage of its formerly extensive coal deposits to process imported raw materials for export. These older industries were in Wallonia, but most coal mines and many of the old factories are now shut. Belgium is still a major steel producer, now using imported coal, but its newer, lighter manufacturing industries, such as chemicals, have been established in Flanders nearer the important ports. The majority of manufacturing companies are foreign owned.

Antwerp is also the world's largest diamond centre—half the world's diamonds pass through the city. This means that Flanders, which used to be a more backward agricultural area, now has a per capita GDP 40% higher than Wallonia—a shift in industrial power that has fuelled resentment between the two communities. The Flemish complain that their taxes are heavily subsidizing the Walloons.

Belgium has run into economic problems in recent years. Though it has many new service industries, unemployment has been high, around 8%, and despite immigration the population is ageing rapidly. Another issue is debt, since governments have frequently attempted to spend their way out of inter-community problems.

Belgium's political landscape is now dominated by the language issue. To deal with this, Belgium has effectively become a federal state,

Belgium is heavily governed

and very heavily governed: in addition to the bicameral federal government there are now separate assemblies not just for Flanders, Wallonia, and Brussels-Capital, but also individual assemblies for the French, Dutch, and German-speaking communities. These various

assemblies have complex overlapping memberships. In total there are around 60 ministers or junior ministers.

In the past, the two main unifying factors have been Catholicism and the monarchy: King Albert II acceded to the throne in 1993 and has helped to serve as a mediator. So far, only around one-fifth of the population vote for parties that want to break up the country, but different groups of Belgians are increasingly leading separate lives.

The language split is also matched among the political parties. In the two main regions, the three largest—the Christian Socialists, the Socialists, and the Liberals—each have autonomous parties. In addition there are two right-wing nationalist parties: for Flanders, the Vlaams Blok; for Wallonia, the National Front.

Belgians' faith in their political system was rocked in the late 1990s by a series of scandals. These included the failure of the justice system to deal with a paedophile ring, allegations of bribery against government officials, brutal police treatment of immigrants, and a number of food scares.

These contributed in 1999 to an end of the 40-year dominance of Christian parties. The right-leaning liberals became the biggest party nationally, while the greens and the Vlaams Blok also made gains. Guy Verhofstadt, of the Flemish liberals, formed a coalition government and instituted a popular series of reforms. As a result he was returned in the 2003 elections at the head of a four-party coalition of Liberals and Socialists. In 2004, after being found guilty of racism, Vlaams Blok was refounded as Vlaams Belang.

Belize

Belize still has a border dispute with Guatemala, but has become steadily more Hispanic

Land area: 23,000 sq. km.
Population: 0.3 million—urban 48%
Capital city: Belmopan, 4,000
People: Mestizo 49%, Creole 25%, Maya 11%, Garifuna 6%, other 9%
Language: English, Spanish, Mayan, Garifuna
Religion: Roman Catholic 50%, Protestant 27%, none 14%, other 9%
Government: Constitutional monarchy
Life expectancy: 72 years
GDP per capita: $PPP 6,950
Currency: Belize dollar
Major exports: Sugar, citrus concentrates, bananas, garments

The northern half of Belize consists largely of lowland swamps, while the main feature in the southern half is the Maya mountain range which rises to 1,112 metres. More than 40% of the territory is covered by rainforests. Just off the coastline there are many small islands, or cays, and one of the world's largest barrier reefs.

As a result of flows of migrants and refugees from El Salvador and Guatemala in the 1980s, probably more than half the population is now Spanish-speaking. Meanwhile English-speaking creoles have emigrated to the USA. In addition, there are smaller numbers of Garifuna in the south.

Around one-quarter of the labour force work in agriculture, mostly on small farms growing maize, beans, and cash crops. The main sources of export income are sugar cane in the north, citrus concentrates in the Stann Creek Valley, and bananas in the south. Forestry is likely to become more important; although there are replanting programmes, the arrival of Asian logging companies raises the prospect of deforestation.

Manufacturing tends to be on a small scale for local consumption. Garment production in export processing zones has fallen as a result of competition from countries with cheaper labour. Tourism is also an increasing source of income; more than 200,000 visitors come each year, many for scuba diving or to visit Mayan archaeological sites. Others visit from cruise ships.

Belize has a border dispute with Guatemala. In 2002, the Organization of American States brokered a new agreement, which neither country has ratified. An increasingly serious problem, however, is the use of Belize for trafficking cocaine from South America, which is leading to rising levels of crime.

Since independence in 1981, Belize has had a stable parliamentary system. The head of state is the British monarch, represented by a governor-general. There are two main parties. The 1998 and 2003 elections were won by the People's United Party (PUP) led by Said Musa. The PUP, previously left-wing, has moved to the centre and tends to be supported by the mestizos. The conservative United Democratic Party has business links and is supported more by creoles. By 2006, economic pressures and public spending cuts were causing unrest and undermining support for the PUP.

Benin

Business has revived, but most is informal so the government gets little tax revenue

Land area: 113,000 sq. km.
Population: 8 million—urban 45%
Capital city: Porto Novo, 238,000
People: Fon, Adja, Yoruba, Barib, and many others
Language: French, Fon, Yoruba, and others
Religion: Indigenous beliefs 50%, Christian 20%, Muslim 20%, other 10%
Government: Republic
Life expectancy: 54 years
GDP per capita: $PPP 1,115
Currency: CFA franc
Major exports: Cotton, oil

Benin has a series of geographical regions that extend northwards from its narrow coastline. Just behind the sandy coast are tidal marshes and lagoons; then there is a flat fertile plateau, the barre. A further series of plateaux lead to the Atakora Mountains in the north-west and to plains that slope down to the Niger River in the north-east.

Benin's people are divided among many ethnic groups, the largest of which are the Fon and the Adja, who make up over half the population. These have an animist religious tradition but there are also significant Muslim influences that come from countries to the north and Christian influences in the south that are a legacy of the French colonial years, so religious beliefs tend to be a mixture of different ideas.

Benin is one of Africa's poorest countries. Around one-third of the population live below the poverty line. Though Benin is self-sufficient in food, more than one-fifth of children under five suffer from malnutrition, and health standards are poor. HIV/AIDS prevalence, however, is still relatively low at around 2%. Malaria remains a serious problem, along with many cases of gastroenteritis linked with contaminated water supplies. Only 34% of adults are literate.

Around half the population make their living from agriculture. Farmers grow yams, cassava, millet, sorghum, beans, maize, and rice. Productivity is low, but in a year of reasonable rainfall Benin is self-sufficient in food. Many people also work on cotton and palm-oil plantations. Benin is a major cotton producer and sales of cotton fibre and seeds make up more than 40% of foreign exchange earnings. Most industry is linked with processing primary products for export.

Benin is also a bustling centre for trade, which makes up more than half of GDP. Its geographical location makes it a key outlet for the landlocked countries to the north. But most of the trade activity is focused on Nigeria, its giant neighbour to the east. Benin is largely an entrepôt state since less than 20% of imports are consumed locally; the rest are re-exported.

Most of this business flows through the informal sector, particularly the Yoruba traders who deal with their counterparts in Nigeria. The fairs and markets along the border with Nigeria are always packed. The Beninois sell rice, cigarettes, and spirits, while

A frenzy of cross-border trade

they buy cars, plastic products, and electrical equipment. But their most important purchase is petrol: every day around 300,000 litres cross the border illegally; the smuggled price is around half the official price.

Benin's commercial capital, Cotonou, is also a major transit point. Its streets are packed with cars and with motorcycle taxis, the zemidjans. In the 1990s, vehicle ownership more than tripled, and the old bridge that crosses the lagoon and links the country with Togo in the west and Nigeria in the east creaks under the daily weight of traffic.

Much of this frenzy of activity is a response to desperation. Many of those now in the informal sector previously worked in larger enterprises or in government services. But the economic crises and government cutbacks in the 1980s and 1990s resulted in heavy job losses. Even government officials need second or third sources of income: many of the zemidjans are owned by civil servants or teachers, who lease them out.

The government largely turns a blind eye to the 'osmosis' in the border region since it provides an important source of funds to so many people. But it does mean that relatively little income is taxed so government finances are weak.

During the early years after independence from France in 1960 political life was punctuated by a series of military coups. In 1972, Major Mathieu Kérékou seized power and from 1974 tried to set the country on a Marxist-Leninist course and changed its name from Dahomey to Benin.

But in 1989 in the wake of an economic collapse Kérékou abandoned socialism and in 1990 organized a national conference which led to a new constitution for a multiparty democracy. Nevertheless, he lost the 1991 presidential election to Nicéphore Soglo. Soglo struggled to build a free-market economy and introduced many austerity measures.

These did not make him popular, and the 1996 presidential election was won by Kérékou, who appropriately campaigned under the symbol of the chameleon. He also won the 2001 presidential election, though amid allegations of fraud and his coalition, the Union pour le Bénin du futur (UBF), followed up with a majority in the 2003 parliamentary elections.

The presidential elections in 2006, however, produced a surprising result. Kérékou was ineligible but neither he nor the UBF could settle on a successor. This left the field open for Boni Yayi, an independent technocrat. After leading in the first round, he formed an ad hoc alliance with Mr Soglo and others and won 75% of the vote in the second round, from an electorate clearly seeking a change in their political options.

Bhutan

Bhutan's development objective is to maximize 'Gross National Happiness'

Land area: 47,000 sq. km.
Population: 2.1 million—urban 9%
Capital city: Thimphu, 70,000
People: Ngalop, Sharchop, Nepali
Language: Dzongkha, Tibetan, and Nepalese dialects
Religion: Buddhism 75%, Hinduism 25%
Government: Buddhist monarchy
Life expectancy: 63 years
GDP per capita: $PPP 1,467
Currency: Ngultrum
Major exports: Electricity, timber

Bhutan has three main geographical zones. The north falls within the high Himalayas which reach altitudes of 7,300 metres. To the south are the 'inner' Himalayas which include a number of fertile and well–cultivated valleys—though travel between them can be arduous. To the far south these descend to the narrow, subtropical Duars Plain that runs along the border with India.

The main ethnic group are the Ngalop, who live mostly in the west and centre. Of the other groups, the largest are the Sharchops in the east and ethnic Nepalese in the south.

Concerned at a largely unauthorized influx of Nepalis, the king from the late 1980s enforced a national language and dress. This prompted an exodus of refugees and 100,000 still live in Nepal. Other security problems include the presence of Indian insurgents who use southern Bhutan as a base.

Bhutan has made remarkable progress in human development since it began to open up to the rest of the world after 1959. Between 1960 and 2001, life expectancy increased from 37 to 66 years. This was based on a uniquely Bhutanese model, balancing modernization with Buddhist values and characterized by the king as the pursuit not of gross national product but of 'Gross National Happiness'.

Around 70% of people work in agriculture, growing subsistence crops such as rice, maize, and potatoes. The best land is in the fertile valleys and is fairly equally distributed though plots are small. Three-quarters of the territory is forested, and timber is an important export.

Bhutan's recent spurt of economic growth has been based on the development of hydroelectric power for export to India. This is not based on dams but mostly uses 'run-of-river' systems to harness the country's fast-flowing rivers.

Development has largely been financed by aid but the completion of the Tala power project should soon eliminate the need for this. Tourism is also a useful source of income, though careful control of numbers makes this an exclusive and expensive destination.

Bhutan is a monarchy, ruled since 1972 by King Jigme Singye Wangchuk. He has however gradually been moving the country towards democracy. In 2005 he presented a new constitution and said that in 2008 he would abdicate in favour of the crown prince, Jigme Khesar Namgyel Wangchuck, and that Bhutan will become a parliamentary democracy.

Bolivia

Bolivia has taken a sharp left turn with the election of its first indigenous president who wants to 'refound' the country

Land area: 1,099,000 sq. km.
Population: 9 million—urban 63%
Capital city: La Paz, 0.8 million
People: Quechua 30%, Aymara 25%, mestizo 30%, European 15%
Language: Spanish, Quechua, Aymara
Religion: Roman Catholic
Government: Republic
Life expectancy: 64 years
GDP per capita: $PPP 2,587
Currency: Boliviano
Major exports: Soya, zinc, gold, tin, natural gas

Bolivia is one of only two landlocked states in Latin America—having lost its access to the sea in 1883 after a war with Chile. It can be divided into three main regions. First, in the south-west and covering about one-tenth of the country there is the bleak, arid expanse of the Altiplano, a plateau some 3,600 metres above sea level. Second, enclosing this plain, are two branches of the Andes with the highest peak at 6,542 metres. Third, to the east and north, and covering around two-thirds of the country, are the lowlands of the Oriente, which includes grasslands and tropical rainforests.

The Bolivians who occupy this often harsh terrain are mostly of Indian origin. The largest groups are the Quechua and Aymara, who live mostly in the highland regions. The women have often retained their traditional dress with colourful petticoats, and in some regions distinctive bowler hats. The mestizo and white population are more likely to be found in the capital, La Paz, and in the richer valleys and lowlands.

This is one of the poorest countries in South America. The rural areas in particular suffer from a lack of safe water and sanitation and poor health services. High levels of infant mortality have depressed population growth. Around two-thirds of Bolivians live below the poverty line. In the face of increasing poverty, many Bolivians, 200,000 or more, have now migrated to work in sweatshops in Argentina.

Agriculture employs around 60% of the labour force, but much of this in the highlands is still primitive cultivation of cereals and potatoes. The more productive land is in the eastern lowlands and is often devoted to commercial farming—including cotton, sugar, and particularly soya which is now the leading official export earner.

Much of Bolivia's wealth, and its potential, lies in its minerals—including tin, silver, zinc, and gold—usually extracted with cheap Indian labour working in harsh conditions.

This reliance on commodities will however continue to leave Bolivia severely exposed to international commodity markets. Other important resources are oil and particularly gas of which there are extensive reserves in the tropical lowlands; 1999 saw the completion of a gas pipeline to Brazil.

Bolivia's largest unofficial export is coca paste. This is derived from coca leaves grown by 35,000 or so peasant farmers in the lowlands. Around 9,000 'laboratories' transform it into coca paste. Bolivians have traditionally chewed the leaf to numb themselves to cold and hunger, but since the 1980s the market for cocaine has transformed this into a lucrative cash crop. Bolivia supplies around 30% of the world's coca paste, with exports worth around $200 million (3% of GDP). With US aid, the government has made efforts, generally unpopular, to eradicate coca cultivation. The area under cultivation, 27,700 hectares, is much less that of a few years ago but is now rising again.

Source of one-third of the world's coca paste

Economic management has often been chaotic. By the mid-1980s, with crashing tin prices, rising international debt, and a world record inflation rate of 20,000%, Bolivia was close to collapse. The subsequent recovery started in 1985 with the shock therapy of the 'New Economic Plan', which dramatically reoriented the economy along free-market lines. From the early 1990s, foreign firms were allowed to acquire 50% shares in the main state companies. These and other changes enabled Bolivia to enjoy investment-led growth which averaged 4% annually in the period 1990–96.

Subsequently, however, the Bolivian economy, like those of its neighbours, suffered a severe slump. And since many people had derived little benefit, there have been frequent protests and demonstrations.

Bolivia's political history has been punctuated by coups and military dictatorships, though since the early 1980s power has generally been transferred constitutionally.

The 1993 presidential election resulted in a coalition led by Gonzalo Sánchez de Lozada, of the Movimiento Nacionalista Revolucionario (MNR). He lost the 1997 presidential election to Hugo Banzer Suarez, formerly a military dictator, but after Banzer stepped down due to ill health, Lozada returned in the 2002 election, just ahead of Evo Morales the radical leader of Bolivia's coca workers.

Sánchez's second term was brief. After violents protests over a proposed gas pipline he was forced out of office in October 2003 to be replaced by a non-party figure, Carlos Mesa. But he too struggled in the face of massive demonstrations and roadblocks.

This opened the path in December 2005 for the election of Evo Morales of the Movimiento al Socialismo (MAS). Morales, Bolivia's first indigenous president, has allied himself closely with Fidel Castro in Cuba and Hugo Chavez in Venezuela. He aims to 'refound' Bolivia and moved swiftly to nationalize the oil and gas industries and launched a programme of land redistribution. In July 2006 a new Constituent Assembly was elected—but the MAS does not have the two-thirds majority needed to change the constitution.

Bosnia and Herzegovina

Bosnia and Herzegovina is held together by aid and diplomatic pressure

Land area: 51,000 sq. km.
Population: 4 million—urban 44%
Capital city: Sarajevo, 526,000
People: Bosniak (Muslim) 48%, Serb 37%, Croat 14%, other 1%
Language: Serbian, Bosnian, Croatian
Religion: Muslim 40%, Orthodox 31%, Catholic 15%, Protestant 4%, other 10%
Government: Republic
Life expectancy: 74 years
GDP per capita: $PPP 5,967
Currency: Marka

The state of Bosnia and Herzegovina, commonly referred to as Bosnia, is a loose federation of two entities created in 1995. One is the Serb Republic (RS) with 49% of the territory and approximately 45% of the population. The remainder is itself a federation: the Federation of Bosnia and Herzegovina, which consists of land controlled by the Bosniak (Muslim) and Croat communities—commonly referred to as 'the Federation'.

The region of Bosnia occupies the north and centre of the country while Herzegovina makes up the south and south-west. The political partition gives most of the lower-lying and better agricultural land in the north to the Serb Republic, and the more mountainous remainder, including the Dinaric Alps, and most of the country's industry to the Federation.

The wars since the early 1990s caused rapid changes in population.

Apart from those people killed there was considerable migration in and out of the country. The war also affected birth and death rates and the population is rapidly ageing. Education too suffered, not just because of physical damage to schools but also because of a brain drain of teachers as well as the fragmentation of the education system along ethnic lines.

Health also deteriorated as a consequence of poverty, stress, and poor diet, as well as alcohol consumption and smoking. Indeed life expectancy started to fall, though it is now increasing again.

The dissolution of Yugoslavia deprived the country of the markets for its industrial goods and the war destroyed many of its factories. As a result, since the late 1980s industry has declined from 50% to 30% of GDP. This has largely been replaced by services, predominantly public administration and defence.

Agriculture has also deteriorated. Many farms have been abandoned and much of the land is contaminated by landmines. Most economic growth has come from aid-financed reconstruction. Even so unemployment remains around 40%.

Bosnia's ethnic strife reflects a long

history of occupation and struggles in the Balkans. By the 20th century, this had left the territory with three main communities who are physically identical but have strong cultural and religious differences: the Croats, who are generally Roman Catholics; the Serbs, who are generally Orthodox; and the Muslims, now officially referred to as Bosniaks, who are a legacy of the long occupation by the Ottoman Empire.

The communist government of Marshal Tito managed to keep these groups together, establishing Bosnia and Herzegovina as a republic within Yugoslavia, and giving the Muslims a distinct ethnic status. The distribution of the groups was complex: some areas had majorities of one group but others were ethnically mixed.

When Yugoslavia started to disintegrate the situation in Bosnia was very difficult. At this point, the Serbs wanted to remain part of Yugoslavia, the Croats wanted to unite with Croatia, while the Muslims preferred a multiethnic independence. They maintain these ambitions to this day. However, they did all make a choice of sorts in March 1992 when a referendum, which the Serbs boycotted, opted for secession and Bosnia duly declared its independence.

After independence, the violent conflicts that had accompanied all these events erupted into full-scale civil war as each group sought to seize territory while driving out the members of the other communities in a bitter and savage process of 'ethnic cleansing'. The war was to kill a

A savage process of ethnic cleansing

quarter of a million people, drive one million people out of the country as refugees, and displace another million or so internally.

In 1994, the USA brokered a ceasefire and in 1995 an agreement reached in Dayton, Ohio, brought the war to an end. This established a central government with a rotating three-member presidency and a two-chamber parliament to deal with foreign affairs and monetary policy. But most of the major functions, including economic policy, taxation, defence, and the police forces, would devolve to the two 'entities'.

The Dayton agreement also provided for the appointment of a High Representative, who in 2006 was a former German cabinet minister, Christian Schwarz-Schilling. The High Representative monitors the agreement, co-ordinates the work of international agencies—and where necessary imposes decisions. In addition there is an EU peacekeeping force, EUFOR, and a small NATO contingent.

Elections in November 2000 resulted in a government coalition of ten moderate parties, the Alliance for Change. But the 2002 elections saw victories for hard-line nationalist parties from each of the three ethnic groups: the Party for Democratic Action, the Croatian Democratic Union of BiH, and the Serb Democratic Party. With the support of more moderate parties they formed new governments at the different levels. They have made some progress, with reforms that give the central government more power. But the administration remains riven by factional disputes.

Botswana

Botswana's diamonds have financed public services but have yet to lift its people out of poverty

Land area: *582,000 sq. km.*
Population: *2 million—urban 52%*
Capital city: *Gaborone, 186,000*
People: *Tswana 79%, Kalanga 11%, Basarwa 3%, other 7%*
Language: *Setswana, English*
Religion: *Indigenous 85%, Christian 15%*
Government: *Republic*
Life expectancy: *36 years*
GDP per capita: *$PPP 8,170*
Currency: *Pula*
Major exports: *Diamonds, copper, nickel*

Apart from hills in the south-east, Botswana is largely a flat plateau at around 1,000 metres above sea level. In the north-west is the extensive Okavango Swamp and in the south the Kalahari Desert. Most people live in the east, where the climate is cooler and the soil more fertile.

Compared with other countries of Sub-Saharan Africa, Botswana has seen little ethnic conflict. Most people are Tswana: a group of eight ethnic clans. The constitution defines these as the 'majority tribe' and they have official representation in an advisory parliamentary chamber, the House of Chiefs. However, there are also around 50,000 Bushmen, who frequently face discrimination —including removal from the Central Kalahari Game Reserve.

Botswana's income from diamond mining has enabled it to invest in social services. Education is universal and free, and the primary and curative health services are extensive. Infrastructure has also improved. An increasing proportion of the population are now urbanized—as people are attracted in particular to the capital Gaborone which has expanded ten-fold in the past 30 years.

But the economy has yet to diversify and to distribute economic opportunities more widely. Diamond mining employs only 5% of those in the formal sector. Unemployment is around 20% and many more are underemployed. One-quarter of the population live in poverty.

One of the most alarming recent developments has been the spread of HIV/AIDS. In 2002, 39% of those aged 15–49 were infected—the highest infection rate in the world. This has had a dramatic impact on average life expectancy: in the early 1990s this was around 65 years but by 2003 it had slumped to 36. As a result population growth has slowed: previously around 3.3% per year, it is now around 2.4%.

Many people still rely on agriculture—through subsistence farming and particularly raising livestock. But even when rainfall is good, they only grow enough food to meet around one-third of national needs. Agriculture by 2002 had fallen to around 2% of GDP.

Nowadays, Botswana's economy is dominated by the diamond industry, which accounts for around 36% of GDP. Diamonds were discovered under the sands of the Kalahari Desert shortly after independence and large-scale extraction was started in 1971 by the Debswana Mining Company—a joint enterprise between the government and the South African company De Beers. Botswana is now one of the world's largest producers of diamonds and new diamond 'pipes' are still being discovered.

Diamonds under the desert

There are also significant deposits of copper-nickel and other minerals, including coal, soda-ash, and gold, but the enterprises that are mining these reserves have proved less successful.

Mining employs few people directly. But it does support a large public sector which employs around half of all those working in the formal sector. The government is anxious to build up manufacturing industry and is offering substantial tax concessions for new investment. An assembly plant established to build Hyundai cars, however, failed in 2000 when its South African owners went into liquidation. Recently efforts have been made to build up a textile industry to take advantage of preferential access to the US market.

Mining has also enabled the government to build up large savings in anticipation of its diamond reserves eventually being exhausted. In 2002, Botswana had foreign reserves of around $5 billion and has been in the unusual position for an African country of lending funds to the IMF.

Botswana has been one of Sub-Saharan Africa's more democratic countries. Since independence in 1966, it has had a 40-member National Assembly which elects an executive president, in addition to its 15-member advisory House of Chiefs. The first president, Seretse Khama, founder of the Botswana Democratic Party (BDP), was determined to build a multiracial society.

Although the country needed to maintain trading links with South Africa in order to survive, Khama opposed apartheid. On his death in 1980, he was succeeded by his vice-president, Quett Ketumile Masire, who also won elections in 1984, 1989, and 1994. When he retired in 1998, he in turn was replaced by his vice-president Festus Mogae.

Until recently, opposition came from the left-wing Botswana National Front (BNF), but in 1998 eleven of the BNF's assembly members formed the new Botswana Congress Party.

With the opposition split, the BDP achieved another convincing victory in the 1999 National Assembly elections. Mogae chose as vice-president Ian Khama, son of the first president. But now he too has been building his own power base, becoming party chairman in 2003, and the BDP has suffered from extensive infighting.

Fortunately for the BDP, so has the BNF, now led by Otsweletse Moupo. Another faction has left to form the New Democratic Front Party. As a result, the BDP was able to win an easy victory in the 2004 National Assembly elections.

Brazil

Destined by size to be the leading country of South America, Brazil has yet to realize its potential

Land area: 8,512,000 sq. km.
Population: 181 million—urban 83%
Capital city: Brasilia, 2.0 million
People: White 54%, mixed white and black 39%, black 6%, other 1%
Language: Portuguese
Religion: Roman Catholic 74%, other 26%
Government: Federal republic
Life expectancy: 71 years
GDP per capita: $PPP 7,790
Currency: Real
Major exports: Manufactures, soya, iron, coffee

Brazil's huge landmass occupies almost half the continent of South America. Geographically, the country's two main features are the Brazilian Highlands and the Amazon Basin. The Brazilian Highlands cover most of the south and east of the country and consist of a vast plateau with an average elevation of 1,000 metres interspersed, particularly in the east, with rugged mountain ranges, some of which rise above 2,800 metres. Much of this area is forested or opens up to extensive prairies.

The Amazon Basin to the north and west covers more than 40% of the country. This is the world's largest river drainage system and most of it is covered with tropical rainforest. While there are still unexplored areas, many parts of the rainforest have now been penetrated by settlers, ranchers, or mining companies, a process of deforestation that has alarming environmental implications globally— in terms of climate change and loss of biological diversity. This area contains around one-fifth of world plant species. Most of Brazil has a humid subtropical climate, but the land to the north-east, known as the sertão, suffers from frequent droughts.

Brazil has long been a racial melting pot. There is little overt discrimination, but people of European origin hold the most powerful positions, followed by those of mixed race (who call themselves 'brown') and blacks, with the small and declining Indian population the most marginalized of all. This mixture of races has generated a vibrant and diverse culture. So although Brazil is unified by the Portuguese language and by Roman Catholicism it also has strong African influences.

This social stratification contributes to severe inequalities. The cities in the south, like Rio de Janeiro or São Paulo, are similar in many respects to those in Europe, though also have desperate shanty towns, called favelas. The north-east is almost another country—deep in the Third World. Brazil has one of the world's most unequal distributions of income. In 2001, the richest 10% of the population got 47% of the income while the poorest 10% got only 0.7%. Around 22% of Brazilians live on less than \$2 per day.

These contrasts are evident in education. Though Brazil has a

The richest 10% of Brazilians get 47% of the income

small proportion of highly educated professionals, most children drop out of school very early. Health services too are skewed. One estimate suggests that 40% of health expenditure is used for sophisticated curative treatments that benefit only 3% of the population. HIV/AIDS is a significant problem but those infected can get free anti-retroviral treatment.

In many respects, Brazil is an advanced industrial economy. Around one-quarter of the labour force work in industry. With such a large population, Brazil was for decades able to direct most of its manufacturing output at its domestic market: not just basic industries, such as food and clothing, but also more sophisticated products from cars to petrochemicals to aircraft—all of which benefited from protection from foreign competition.

By the late 1980s, however, the limitations of this strategy were becoming clear. Brazilian personal computers, for example, were much more expensive than foreign equivalents. From 1990 therefore, the government opened up the markets. This helped stimulate greater efficiency in some areas—though others such as garments or shoes have shrunk in the face of Asian competition.

Brazil's industry has benefited from its wealth of mineral resources. It has around one-third of the world's reserves of iron ore, as well as large deposits of bauxite, coal, zinc, gold, and tin. It also has extensive offshore oil reserves that have enabled the state oil company, Petrobrás, to supply more than half the country's needs. Most electricity, however, comes from hydroelectric plants—one of the largest in the world being the Itaipú dam shared with Paraguay on the Paraná River.

With such a huge landmass, Brazil might also be expected to be a major agricultural producer. Brazil is indeed largely self-sufficient in basic foods. It also has a large livestock herd, and is an exporter of cash crops like soya and coffee. But output is less than might be expected. This is partly because much of the land area, particularly in the Amazon Basin, is unsuitable for farming. But even the better farming land is often used very inefficiently. In fact much of the land held by the largest landowners is scarcely used at all. Some 58,000 large landowners hold half the country's farmland, while three million small farmers make do on 2%, and millions more have no land at all. Many governments have promised land reform; few have delivered.

Brazil has often been plagued by inflation—usually in at least two digits and sometimes four or five. This pattern appeared finally to have been broken early in 1994 with the introduction of a new currency, the 'real', which was pegged to the US dollar, a move which was strikingly successful at reducing inflation.

Any Brazilian president's task is

complicated by the dispersed and fragmented nature of the political system. Brazil is a federal democracy and each of the 26 states has its own legislature and administration. States and their municipalities control over two-fifths of total tax revenues and have considerable freedom.

Federalism has the merit of permitting decentralization, but it also produces confusing overlaps. Universities and hospitals, for example, may be controlled either by the federal or state governments. There are also huge disparities between the states: in 1996 the average per capita income in the state of São Paulo was more than ten times that of the north-eastern state of Piaui or of Amazonas. Governments have tried to narrow these gaps by building infrastructure and offering tax breaks. As a result, many new electronics factories have sprung up in Manaus in the middle of the rainforest.

The federal structure also complicates political manoeuvres in the capital, Brasilia. The president **Brazilian states have considerable freedom** does in theory have considerable power of patronage. But party allegiances are notoriously weak and temporary—over the period 1994–98, almost half the 513 members of the lower house switched allegiance. Most have greater loyalty to their local power base and concentrate on extracting federal largesse for their state or municipality. Party organization is also blurred by powerful cross-party interest groups such as the 'ruralistas', who lobby for large landowners.

For much of the country's history

Brazil has been under centralized, authoritarian rule, including a military government from 1964 to 1985. In 1988 the country adopted a new constitution that provided for a directly elected executive president in addition to elections for the two houses of Congress. The 1988 election was won by the conservative Fernando Collor de Mellor. He resigned in 1992, having been impeached for corruption, and Vice-President Itamar Franco took over.

The 1994 election was won by Fernando Henrique Cardoso who as Franco's finance minister had been the architect of the 'real plan'. The success of this also helped him to amend the constitution to be allowed to have a second term in 1988.

By the 2002 elections, however, the situation had deteriorated and the currency came under pressure. Cardoso's chosen successor José Serra proved a poor candidate to face up to Luis Inácio ('Lula') da Silva, the charismatic leader of the left-wing Workers Party (PT), who in a second round run-off against Serra gained 61% of the votes.

President da Silva was quick to reassure the international markets that, while promising radical social change, he would pursue fairly orthodox economic policies. He maintained the confidence of the business sector, appointed a fairly centrist cabinet, and remains popular among the public. In 2005 his popularity sank as the PT became embroiled in financial scandals, but thanks to a healthy economy and a new public welfare programme it now appears to have recovered and he seems likely to win the 2006 election.

Brunei

A high-income sultanate floating on oil, but subjected to autocratic rule

Land area: 6,000 sq. km.
Population: 0.4 million—urban 76%
Capital city: Bandar Seri Begawan, 27,000
People: Malay 67%, Chinese 15%, indigenous 6%, other 12%
Language: Malay, English, Chinese
Religion: Muslim 67%, Buddhism 13%, Christian 10%, indigenous and other 10%
Government: Constitutional sultanate
Life expectancy: 76 years
GDP per capita: $PPP 23,600
Currency: Bruneian dollar
Major exports: Oil, gas

Brunei occupies a small section of the north coast of Borneo, surrounded by the Malaysian state of Sarawak. A narrow coastal plain rises to low hills in its eastern section and to mountains in the west. Unusually for this part of Asia, forests still cover around 85% of the territory—thanks to its other main natural resource: oil.

Since 1929, oil and gas have been the main sources of wealth. Oil not only gives Brunei's citizens a high per capita income; it also funds extensive free, or highly subsidized, health and education services. The major health concerns are those of affluence: a high-fat diet and a sedentary lifestyle.

Most Brunei citizens are of Malay extraction, and tend to seek work either in the public sector, or in large prestigious companies such as Brunei Shell Petroleum. Most of the entrepreneurial energy comes from the less privileged Chinese minority, who are considered non-citizens, and also from temporary foreign workers who make up around 45% of the labour force. Officially the unemployment rate among Bruneians in 2003 was 5% but was probably nearer 10%.

Brunei's oil and gas reserves are expected to last 40 years, and, given the country's extensive foreign investments, should provide an income beyond that. Even so, oil and gas represent a shrinking proportion of GDP—down from 80% in the early 1980s to 50% in 2002, as a result of falling oil prices and an extension of government services. Moreover, there are doubts about the value of the holdings of the Brunei Investment Agency.

As a result the government has concluded that the economy is unsustainable and has proposed a series of measures to streamline expenditure, diversify the economy, and reduce dependence on foreign workers—though has yet to act on most of these proposals.

In theory, Brunei is a constitutional sultanate; in practice it is an absolute monarchy. When the country became independent in 1985, Sultan Hassanal Bokiah (whose family have ruled the country for 600 years) dissolved the existing legislative assembly and has since ruled by decree in a biennially renewed 'state of emergency'.

The sultan occupies numerous government posts, including prime minister, head of the police force, and leader of the Islamic faith. Political parties are proscribed, the press is strictly censored, and no public criticism is permitted. Fundamental change seems remote.

Bulgaria

Bulgaria is now politically more stable and is about to join the European Union

Land area: 111,000 sq. km.
Population: 8 million—urban 70%
Capital city: Sofia, 1.1 million
People: Bulgarian 84%, Turk 9%, Roma 5%, other 2%
Language: Bulgarian, Turkish
Religion: Bulgarian Orthodox 83%, Muslim 12%, other 5%
Government: Republic
Life expectancy: 72 years
GDP per capita: $PPP 7,731
Currency: Lev
Major exports: Base metals, chemicals, textiles, machinery, agricultural products

Bulgaria has three main natural regions, each of which extends across the country from west to east. The most northerly, starting from the banks of the Danube, which marks the border with Romania, is a plain with low hills that takes up around one-third of the country. To the south of this plain is the second main region, the Balkan Mountains. Further south still, separated from these mountains by the narrow Thracian Plain, is the third region, the Rhodope Mountains, which form the border with Greece.

Bulgaria has a fairly homogenous population and has so far avoided serious ethnic conflict. The largest minority are around 800,000 ethnic Turks, who live largely in the north-east or in the east of the Rhodope Mountains. The Turks tend to be worse off than most Bulgarians, and in the communist era suffered legal discrimination. They have their own political party but are not very assertive and many have been emigrating to Turkey. Even worse off are the Roma, or gypsies, who are economically marginalized and frequently the victims of heavy-handed policing.

Like many other countries of Eastern Europe, Bulgaria now has a shrinking population—a result of a falling birth rate, a higher death rate, and emigration, particularly of ethnic Turks, around 200,000 of whom left during the 1990s. From 1990 to 2004 the total population fell from 8.7 million to 7.8 million.

Public services have deteriorated since the collapse of communism. Education had been one of Bulgaria's strong points but spending has fallen and the quality has declined. Health standards too suffered: infant mortality rose, and although it has now fallen again is still fairly high.

During the communist era, Bulgaria's economy was heavily industrialized, but production fell steeply during the transition. The industries that held up better were basic metallurgy and chemicals, which along with textiles have been the major exports. By 2004 the GDP had returned to 92% of its level in 1989.

Part of the problem has been an erratic and often slow process of privatization. Even so, by the end of 2003 more than half of state enterprise assets had been sold. Foreign enterprises have tended to stay away—dissuaded by economic and political turmoil and a weak judiciary.

One-quarter of the population still make their living from agriculture, growing basic grains. Most output is now in private hands, including a number of western-style co-operatives. The late 1990s saw some good harvests and an expansion of cultivation of crops such as grapes for the production of wine—with most exports going to the UK.

A leading wine exporter to the UK

Unemployment remains a major problem, although the rate has come down; early in 2005 it was still 11%. Future hopes are pinned on the EU which Bulgaria is set to join in 2007.

Bulgaria's post-communist political development has often been conflictive. The former communists have regrouped as the Bulgarian Socialist Party (BSP), which has support among older people in the rural areas and among those with strong nationalist sentiments.

Until the last election their main rivals were the Union of Democratic Forces (UDF), a coalition of centre-right organizations that has greater support in the cities and among the younger and more educated. But neither of these groups has followed a clear political direction and both are prone to shifts and splits based more on personalities than policies. Other important parties include the party

for ethnic Turks, the Movement for Rights and Freedom (MRF), and the Bulgarian Business Bloc.

The BSP won the first post-communist election in 1990. But this government did not last and was followed by a number of short-lived administrations led either by the BSP or the UDF. In 1994, the BSP won an absolute majority but its poor performance again opened the way for the UDF, whose candidate, Petar Stoyanov, won the 1996 election for the ceremonial role of president.

Then in April 1997, an alliance led by the UDF won the parliamentary elections choosing Ivan Kostov as prime minister. Kostov brought more coherence to his own party and faced up to the conglomerates but was undermined by corruption and became increasingly unpopular. The 2001 presidential election was won by the BSP's leader, Georgi Purvanov.

Such was the disillusionment with the two main parties that both were swept aside in the June 2001 parliamentary elections by a coalition of the MRF and new right-wing party, the Simeon II National Movement (SNM) headed by Simeon Saxe-Coburg, the former king of Bulgaria, who became prime minister. He was successful in promoting stable economic growth, but less so in tackling social issues such as crime.

The 2005 election was complicated by the emergence of a new nationalist coalition, Ataka, hostile to the Turks and the Roma. The outcome was a coalition between the BSP, the SNM, and the MRF, with BSP leader Sergei Stanishev as prime minister. The coalition seems likely to hold, at least until successful accession to the EU.

Burkina Faso

Burkina Faso has few natural resources, a fragile environment, and a rapidly growing population

Land area: 274,000 sq. km.
Population: 12 million—urban 18%
Capital city: Ouagadougou, 1.1 million
People: Mossi over 40%, also Gurunsi, Senufo, Lobi, Bobo, Mande, Fulani
Language: French and local languages
Religion: Muslim 50%, indigenous beliefs 40%, Christian 10%
Government: Republic
Life expectancy: 48 years
GDP per capita: $PPP 1,174
Currency: CFA franc
Major exports: Cotton, livestock, gold

Burkina Faso extends over a broad plateau in the Sahel—the fringe of the Sahara Desert. Most of the territory is savannah grassland scattered with small bushes and trees. The climate is generally arid with a short rainy season and a long dry season. But the rains are unpredictable and the land is vulnerable to drought. The country does have major rivers, including the Red, White, and Black Voltas, though these frequently dry up.

Burkina Faso has numerous ethnic groups, of whom the largest, the Mossi, make up more than half the population. Most people live in the rural areas in the east and centre of the country. Half live below the poverty line. One-third of pre-school children are chronically malnourished and the population is largely uneducated: only around one-third of children enrol in primary school—one of the lowest proportions in the world. Health standards are undermined by endemic diseases like malaria and bilharzia, as well as by sleeping sickness transmitted by the tsetse fly, which makes land alongside the rivers virtually uninhabitable. Around 4% of people are infected with HIV/AIDS.

Most people make their living from agriculture and livestock—which provides around 90% of jobs and 30% of GDP. Farmers grow a number of subsistence food crops such as millet and sorghum. Food production has increased in recent years—though this was more the result of bringing additional land under cultivation than achieving increases in productivity.

The other main crop is cotton. Burkina Faso is one of the region's leading producers, and cotton brings in around 70% of export income. Other cash crops include shea nuts, sesame seeds, and sugar cane. Even so, only around 13% of the land is under cultivation: much of the rest of the territory, particularly in the north, is given over to livestock, largely for export to neighbouring countries such as Ghana and Côte d'Ivoire.

Burkina Faso also has mineral potential. At present, most extraction is of gold by small-scale panners.

But the government has attracted a number of foreign mining companies which are looking at the possibilities for zinc, manganese, limestone, phosphates, and diamonds.

The population continues to grow rapidly, by 3% annually, which is putting increasing pressure on the environment. Many of Burkina Faso's traditional systems for managing the fragile ecology are breaking down. Much of the thin topsoil has been lost and in the past 15 years or so a combination of drought, over-grazing,

The ecology is breaking down
brush fires, and unregulated felling has removed nearly 60% of the trees.

Over a similar period the water table has fallen by around 20 metres. Many parts of the country have already gone beyond what is considered to be a critical limit of 50 inhabitants per square kilometre.

Bringing the country's population and resources into balance will be an increasingly difficult task. Though economic growth and population growth have roughly kept pace, already millions of people migrate seasonally to neighbouring countries in search of work. However the recent conflict in Côte d'Ivoire has reduced these opportunities.

As one of the world's poorest countries, Burkina Faso has been a major recipient of foreign aid. Its high levels of poverty and its participation in an IMF-backed structural adjustment programme have also qualified it for debt cancellation. But donors are becoming increasingly concerned about the quality of governance, particularly repression and corruption.

Since independence from France in 1960, governments in Burkina Faso have rarely been free of military influence. One of the most distinctive periods was between 1983 and 1987, when a group of army officers led by Thomas Sankara seized power and embarked on a left-wing agenda determined to redistribute resources to peasant farmers and to free the country from dependence on foreign aid. He also changed the country's name from Upper Volta to Burkina Faso—which means 'land of the dignified'.

This radical experiment ended in 1987 with Sankara's assassination in a violent coup by Captain Blaise Compaoré, who in 1989 founded a new party which later merged with others to become the Congrès pour le démocratie et le progrès (CDP). In 1991, under pressure from donors, he introduced a new constitution permitting opposition parties.

The opposition boycotted the presidential elections in 1991 and again in 1998 allowing Compaoré to be re-elected with a large majority. In 1998, however, there were widespread protests over the murder of a newspaper editor, Norbert Zongo, which led to further constitutional reforms.

For the presidential election in 2005, 16 opposition parties formed a new coalition, Alternance. But they failed to unite around a single candidate and with the support of the official opposition, Alliance pour la démocratie et la fédération-Rassemblement démocratique africain (ADF-RDA), Compaoré was comfortably returned to office in a relatively free and fair election.

Burma (Myanmar)

Burma has suffered years of repression in the grip of a military regime that renamed it Myanmar

Land area: 677,000 sq. km.
Population: 50 million—urban 30%
Capital city: Pyinmana, 100,000
People: Burman 68%, Shan 9%, Karen 7%, Rakhine 4%, Chinese 3%, Mon 2%, Indian 2%, other 5%
Language: Burmese and otherss
Religion: Buddhist 89%, Christian 4%, Muslim 4%, other 3%
Government: Military dictatorship
Life expectancy: 60 years
GDP per capita: $PPP 1,700
Currency: Kyat
Major exports: Gas, food, timber, prawns

Burma consists largely of the central fertile valley of the Irrawaddy River, encircled by a horseshoe-shaped mountain system running north to south. The highest mountains are to the north, while to the west are two lower ranges, the Chin Hills and the Arakan range. To the east is the extensive Shan Plateau which consists of mountains that intersect with broken ranges of hills and river valleys.

Three-quarters of the population, mostly Burmans, live in the central valley and the coastal strips. Although sparsely settled, Burma has a complex ethnic mix, with 21 major groups and over 100 languages. The second largest group are the Shan, who live on the Shan plateau. The Karen live in the delta, the coastal areas to the south, and the hills bordering on Thailand, while the Rakhine live in the south-west. These groups have engaged in long struggles with the military government. Around 120,000 live in refugee camps in Thailand.

Education standards are low and more than one-third of children are malnourished. Intravenous drug use is growing and half a million people are now HIV-positive. Over one-third of public spending is used to finance the army.

Two-thirds of households make their living from agriculture, which accounts for more than half of GDP. But they are short of fertilizer and other inputs so productivity is low. Another source of rural income is forestry. Burma has around 75% of the world's teak reserves, but over-logging, often through government contracts with Thai companies, is rapidly stripping the forests. The most profitable crop is opium poppies, which enable Burma to supply around 60% of the world's heroin. Burma is also a major source of amphetamines.

Burma has made slow progress in industrial development. However, recent offshore discoveries have made natural gas the leading export earner. Foreign investment is limited. A number of Hong Kong and

Korean companies have established a garments industry, though in 2003 the US banned imports. Faced with consumer boycotts, Levi-Strauss, Reebok, and British Home Stores have pulled out and Texaco has withdrawn from oil and gas.

The present regime dates to a coup in 1988. Millions of people had taken to the streets to protest against military rule. The response was brutal. Soldiers sprayed bullets indiscriminately into the crowds and during this incident and the subsequent repression 10,000 people may have died. At this point,

Aung San Suu Kyi arrested
a new group of senior military officers seized power. They called themselves the State Law and Order Restoration Council (SLORC), and renamed the country Myanmar. In 1989, they placed under house arrest the leading opposition politician Aung San Suu Kyi, daughter of Aung San, a hero of Burma's independence struggle.

In 1990 SLORC, presuming wrongly that their grip on power would intimidate people into voting for military-backed parties, held multiparty elections for representatives to design a new constitution. In the event, 80% of the seats were won by Aung San Suu Kyi's party, the National League for Democracy. SLORC refused to accept the results. The assembly never met.

This marked the onset of one of the world's most repressive regimes. The UN Human Rights Commission has accused the government of torture, summary executions, and forced displacement and oppression of religious and ethnic minorities. SLORC also overturned the previous

regime's socialist model—opening the country up to foreign companies, particularly for timber extraction and the export of natural gas. To build the country's infrastructure it has resorted to forced labour, displaying what an ILO report called 'a total disregard for the human dignity, safety and health of the people'.

In late 1997, SLORC reformed as the State Peace and Development Council—with most of the same personnel but a less forbidding acronym.

Many ethnic groups have stubbornly resisted the regime. However, the army, largely staffed by Burmans, has steadily subdued them and most have signed ceasefires. Only the Karen National Liberation Army and two others are still holding out.

The most potent focus of non-violent opposition remains Aung San Suu Kyi, who in 1991 won the Nobel Peace Prize and has now had several periods of house arrest, renewed in 2006 for yet another year.

Most power rests with SPDC chairman General Than Shwe. He has supposedly established Burma on a 'roadmap' to democracy and reconvened the National Convention to draft a new constitution. But this is not allowed to challenge the SPDC in any way and meets only sporadically. Meanwhile the junta is preparing a front body, the Union Solidarity Development Association, to act as a potential political wing and has been moving the capital to a remote jungle location at Pyinmana.

With an incompetent and brutal government indifferent to international opinion, the outlook for the people of Burma is bleak.

Burundi

Burundi has been plagued by ethnic violence, though has avoided genocide on the scale of neighbouring Rwanda

Land area: 28,000 sq. km.
Population: 7 million—urban 10%
Capital city: Bujumbura, 140,000
People: Hutu 85%, Tutsi 14%, other 1%
Language: Kirundi, French, Swahili
Religion: Christian 67%, indigenous beliefs 32%, other 1%
Government: Republic
Life expectancy: 44 years
GDP per capita: $PPP 648
Currency: Burundi franc
Major exports: Coffee, tea, cotton

The west of Burundi lies along the Great Rift Valley, with Lake Tanganyika forming the southern two-thirds of the border. To the east the land rises first to high mountains and then descends across a hilly plateau to the border with Tanzania. Much of this area was originally forested, but most of the land has now been cleared for cultivation. This has resulted in extensive soil erosion.

Burundi's population remains sharply divided between two main ethnic groups. The majority are the Hutu, who are agriculturalists. For centuries they have been dominated by the minority Tutsi, most of whom raise cattle.

In recent decades, there have been a series of Hutu challenges to Tutsi domination, following which the security forces have taken revenge on the Hutu. Waves of killings occurred

in 1965, 1969, 1972, 1988, and 1991. In 1993 it was Hutu militias, however, who killed more than 100,000 Tutsi. And from 1994 onwards, a number of Hutu armed opposition groups fought the Tutsi-dominated armed forces. Half a million people, mainly Hutu, fled into Tanzania and the Democratic Republic of the Congo—and similar numbers, mainly Tutsi, were displaced within Burundi.

The civil war has largely brought development to a halt. Only around half the population are literate and the proportion is probably falling. Schools have frequently been a target of the militias and many teachers and pupils have been killed. Health standards too are falling. More than 40% of children are malnourished and health services have deteriorated as medical staff have been caught up in the fighting. Around 11% of people are HIV-positive.

More than 90% of people live in the rural areas, mostly growing subsistence crops such as cassava, sweet potatoes, bananas, beans, and maize. Rapid population growth in recent decades has intensified pressure on the land. Even so, the soil is relatively fertile and until the mid-1990s the country was mostly self-sufficient in food. Since then, however, Burundi has become reliant on international food aid. The main cash crops are coffee, tea, and cotton, which are also grown on peasant smallholdings.

Until independence in 1962, Belgium administered Burundi and neighbouring Rwanda as one country. In fact, they have always been fairly distinct territories, naturally divided by rivers, and with their people speaking different languages.

Burundi achieved independence as a monarchy but in 1966 the king was deposed by a Tutsi police commander who declared himself president and the country a republic. The decades that followed saw regular purges, massacres, and reprisals, one of which

Regular purges and massacres
in 1972 resulted in the death of at least 100,000 Hutu.

Another coup by another Tutsi in 1974 brought Colonel Jean-Baptiste Bagaza to power. He continued to exert Tutsi dominance but was overthrown in another coup by Major Pierre Buyoya. He was a more conciliatory figure and deliberately included Hutu leaders in the government.

In 1993 Burundi had its first free elections. These were won by a new party, the Front pour la démocratie au Burundi (Frodebu). Although ethnically mixed, it had a large Hutu following. A Hutu president, Melchior Ndadaye, took office with a Tutsi prime minister.

National unity was short-lived. Attempts to promote more Hutu within the civil service alarmed the military, who within five months staged another coup—killing Ndadaye and other ministers. An estimated 100,000 people died in the ensuing fighting and up to 700,000 refugees fled to Tanzania.

Attempts to re-establish a legitimate Frodebu government were weakened when the interim president was killed in a plane crash.

Following another coup in 1996, Buyoya returned, though soon entered into a power-sharing agreement with Frodebu. From 1998 to 2001 Burundi was also involved in the war in neighbouring Congo.

External mediation started in 1998, steered initially by former Tanzanian President Julius Nyerere then by Nelson Mandela resulting in a new constitution and a transitional administration.

According to the new constitution 60% of seats in the National Assembly and the executive must be filled by Hutus, while the two groups share the most senior posts in the army and the Senate.

Elections in 2005 resulted in a victory for Pierre Nkurunzizawgi, who had a Hutu father and a Tutsi mother, and his party the Conseil national pour la défense de la démocratie–Forces pour la défense de la démocratie (CNDD–FDD). Although the CNDD–FDD was formerly a Hutu militia it has successfully drawn in some Tutsis.

Nkurunzizawgi may struggle to keep his party together and fend off disgruntled Tutsis in the army. But the main problem is that not all the rebel groups accepted the peace agreement. The main opposition comes from the Parti pour la libération du peuple Hutu–Forces nationales pour la libération (Palipehutu–FNL), led by Agathon Rwasa. The negotiations, and the fighting, continue.

Cambodia

Decades of war and political upheaval tore Cambodia apart. Democracy has been established but is fragile

Land area: 181,000 sq. km.
Population: 14 million—urban 19%
Capital city: Phnom Penh, 1.0 million
People: Khmer 90%, Vietnamese 5%, Chinese 1%, other 4%
Language: Khmer, French
Religion: Theravada Buddhism
Government: Constitutional monarchy
Life expectancy: 56 years
GDP per capita: $PPP 2,078
Currency: Riel
Major exports: Logs, timber, rubber, manufactured goods

Three-quarters of Cambodia consists of a large central plain, through which passes the Mekong River on its way from Laos in the north-east to Vietnam in the south. The plain also includes a large lake, the Tonle Sap, which drains into the Mekong—though during the rainy season, while the Mekong is flooding it also sends water back into the lake. Cambodia has a number of mountainous regions—including the Dangrek Mountains on the northern border with Thailand, as well as ranges to the east and south-west.

Cambodians are mostly Khmers, but there are also significant Vietnamese and Chinese minorities. The vast majority live and work in the rural areas. This is one of the poorest countries in Asia. Standards of literacy and health are low; only one-quarter of the population has access to safe drinking water. The

health system is weak, offering care only to about half the population. Around half of Cambodia's children are malnourished, more than 40% are working, and many girls are trafficked into the sex trade to Thailand and elsewhere. Cambodia has however been successfully addressing HIV/AIDS: adult infection rates are now down to 2.6%.

Three-quarters of the labour force work in agriculture and fishing. The main crop is rice but yields are relatively low. This is partly because production is generally on a very small scale and many farmers lack irrigation systems. Relying on seasonal rains, they can only get one crop per year. Even so, Cambodia is self-sufficient in rice and can even produce a surplus. The government is keen for farmers to diversify into other crops but they are hampered by weak infrastructure and underdeveloped marketing systems. The most important cash crop is rubber which is produced on government-owned plantations.

Forestry is another source of export income. But logging companies are cutting down large numbers of teak trees, often illegally, and deforesting vast areas. Over the past 30 years the proportion of land covered by forests has fallen from 70% to 30%.

Industrial development has been slow. One of the main growth areas has been in garment manufacturing. By 2004 there were around 200 factories which employed 150,000 workers. Participation in a labour inspection programme has made exports more acceptable to foreign buyers. Tourism is another important source of foreign exchange. Around 1.4 million visitors arrived in 2005, many heading for the extraordinary complex of temples at Ankor Wat.

Cambodia's recent political history has been both tragic and complex. One constant figure was Norodom Sihanouk. Between 1941 and 1970, he ruled either as king or prime minister, attempting to steer a neutral course

The constant figure of Sihanouk

between right and left. In 1970, he was deposed by more conservative forces in a US-backed coup, though this republic only lasted until 1975 when it fell to the communist Khmer Rouge.

Headed by Pol Pot, the Khmer Rouge unleashed a regime of extraordinary ruthlessness that killed two million people. Sihanouk returned—to house arrest.

But the Khmer Rouge made the mistake of antagonizing neighbouring Vietnam. In 1979, the Vietnamese responded by invading Cambodia and installing a new regime, forcing the Khmer Rouge back underground. In 1989 the Vietnamese finally withdrew, and in 1991 a UN-brokered peace process led to elections in 1993.

The outcome was evenly balanced between a royalist party, Funcipec, led by Sihanouk's son Prince Ranariddh,

and the Cambodian People's Party (CPP), led by a political strongman, Hun Sen, a former Khmer Rouge defector who had been prime minister in the Vietnamese-installed regime. In the same year, a new constitution established a constitutional monarchy with Sihanouk once again as king.

Hun Sen worked with Prince Ranariddh as joint prime minister until 1997 when he ousted him in a violent coup. Ranariddh fled, but returned for new elections in 1998. Again, the CPP came out ahead, but without the two-thirds majority required to rule outright. Funcipec came second, and the 'self-named' party of Sam Rainsy came third. Ranariddh subsequently agreed to join Hun Sen's uneasy coalition.

The elections in July 2003 were similar to those in 1998—violent with no clear outcome—though the Sam Rainsy party strengthened its position. The CPP won 73 seats, Funcipec 26, and Sam Rainsy 24. This time there was an inordinate delay in forming a coalition. In July 2004 Hun Sen and Prince Ranariddh finally agreed to share power with the former as prime minister and the latter as speaker of the national assembly. In October 2004, Sihanouk abdicated and was succeeded by another of his sons, Norodom Sihamoni.

Another long-running issue is that of a tribunal to try Khmer Rouge leaders for genocide. The government and the UN in 2003 agreed on the arrangements. The prosecution will be directed jointly by the UN and Cambodia. Some preparations have been made but trials have yet to start, and will be too late for Pol Pot who died in 1998, unrepentant to the last.

Cameroon

Potentially one of Africa's richer countries, Cameroon has squandered its oil wealth

Land area: *475,000 sq. km.*
Population: *16 million—urban 51%*
Capital city: *Yaoundé, 1.4 million*
People: *Fang 20%, Bamiléké 20%, Duala 15%, Kirdi 11%, Fulani 10%, and numerous other groups*
Language: *24 African language groups, English, French*
Religion: *Indigenous beliefs 40%, Christian 40%, Muslim 20%*
Government: *Republic*
Life expectancy: *46 years*
GDP per capita: *$PPP 2,118*
Currency: *CFA franc*
Major exports: *Oil, timber, cocoa, cotton*

Cameroon can be divided into four main regions. First, there is a coastal belt with mangrove forests and swamps that stretch up to 60 kilometres inland. These give way to the east to rocky plateaux. Much of the far north is a broad savannah plain with occasional hills leading to the shores of Lake Chad. The highest part of the country is along the border with Nigeria, a region that includes Mount Cameroon, the highest point in West Africa.

Cameroon has a diverse population, fractured along lines of ethnicity and language. There are thought to be around 200 ethnic groups. In the west the largest include the Bamiléké, who are one of the main commercial influences in the larger cities. In the north there are the Fulani and the Kirdi. And in the south there are the politically more powerful Fang and Beti. These differences open up a number of potential divisions: between the Islamic north and the Christian south, for example, and between pastoralists and farmers. But the most significant political differences arise from the colonial experience. The east of Cameroon was colonized by the French and the west by the British. Today the country is around 80% Francophone and 20% Anglophone—with the latter tending to be marginalized.

Although they had been doing well by the standards of Sub-Saharan Africa, most Cameroonians have seen a fall in living standards since the mid-1980s and around 40% live below the poverty line. Primary school enrolment increased to around 95% after school fees were abolished in 2000, but many children drop out. Health has been hit by reductions in public expenditure and around 7% of adults are HIV-positive. Only half the population have access to clean water and sanitation.

More than half of Cameroonians still rely directly or indirectly on agriculture. The land is fertile and the country is usually self-sufficient in basic food crops like cassava, corn, millet, and plantains. The most important cash crops are cocoa, coffee, bananas, and cotton, which are mostly grown on smallholdings, though they are marketed through state corporations. In the south, there are also plantations growing palm oil and rubber. The country's extensive grasslands also offer grazing for livestock, which provide meat both for local consumption and export.

Cameroon's dense forests in the centre and south, as well as in the coastal belt, have also been an important source of income and the largest source of foreign exchange after oil. Around one-third of the

Rainforests threatened by logging and the oil pipeline
forest area has been exploited, chiefly for the export of mahogany, teak, and ebony. This activity is largely in the hands of multinational companies, but there have been increasing concerns about the over-exploitation of the forests as well as about the effects on the country's oldest inhabitants, the pygmies—hunter-gatherers who live in the southern forests.

Economic prospects brightened with the discovery of offshore oil in 1976. Production started in 1978 and peaked in 1985. Since then, some of the better fields have matured and companies are having to exploit more marginal deposits. Oil flows provided a sudden injection of wealth into the economy, but much of this was dissipated in unwise public expenditure, and through corruption. When the oil price crashed in the mid-1980s, Cameroon crashed with it, suffering one of Sub-Saharan Africa's steepest economic declines. Over the

same period, Cameroon also suffered from the overvaluation of the CFA franc. The 50% devaluation in 1994 caused a surge in inflation, though the economy subsequently recovered and growth has been 4% to 5% per year.

One of the most contentious development projects has been the 1,100-kilometre oil pipeline from Chad through Cameroon's rainforests to the Atlantic coast. This project, completed in July 2003, is likely to result in major ecological damage.

Since 1982, political power has been in the hands of President Paul Biya. He is from the southern Beti group, but has adroitly manipulated the country's various ethnic divides. Initially, he headed a one-party state, but a wave of social agitation in 1990 forced Biya to permit multiparty politics. In a rigged 1992 election Biya was elected president at the head of his own party, the Rassemblement démocratique du peuple Camerounais (RDPC)—narrowly defeating John Fru Ndi of the main Anglophone opposition party, the Social Democratic Front (SDF).

Biya won again in 1997, but this time all three opposition parties boycotted the election. The SDF, whose power base is among the Bamiléké, wants constitutional reform, and its more radical elements demand secession for the English-speaking provinces. For the December 2004 elections the SDF and the other main opposition party, Union démocratique du Cameroun (UDC), ran a joint candidate but he made little impression and Biya was returned for a fifth term with 71% of the vote. The opposition claimed that they were victims of intimidation and fraud.

Canada

An increasingly multicultural population largely free from racial tension

Land area: 9,976,000 sq. km.
Population: 32 million—urban 80%
Capital city: Ottawa, 1.1 million
People: British origin 28%, French origin 23%, other European 15%, Amerindian 2%, other 32%
Language: English, French
Religion: Roman Catholic 43%, Protestant 23%, other 34%
Government: Constitutional monarchy
Life expectancy: 80 years
GDP per capita: $PPP 30,677
Currency: Canadian dollar
Major exports: Newsprint, pulp, timber, oil, machinery, vehicles, gas

Canada is, after Russia, the world's second largest country, though much is barren and sparsely populated. The more remote areas include the icy wastes of the Arctic archipelago in the north, the splendour of the Rocky Mountains in the west, and the stormy Newfoundland coast in the east. The largest region, around Hudson Bay, is the flat, rocky Canadian Shield which is studded with thousands of lakes. To the west are the lowlands of the interior plains, and to the south-east the Great Lakes-St Lawrence region that borders on the USA. Around 85% of Canadians live within 350 kilometres of the border with the US.

Canada's original population, the 'first nations', now make up less than 2% of the population, and are among the poorest people. Many are now claiming land and other rights from provincial governments. In 1997, the Supreme Court ruled that the government has 'a moral, if not legal, duty' to settle their claims.

Everyone else is of immigrant descent, and Canada remains a country of immigration. In 2001, 18% of the population were foreign-born. Immigration is strictly limited to around 250,000 per year, but the pattern of arrivals altered dramatically from the 1960s, following changes that removed the privileges of European immigrants. Now more than half of immigrants come from Asia, many of them of Chinese origin who settled on the Pacific coast. Despite this influx, Canada remains relatively free of racial tension.

The wealth of natural resources and a dynamic immigrant population have made Canada one of the world's richest countries. It also has strong health and education services. Nevertheless there have been some economic problems in recent years, notably unemployment, though by mid-2006 this had dropped to 6%.

Three-quarters of Canadians are employed in service industries, with the most dynamic growth in larger cities like Toronto and Vancouver. Manufacturing, particularly in high-technology goods, has also expanded, boosted by exports to the US.

Mining now employs only around 1% of the workforce but still makes a vital contribution to the economy. Canada is the world's largest producer of uranium and potash and is an important source of many other metals, including nickel, zinc, platinum, copper, and titanium. It also has, 95% in the form of 'tar sands' in Alberta, the world's second largest reserves of oil and is the third largest producer of natural gas.

The Canadian prairies produce vast quantities of wheat, oats, barley, and many other crops, as well as livestock. But agriculture employs only 2% of the workforce. Canada's forests also enable it to be the world's leading exporter of wood pulp and newsprint. And its rich fishing grounds on both Pacific and Atlantic coasts also make it a leading fish exporter though the industry has suffered from over-fishing.

Primary products remain important sources of export earnings, although they are now outpaced by manufactured exports. Around 80% of exports go to the USA, and 70% of imports come from the USA—trade which was boosted by the 1994 establishment with the USA and Mexico of the North American Free Trade Agreement (NAFTA). There have, however, often been strained trade relations with the USA, notably over exports of timber.

Trade is mainly with the USA

Canada's head of state is the British monarch—a constitutional anomaly that seems unlikely to change since Canada has a strong federal system with semi-autonomous provincial governments—each of which would have a right of veto on constitutional change. In 2005, a former refugee from Haiti, Michaëlle Jean, was appointed governor-general.

For many years, Canadian politics at the federal level was the domain of the centre-left Liberal Party or the centre-right Progressive Conservative Party. The smaller left-wing New Democratic Party has traditionally been stronger in some provincial parliaments. But the 1990s saw dramatic changes. These included the collapse in 1993 of the Conservative Party, and the emergence of the right-wing Canadian Alliance.

The parliament also has a strong federal representation for the Bloc Québécois which wants independence for the French-speaking province of Quebec. The province rejected this in a 1995 referendum, but in polls in 2005 more than half of respondents said they would vote for sovereignty.

The 1997 and 2000 federal elections resulted in overall majorities for the Liberal Party led by Prime Minister Jean Chrétien. In December 2003 he resigned making way for Paul Martin. Then the Canadian Alliance and the Progressive Conservatives merged into a single Conservative Party led by Stephen Harper. The June 2004 election was one of the most closely fought for decades and the Liberals only just scraped through to form a minority government.

In 2006, however, a further election allowed the Conservatives and Harper to gain 125 of the 308 seats in parliament and form a minority administration—which has been cutting taxes and trying to repair relations with the USA.

Cape Verde

One of Africa's smallest countries—heavily dependent on emigrants' remittances

Cape Verde is an archipelago of ten islands off the coast of West Africa. Much of the territory is mountainous with little arable land. Combined with frequent droughts, this has made agriculture difficult, and has resulted in serious environmental degradation.

Two-thirds of Cape Verdeans are creole, of mixed Portuguese and African descent, and half live on the island of Santiago. Because of the harshness of the terrain and the resultant poverty, they have had a strong tradition of emigration. Today at least half a million Cape Verdeans live abroad, either in the USA or Europe.

Aid-financed public investment in education and health services in the decades following independence in 1975 helped to give Cape Verde one of Sub-Saharan Africa's higher levels of human development.

However this also made the country very dependent on aid—which in 2004 amounted to $280 per capita and 14% of GDP. The country has also benefited from emigrants' remittances which in 2003 were equivalent to 10% of GDP.

Agriculture still employs more than half the workforce but only produces around 15% of the country's food needs. And as it does not offer much prospect of future employment the

> **Land area:** 4,000 sq. km.
> **Population:** 0.5 million—urban 56%
> **Capital city:** Praia, 106,000
> **People:** Creole 71%, African 28%, other 1%
> **Language:** Portuguese, Crioulo
> **Religion:** Roman Catholicism and indigenous beliefs
> **Government:** Republic
> **Life expectancy:** 70 years
> **GDP per capita:** $PPP 5,000
> **Currency:** Cape Verde escudo
> **Major exports:** Footwear, garments, fish

government has been liberalizing the economy in an attempt to attract foreign investment for light industry, such as textiles, shoe making and food processing. This may help, but seems unlikely to be sufficient to absorb the unemployed, who in 2001 were 19% of the workforce.

Tourism has also become more important and now accounts for around 11% of GDP with around 185,000 visitors a year.

Cape Verde, which was in union with Guinea-Bissau until 1980, was one of the first African one-party states to democratize. The first multiparty elections in 1991 saw a defeat for the long-ruling socialist Partido Africano da Independência de Cabo Verde (PAICV) and victory for the more market-oriented Movimento para a Democracia (MPD)—a victory repeated in 1995.

From 1999 however, the MPD became caught up in faction fighting. In 2001 the PAICV took advantage of this to win both legislative and presidential elections. The strong performance of both Prime Minister José Maria Neves and President Pedro Pires gave them another convincing victory in 2006—though the MPD contested the results.

Central African Republic

This coup-stricken country is once again in military hands

Land area: 623,000 sq. km.
Population: 4 million—urban 43%
Capital city: Bangui, 687,000
People: Baya 33%, Banda 27%, Sara 10%, Mandjia 13%, Mboum 7%, M'baka 4%, other 6%
Language: French, Sangho, Arabic, Hunsa, Swahili
Religion: Indigenous beliefs 35%, Protestant 25%, Roman Catholic 25%, Muslim 15%
Government: Republic
Life expectancy: 39 years
GDP per capita: $PPP 1,089
Currency: CFA franc
Major exports: Diamonds, timber, cotton

The Central African Republic is a landlocked country that sits on a plateau around 700 metres above sea level. The north of the country is savannah with some extensive grasslands. The south has more luxuriant vegetation, particularly along the river valleys and in the dense tropical rainforests.

The country is sparsely populated but ethnically complex. The main division is between savannah dwellers such as the Sara and Mandjia, and the more dominant 'riverines': those such as the M'baka who live in the south along the Ubangi River and who had greater contact with the French colonists. There are many languages, though the national lingua franca is Sangho, and the official language in the education system is French.

The most important transport artery is the Ubangi River along the southern border with the Democratic Republic of Congo.

Years of unrest and ethnic strife have taken their toll on human development. The education system is weak and has been further disrupted by fighting; only around 48% of the population are literate. Health standards too are low: in the rural areas where more of the people live few have access to safe water or sanitation, and around one-quarter of all children are malnourished. Only half the population have access to basic health care and 14% are HIV-positive. Two-thirds of the population live below the poverty line.

Two-thirds of the population make their living through subsistence agriculture, growing such crops as cassava, peanuts, corn, and millet. They also grow cash crops, including coffee and cotton, but the poor state of communications means that many households effectively live outside the cash economy. Although food production has increased in recent years, and has not been too disrupted by the fighting, the country is still not self-sufficient, and many people have relied on distribution of food aid.

One-fifth of the country is covered by tropical rainforests and there are more than 50 species of commercially viable trees. Transport difficulties have hampered the development of forestry but logging has now made timber the most important export earner. The government has attempted to conserve the rainforest by reducing the proportion of wood exported as timber and encouraging the production of veneers and plywood.

The Central African Republic also has a number of mineral resources, including iron ore, uranium, and gold, but these have not been exploited because of low prices and high transport costs.

More important are alluvial diamonds. Found in the west of the country, these are often mined by individuals and co-operatives and offer a significant source of rural employment. Diamonds are also the second largest export earner, though at least two-thirds of the output is smuggled out of the country. Official export and cutting operations are in the hands of a joint venture between the state and South Korean and Belgian companies.

The Central African Republic achieved independence in 1960, but from 1965 the country's development **One of Africa's** was overshadowed **most notorious** by the bizarre, **dictators** tyrannical rule of Jean-Bedel Bokassa, who in 1977 in an extravagant ceremony had himself crowned Emperor Bokassa I. France sustained him in power but in 1979 finally lost patience and sent in troops to remove him. However the chosen successor was himself removed (again with French support) in a coup in 1981 by General André Kolingba.

Kolingba offered a degree of stability and put more emphasis on economic development. He also attempted to legitimize his rule by creating a single political party, the Rassemblement démocratique centrafricain (RDC). But in the face of popular and international pressure in 1992, he had to legalize other parties.

In 1993 Kolingba lost the ensuing presidential election to Ange-Félix Patassé, who had been one of Bokassa's prime ministers. His party, the Mouvement pour la libération du peuple centrafricain (MLPC), gets most of its support from the north-west of the country, where Patassé himself comes from. Kolingba's RDC has its support in the south.

Patassé's government soon proved inept and corrupt and in 1996 the army staged a mutiny, demanding back pay for themselves and other government workers. The French army, which maintained a garrison in the country, came to his rescue.

A further coup attempt in May 2000 was also seen off, this time with the help of troops from Libya. However the rebellion continued with the sacked commander of the armed forces, General Bozizé, launching attacks from bases in southern Chad.

In March 2003, Bozizé finally took the capital and seized power, dismissing the National Assembly. Bozizé promised multiparty elections and said he would not be a candidate. But to nobody's surprise in March 2005 he did run again. He failed to win outright but won a subsequent run-off in May against Kolingba, possibly because voters suspected that if defeated he would simply launch another coup.

Chad

Chad has had decades of civil war, but now seems more stable, and its future will be transformed by oil

Land area: *1,284,000 sq. km.*
Population: *9 million—urban 25%*
Capital city: *Ndjamena, 737,000*
People: *Numerous Muslim and non-Muslim groups*
Language: *French, Arabic, many others*
Religion: *Muslim 51%, Christian 35%, animist 7%, other 7%*
Government: *Republic*
Life expectancy: *44 years*
GDP per capita: *$PPP 1,210*
Currency: *CFA franc*
Major exports: *Cotton, livestock, textiles*

Chad's territory forms a basin centred on Lake Chad on the western border. The terrain rises steadily, eventually reaching mountains in the north and east between which, in the north-east, lies a sandstone plateau. The climate varies from hot and dry in the north to tropical in the south, where the country's small amount of arable land is to be found.

Chad is ethnically very diverse, with more than 200 distinct groups. Conventionally, these are divided into the Islamic north, where most people raise cattle and the common language is Arabic, and the Christian or animist south, where people are farmers and the common language of the élite is French. The dividing line is taken to be the Shari River. Though the two parts have roughly the same population, the south represents less than one-tenth of the territory. This north–south divide is something of a simplification, since each part has numerous internal divisions, but generally the north has been more traditional, and the south, which was more effectively colonized by the French, is economically more modern.

Throughout the country, poverty is severe and widespread. Infant mortality is high, life expectancy is low, and the literacy rate is only 26%. Over the past 30 years up to 400,000 people have died as a consequence of warfare.

Until very recently Chad's economy has seen little progress since independence, hampered by poor infrastructure and endemic political conflict. Around 25% of GDP is generated by agriculture, which in 2004 employed 72% of the workforce. Sedentary agriculture is almost entirely in the south, on the lake shore, and along the river valleys.

In a good year, production of rice, sorghum, millet, wheat, and other foodstuffs is sufficient for local consumption, but output has frequently been constrained by drought and civil conflict. Cash crops include sugar and peanuts, but the most important, and the major export earner, is cotton—which directly or indirectly is thought to employ about one million people. In the centre and north of the country people rely for income on their livestock—around five million cattle and six million sheep and goats. Most of Chad's industry has also involved processing agricultural products—ginning cotton and producing meat and hides.

But things are about to change radically now that oil production is underway. Following oil discoveries in 1994, ExxonMobil, Petronas, and Chevron have developed oilfields in Doba in the south and built a 1,100-kilometre pipeline to the port of Kribi in neighbouring Cameroon. The first oil was pumped in July 2003.

Oil income is expected to boost Chad's budget by 50% over the next 25 years. How will this windfall be used? Chad, under pressure from the World Bank, which helped finance the pipeline, has at least set up an independent watchdog, the Revenue Oversight Committee, to scrutinize the use of the oil income and to try to increase transparency. Local people will, however, see very little extra direct income since the plant will employ only around 1,000 people. Those who have gained the most so far seem to be traders and others catering to expatriates.

Watchdog scrutinizes oil revenues

The oil companies will also be hoping that Chad's multiple conflicts die down. These became evident soon after independence in 1960, and have usually involved direct or indirect participation by neighbouring countries. The first, one-party, regime, supported by the French, came to be seen as a southern dictatorship. This prompted the emergence in the north of a guerrilla movement, the Front de libération du Tchad (Frolinat).

Following a military coup in 1975, the new regime, although again composed mostly of southerners, tried to make peace with Frolinat. This failed, not least because the rebellion was being funded by Libya. In 1982, however, Hissène Habré, a northern leader, seized power, backed by the USA, Sudan, and Egypt, but opposed by Libya. This provoked a rebellion in the south and warfare that was to claim 40,000 victims.

In 1990, Habré's former army head, Idriss Déby, launched the Mouvement patriotique du salut (MPS). With Libyan support, this rapidly ousted Habré and installed Déby as president.

So far Déby seems to have brought a measure of democracy and stability—by Chad's standards. In 1996, a national referendum approved a new constitution that introduced multiparty politics. In the ensuing elections, President Déby won with 67% of the vote, and the MPS obtained a majority of the seats in the National Assembly. He was re-elected in 2001 and the MPS also gained a large majority in the 2002 legislative election. In 2005 Déby introduced a constitutional amendment to remove the two-term limit on the presidency.

He faces fierce opposition both within his own ethnic group and from armed rebels in the north, the Front uni pour le changement, which staged a large-scale attack on Ndjamena in March 2006. With French support, however, Déby managed to repel them and in May was duly re-elected for a third term.

Channel Islands

Dependencies of the British crown—constitutionally odd and financially rich

Land area: 300 sq. km.
Population: 156,000—urban 29%
Main towns: St Helier, St Peter Port
People: British and Norman French
Language: English, French
Religion: Christian
Government: Dependencies of the British crown
Life expectancy: 79 years
GDP per capita: $PPP 40,000
Currency: Jersey pound, Guernsey pound
Major exports: Vegetables, fruit, flowers

The Channel Islands consist of four main islands and numerous other islets. They are governed as two separate 'bailiwicks' based on the two largest islands: Jersey, with 58% of the population; and Guernsey, with 42%, which also covers the two smaller islands of Alderney and Sark.

Natives of the islands generally speak either English or French, though there are vestiges of a Norman-French dialect. There are also large numbers of settlers: in Jersey only half the resident population were born on the island; most of the rest come from elsewhere in the British Isles, often as tax exiles, and there are also many immigrant workers from Portugal.

The islands have traditionally exported horticultural produce to the UK—fruit, flowers, tomatoes, and potatoes. But agriculture now accounts for less than 5% of GDP. Tourism is more significant, accounting for 24% of GDP in Jersey and 14% in Guernsey. Around 80% of visitors come from the UK, attracted by the mild climate and pleasant scenery, though most come only for short breaks.

More recently, the islands' economies have become dominated by financial services—60% of GDP in Jersey, and 55% in Guernsey. With little or no corporation tax, they became very attractive to British companies in 1979 after the abolition of exchange controls. By 1998, the two had between them around $400 billion in bank deposits. The business is fairly clean, though a British government report concluded that 'the extent of disreputable business is hard to judge' and the scale of tax evasion and fraud 'unusually hazardous' to assess. The report also discovered a strangely high level of entrepreneurial activity on Sark, whose 575 inhabitants between them held 15,000 company directorships.

These opportunities are created by the islands' unusual constitutional position, with respect to both the UK and the EU. They are neither sovereign states nor part of the UK, but remnants of the duchy of Normandy, a legacy of William the Conqueror, and come under the jurisdiction of the British monarch rather than of the state. They have autonomy in domestic policy, including fiscal policy, and Jersey issues its own coins and notes. The monarch appoints a lieutenant-governor to each of the bailiwicks. Each also has an appointed bailiff who is president of the local Assembly of the States, which has both elected and appointed members.

Chile

Chile has set aside the Pinochet years, and now has a stable democracy and a successful economy

Land area: 757,000 sq. km.
Population: 16 million—urban 87%
Capital city: Santiago, 6 million
People: European and mestizo 95%, Indian 3%, other 2%
Language: Spanish
Religion: Roman Catholic 89%, Protestant 11%
Government: Republic
Life expectancy: 78 years
GDP per capita: $PPP 10,274
Currency: Peso
Major exports: Copper, fruit, cellulose, fishmeal

Chile is remarkably narrow: 4,329 kilometres long and on average no more than 180 kilometres wide. From north to south run the rolling hills of the coastal highlands in parallel with the massive ranges of the Andes to the east. Between them lies a central valley that is most evident in the middle third of the country. Chile's length also offers striking climatic variations, from the heat of the Atacama Desert in the north, through a temperate centre, then a cool, wet south extending through the lakes and fjords and on to the stormy Straits of Magellan.

Most of Chile's people are concentrated in the central valley and around the middle of the country. Around 40% live in the capital, Santiago. Chileans are almost entirely of mestizo or of European descent. The small Mapuche Indian population is concentrated in an area 700 kilometres south of Santiago.

Chile has at times viewed itself as more European than South American, and has higher standards of education and health than its neighbours. All wage workers contribute to a health insurance system. However the country is steadily becoming more unequal, with the richest 20% getting 62% of the income and the poorest 20% only 3%.

Although the economy is now more broadly based it remains very dependent on natural resources. One of the economic mainstays is mining in the northern deserts. Chile is the world's largest producer of copper and has one-third of global reserves. The country is also rich in other minerals, including gold, molybdenum, silver, and iron; and it can count on substantial reserves of oil and gas.

Chile's climatic diversity also permits a diverse agricultural output. Agriculture and fisheries employ around 12% of the workforce. Chile is the world's second largest fish producer. The climate in the central zone is ideal for multinational companies to grow apples, pears, and grapes for export. Although the industry is profitable, the workforce is often poorly paid. Also doing less well are the 200,000 wheat-growing campesino farmers who in a very liberal trade regime are suffering from cheaper imports. Land ownership is also becoming ever more concentrated into larger farms.

Chilean manufacturing has been diverse. In the past, it was designed for import substitution. But Chile's open economy now means that it tends to operate in areas where it has the greatest competitive advantage. Thus, leading manufactured exports are usually closely linked to agriculture, including cellulose and fishmeal. Chile is also now the world's fifth largest wine exporter.

Chile's economic and political history was transformed in 1973. Previously, the country had enjoyed a long sequence of democratic governments, but this pattern was shattered by a CIA-supported military coup which ousted socialist President Salvador Allende and ushered in the era of General Augusto Pinochet. For the next 17 years, Chile became a byword for torture, repression, and the abuse of human rights: within three years around 130,000 people had been arrested. Ultimately 3,197 people were to die for political reasons, including 1,102 who 'disappeared'.

Pinochet made Chile a byword for torture and repression

Pinochet's political ferocity had been combined with radical economic liberalism—and subsequent governments followed the same economic path. But the 'Chilean model' is actually a mixture of policies. It does espouse open markets, low trade tariffs, deregulation, and privatization, including a privatized pension scheme. But it also has had significant elements of regulation—including controls on capital flows and some indexing of wages and prices.

Democracy was restored only in 1990, and a government was formed by the Concertación de Partidos por la Democracia—a 13-party coalition that included parties from moderate conservative to the far left. Concertación won again in 1993 led by Christian Democrat Eduardo Frei. For most of the 1990s the right-wing opposition had been hampered by its association with Pinochet.

For the 2000 presidential election, however, it seemed that voters were more interested in issues of employment. In a close result, the Concertación candidate, Ricardo Lagos, won a six-year term, defeating Joaquín Lavín of the right-wing coalition, the Alianza por Chile.

Lagos appeared successfully to have balanced the different wings of Concertación, achieved strong economic growth, and worked with the opposition on constitutional reforms, which included reducing the presidential term to four years.

For the 2006 presidential election the Concertación candidate was former health minister Michelle Bachelet of the Partido Socialista who became Chile's first woman president. Although on the left wing of her own party she appears to be taking a pragmatic approach while vowing to address rising inequality.

The issue of what do about Pinochet lingers on. In 2002 the Supreme Court granted him immunity but in 2005 reversed that decision.

China

China is reforming into an economic superpower, but the Communist Party retains its political grip

Land area: 9,561,000 sq. km.
Population: 1,300 million—urban 39%
Capital city: Beijing, 8.0 million
People: Han Chinese 92%, Zhuang, Uygur, Hui, Yi, Tibetan, Miao, Manchu, Mongol, Buyi, Korean, and other nationalities 8%
Language: Mandarin, Yue (Cantonese), Wu (Shanghaiese), Minbei (Fuzhou), Minnan (Hokkien-Taiwanese), Xiang, Gan, Hakka dialects, minority languages
Religion: Taoism, Buddhism, Muslim, Christian
Government: Communist state
Life expectancy: 72 years
GDP per capita: $PPP 5,003
Currency: Renminbi (yuan)
Major exports: Electrical machinery, clothing, footwear, coal

China's vast landmass incorporates immense topographical and climatic diversity. But it can be divided into three broad areas. First, there is the north-western region, which is predominantly mountainous, including the Tien Shan ranges that average around 4,000 metres—though this area also includes the Tarim and Dzungarian basins and the Takla Makan Desert. A second main region lies to the south-west and includes the Tibetan Plateau and other highlands. The remainder of the country, extending east to the coast, has some mountains but is mostly low-lying and includes extensive river systems flowing west to east: the Huang Ho (Yellow River) to the north and the Yangtze to the south.

This huge territory also has considerable climatic variation—colder and drier to the north and on the mountains and steppes in the interior, and warmer and wetter to the south and east.

Most people live in the eastern part of the country and are ethnically fairly homogenous—more than 90% are Han Chinese. Nevertheless the Han differ regionally, particularly in language. The standardized common language is based on the Mandarin dialect and is used in government as well as in schools and universities, but there are also a number of Han dialects that share most of the same characters though are mutually incomprehensible.

In addition to the Han there are around 75 million Chinese who belong to any of 55 or more different ethnic groups that are settled across more than half the territory. These include the Zhuang, the Hui (Muslims), the Miao, the Mongols, and the Tibetans. Where they are concentrated in specific areas, they theoretically have some autonomy, though this is fairly limited.

One of China's central preoccupations has been population

growth. Population control efforts intensified after 1979 with the introduction of the 'one-child family' policy—with fines for parents who had two or more children. Although subsequently relaxed somewhat, this has had a dramatic effect. Birth rates are now well below replacement level. This has slowed the growth of the population, which is expected to

China has more men than women

stabilize at around 1.6 billion by 2050. But it has also increased the average age: by 2030 on current trends, one in five people will be over 60. An unusually high proportion will be men. By 2000, there were 107 men for every 100 women.

Although poverty has fallen, 16% of the population still live below the poverty line. Although the cities have seen the emergence of a property-owning middle class, life in the countryside is often harsh. The income of rural families is less than half of those in the cities, and services there remain poor. Few have access to safe sanitation and their health services are also inferior to those in the cities. Another pressing health issue for the country as a whole is HIV/AIDS. China currently has around one million people who are HIV-positive.

As a result of economic reforms, China is becoming a steadily more unequal society. The poorest 20% of the population receive only 5% of national income. The imbalances between rural and urban areas have also generated up to 150 million unregistered migrant workers who have travelled to the cities, many of whom are found on construction sites.

The first round of market reforms started after 1978, when the government dismantled the collective farms and passed responsibility to households. In response, farm output increased by 50%. But in the past few years agriculture has stagnated. Farmers do not have title to their land, so they are reluctant to make improvements.

China is still able to feed itself—indeed in the late 1990s there were often food surpluses. But the long-term prospects are more worrying. The country already has a relatively small proportion of agricultural land—only around 16% is cultivable—and this continues to shrink as a result of environmental degradation and industrial and housing development. In 2004, agriculture, although it employed 45% of the labour force, only accounted for 15% of GDP.

To feed its industries, China can rely on a wide variety of mineral deposits. It has enough iron ore for

The Three Gorges dam displaced one million people

its steel industry. It also has extensive coal reserves, particularly in the north. In addition, a number of provinces have small onshore oil deposits, and there are offshore possibilities. Hydroelectric power is also important and has benefited from extensive investment, most recently in the controversial $28-billion 'Three Gorges' dam on the Yangtze, a project that has already displaced more than one million people.

In recent years, China's industry has been transformed. Until the late 1970s, industry was dominated by state-owned industrial enterprises. Since then, there has been a dramatic shift. Many state owned enterprises have been reformed and between 1998 and 2002 laid off more than 26 million workers. Now more than half of industrial output is in the private sector. The government from 1978 also promoted township and village enterprises (TVEs). This boosted rural incomes and employment, and around one-quarter of the rural labour force are now non-agricultural. By 2004, TVEs were employing 139 million rural workers.

Many foreign businesses have also arrived, tempted by cheap labour and the world's most populous market. Most have invested in coastal areas— one-quarter in Guangdong alone. By

Foreign investors flock in

2004 foreign-invested enterprises accounted for 31% of industrial output. In 2005 foreign investment was $55 billion. This investment has also changed China's pattern of exports. Much is still in the form of simple manufactured goods: China manufactures one-third of the world's suitcases, and one-quarter of its toys. However there has been a shift away from garments and textiles and towards electronic goods. Around one-fifth of exports go to the United States, which has become increasingly concerned about China's economic clout. China's importance as a global manufacturing base was further increased by its entry in 2001 into the World Trade Organization.

China's constitution of 1982, as amended in 1988, declares the country to be a socialist dictatorship led by the working class. In practice, it continues to be run by the Communist Party, which with 70 million members, 5% of the population, is the world's largest political party. The party and

the government are more or less the same thing and the most important decision-making body is the party's Politburo Standing Committee.

The army has also been a major political and economic force. Its influence has receded, especially since it has been forced to give up all its lucrative commercial interests, but it remains the guarantor of party rule.

China's economic transformation started with Deng Xiaoping who emerged as the leading figure after the death of Chairman Mao Zedong in 1976. In economic policy, Deng was fairly pragmatic: politically, he was more orthodox. Though Deng opened the labour camps and reduced government interference in people's private lives he was less tolerant of political dissent. This became chillingly clear in June 1989 when troops opened fire on pro-democracy demonstrators in Beijing's Tiananmen Square, killing hundreds of people.

When Deng died in 1997 he was replaced by his chosen successor Jiang Zemin—though Jiang enjoyed less authority than Deng and China's leadership has subsequently been more collective. In March 2003, China then moved on to its fourth generation of leadership when the National People's Congress—a 3,000-member rubber-stamp parliament—appointed a new president, Hu Jintao, and premier, Wen Jiabao.

China remains a repressive society, with no challenge permitted to party rule: the religious cult Falun Gong, for example, has been under strong attack. In the long term, however, the authority of the Communist Party is likely to be eroded as the market economy makes steady inroads.

Colombia

Colombia is at last making progress against paramilitary forces and armed insurgents but is still plagued by drugs

Land area: 1,139,000 sq. km.
Population: 44 million—urban 76%
Capital city: Bogotá, 6.6 million
People: Mestizo 58%, white 20%, mulatto 14%, black 4%, mixed black–Indian 3%, Indian 1%
Language: Spanish
Religion: Roman Catholic 90%, other 10%
Government: Republic
Life expectancy: 72 years
GDP per capita: $PPP 6,702
Currency: Peso
Major exports: Oil, coffee, chemicals, coal

The western half of Colombia is dominated by three Andean mountain chains. To the west, they descend to the Pacific and Caribbean coasts. To the east, their foothills lead from plains on to savannah and to rainforests that cover around two-thirds of the country.

Colombia's diverse ethnic make-up changes from one region to another. The highest proportions of blacks and mulattos—mixtures of black and white—are to be found in the coastal regions. The whites and mestizos—mixtures of white and Indian—who make up most of the population, live in the valleys and basins between the mountain ranges. The Indian population are mostly to be found in the isolated lowlands. Though now only 1% of the population they are still very diverse, with more than 180 languages and dialects. Colombia is a very polarized society with wide disparities in income—the richest 10% of the population, predominantly white, get 47% of national income.

Around one-fifth of the workforce are employed in agriculture. They grow a range of food crops such as maize and rice, as well as cash crops for export, of which the most important is coffee; Colombia is the world's leading producer of the mild arabica variety. Colombia also exports large quantities of cut flowers as well as bananas and sugar. However, land distribution is very unequal with around 40% in the hands of large landowners.

Colombia has rich mineral resources. The most important is oil, which accounts for around one-quarter of exports, though production is declining and eventually Colombia could become a net importer. In addition, Colombia has Latin America's largest coal reserves and is a significant producer of gold, emeralds, and nickel. The country also has a diverse manufacturing sector—in areas such as food processing, chemicals, textiles, and clothing, most of which is controlled by large conglomerates.

The great unknown element in

Colombia's economy is the drugs trade which is thought to represent around 3% of GDP. Colombia processes its own heroin poppies and coca leaf, as well as coca paste imported from Peru and Bolivia. With substantial support from the USA, through 'Plan Colombia' the government is reported to be spending more than $1 billion per year combating drugs, but the traffic continues unabated, and it supplies around 80% of the cocaine that reaches the USA.

Blessed with resources legal and illegal, Colombia has enjoyed steady economic growth. Its economic management has also been fairly steady by Latin American standards, avoiding the cycles of boom and bust as well as the debt crises. Nevertheless, unemployment in 2005 was 17%. The government has had to keep public spending high to pay for the war against the guerrillas.

Colombia is also distinctive in South America for having had a series of democratically elected non-military governments. Even so, the political system has long been steeped in violence. Between 1948 and 1957 this involved warfare between the Liberal and Conservative parties—a period known as 'la violencia' that cost 300,000 lives. Then the 1960s saw the rise of left-wing guerrilla movements, including the Revolutionary Armed Forces of Colombia (FARC), the National Liberation Army (ELN), and M-19.

In recent years there have been a series of ceasefires and amnesties. Eventually, M-19 laid down its arms and entered politics, ultimately unsuccessfully. But political resolution is more difficult nowadays since the guerrillas are less concerned about revolutionary ideals and more preoccupied with making money through drugs, kidnapping, and extortion. Add to these the private armies of drug dealers and right-wing militias and Colombia at times has seemed on the brink of anarchy and civil war. This has taken a terrible human toll. Over the past ten years, around 35,000 people are thought to have died.

Colombia on the brink of anarchy

When Andres Pastrana of the Conservative Party was elected president in 1998, he promised to make peace with the rebels and reform the army but had little success.

The elections of 2002 reflected public concern about the lack of progress and resulted in the election of an independent candidate, Álvaro Uribe. His policy of 'democratic security' seems to have been surprisingly successful at combating the violence. Uribe increased the number of security personnel, ensured that every town had a strong police presence, and launched a military campaign against the guerillas while building a network of one million informants.

As a result, some 30,000 right-wing paramilitaries have laid down their arms and the ELN may follow suit. The murder rate is lower than it has been for decades. The FARC, however, though weakened somewhat is still rejecting Uribe's demands for a ceasefire.

The voters were impressed. In the 2006 elections they awarded him a second presidential term with almost two-thirds of the votes.

Comoros

The Comoros have suffered from endless coups and inter-island tensions

The Comoros consist of three volcanic islands in the Indian Ocean. The largest is Grande Comore (also called Ngazidja), which includes Mount Karthala, an active volcano. To the south-east is the island of Anjouan, and between them is the smallest island, Mohéli. The terrain includes steep hills and some fertile valleys, but the land has been over-exploited and the soil has been eroded.

Comorans are descended from immigrants from Africa, Asia, and beyond, so the population is very diverse. The majority speak Comoran, which is a Bantu language, though the official languages are French and Arabic. Comorans are as poor as the poorest people in other African countries, and their country's limited resources offer a narrow range of opportunities.

Most people still make their living from agriculture which accounts for 50% of GDP. They do grow food but not enough, so much has to be imported. They also grow cash crops, notably vanilla, cloves, and ylang-ylang (a perfume essence), but the country remains reliant on foreign aid. Population density and land shortage are forcing migration to the towns, or overseas; around 150,000 now live abroad. Economic activity is dominated by the government.

Land area: 2,000 sq. km.
Population: 0.8 million—urban 35%
Capital city: Moroni, 16,000
People: Antalote, Cafre, Makoa, Oimatsaha, Sakalava
Language: Arabic, French, Comoran
Religion: Sunni Muslim 98%, other 2%
Government: Republic
Life expectancy: 63 years
GDP per capita: $PPP 1,714
Currency: Comoran franc
Major exports: Cloves, vanilla, ylang-ylang

The islands were a French colony grouped with the nearby island of Mayotte. Following a referendum in 1974, the three islands voted for independence while Mayotte chose to remain with France. Even so, France has remained a significant influence—and French and other foreign mercenaries have regularly provided the personnel for a series of coups and counter-coups, of which there have been 21 since independence.

In 1997, eyeing the subsidies and aid showered on Mayotte, the islands of Anjouan and Mohéli seceded from the federation, demanding to be returned to French control. France declined the offer.

In April 1999 the army took over 'to preserve national unity' and subsequently reached an agreement with the separatists in the 'Fomboni Accord' which established a federal structure. In April 2002, the army chief, Colonel Azali Assoumane, was elected federal president.

The 2006 presidential election, which was largely peaceful, was won by Ahmed Abdallah Sambi, leader of an Islamist party, the Front National de la Justice. Sambi is from Anjouan and will face opposition from the Grande Comore-based bureaucracy.

Congo

The smaller of the two Congos is potentially oil-rich, but its recurring civil war has held back development

Land area: 342,000 sq. km.
Population: 4 million—urban 54%
Capital city: Brazzaville, 1.2 million
People: Kongo 48%, Sangho 20%, M'Bochi 12%, Teke 17%, other 3%
Language: French, Lingala, Monokutuba, and many local languages
Religion: Christian 50%, animist 48%, other 2%
Government: Republic
Life expectancy: 52 years
GDP per capita: $PPP 965
Currency: CFA franc
Major exports: Oil, timber

The country's official name is the Republic of the Congo, though it is also known as Congo (Brazzaville). Its territory consists largely of two river basins separated by mountains and plateaux. The smaller of the two is the Niari basin in the south and south-west. To the north-east of this the land rises to the Batéké and Bembe plateaux, and then descends to the basin of the Congo River, which also serves as the border with the Democratic Republic of the Congo to the east. More than half the land area is covered by dense tropical forests.

Ethnically, Congo is diverse, though around half the population belong to the Kongo ethnic group, who are concentrated in the capital, Brazzaville, and in the south. The Sangho live in the remote forests of the north, the M'Bochi are in the centre-north, while the Teke are in the Batéké plateau. Congo is the most urbanized country in Sub-Saharan Africa and prior to its civil war from 1997 the concentration of people in Brazzaville helped to ensure relatively good educational standards. But the war has undermined the education system and has also taken a toll on health—not only killing more than 10,000 people and destroying clinics but also enabling HIV/AIDS to spread more rapidly: up to 5% of the adult population could now be infected.

Since the late 1970s, the Congo's main source of income has been oil. Oil makes up most of export earnings and over 40% of GDP. Reserves are largely offshore and serviced by the port of Pointe Noire, with Total and Agip as the leading producers.

Though this income has given the country a relatively high per capita GDP, the wealth has not percolated through to the rest of the economy. There is little other industry and most people in Brazzaville who do not work for the government are employed in the informal sector.

The country's infrastructure is also weak—even Pointe Noire and Brazzaville are not connected by an all-weather road, though there are a number of railways.

People outside the cities generally survive through farming. The country has good soil and ample rainfall, but agriculture is underdeveloped. Most farmers grow crops like cassava, plantains, and groundnuts for their own consumption. Because road links are so poor they find it difficult to market their crops, so most of the food consumed in Brazzaville is actually imported, usually across the Zaire River from Kinshasa.

The country's vast forest resources, which include around 600 species of timber, have also been under-exploited. For loggers too the main handicap is a lack of transport. The only forests that have been felled are near the coast. Environmentalists are now, however, worried about the arrival of Asian logging companies looking for new opportunities.

Congo's current political rivalries have their roots in the 1960s and 1970s when this was the Marxist 'People's Republic of the Congo' run by the Parti congolais du travais (PCT). In 1979, Colonel Denis Sassou-Nguesso, a northerner from the M'Bochi ethnic group—from which the PCT drew most of its support—took over the leadership and was elected president in 1984, and again in 1989. But by the early 1990s, popular pressure forced the PCT to legalize other parties, and in 1991 a national conference approved a new constitution and changed the country's name to the Republic of the Congo.

The PCT lost the democratic parliamentary elections in 1992 when the two leading parties were the Union panafricaine pour la démocratie sociale (UPADS), led by Pascal Lissouba whose support came from the centre-south, and the Mouvement congolais pour la démocratie et le developpement intégral (MCDDI), led by Bernard Kolélas who had strong support in Brazzaville. The presidential election in the same year was won by Lissouba.

The country began its slide into chaos in 1993 following parliamentary elections that also gave a victory to Lissouba but were disputed by Kolélas. The three main parties had recruited their own private militias. The PCT had the 'Cobras', the MCDDI had the 'Ninjas', and UPADS the 'Zulus'. Each seized areas of Brazzaville and more than 2,000 died in the ensuing battles. Order was eventually restored when UPADS and the MCDDI agreed to share power while Sassou-Nguesso retired to his northern power base.

Fighting between Cobras, Ninjas, and Zulus

When Sassou-Nguesso returned to Brazzaville with his Cobras for the next election in 1997, Lissouba tried to arrest him and disband his militia. He failed and, following a six-month civil war that destroyed much of Brazzaville, Sassou-Nguesso regained power and Lissouba fled. Fighting resumed in 1999, but early in 2000 the president of Gabon helped broker a ceasefire. In 2002 a referendum approved a new constitution. But the opposition claimed this was biased and boycotted the subsequent 2002 presidential election which Sassou-Nguesso won. Fighting resumed yet again, with a rebel faction led by Pasteur Tata Ntoumi. This was ended by a further peace deal in 2003 which seems to be holding.

Congo, Democratic Republic

Formerly Zaire, the country became a regional battlefield, but has achieved a ceasefire

Land area: 2,345,000 sq. km.
Population: 54 million—urban 32%
Capital city: Kinshasa, 5.0 million
People: Mongo, Luba, Kongo, Mangbetu-Azande, and many others
Language: French, Lingala, Kingwana, Kikongo, Tshiluba
Religion: Roman Catholic 50%, Protestant 20%, Kimbanguist 10%, Muslim 10%, other 10%
Government: Republic
Life expectancy: 43 years
GDP per capita: $PPP 700
Currency: Congolese franc
Major exports: Diamonds, cobalt, coffee, copper, oil

The dominant feature of this vast country, accounting for around 60% of its territory, is the Congo River basin centred on the north-west. This includes some of the world's most dense tropical rainforests. From this depression the land rises to the south and west to plateaux, and then to mountains along the country's borders.

With more than 200 ethnic groups, the population is very diverse. The majority, including the Mongo, Luba, Lunda, and Kongo are Bantu speakers. But in the north there are major Sudanese groups such as the Mangbetu-Azande. There are also pygmy people in the forests. Relations between these groups have often been tense and violent—divisions that politicians have exploited.

Standards of human development have been falling steadily—a result of government neglect, withdrawal of international aid, and civil war. Incomes have declined steeply: GDP per capita in 2006 was lower than at independence, though has started to recover. Most education is provided by Catholic missionaries but fewer children are now attending school and the literacy rate is thought to be falling. Because children are less likely to be immunized, infant mortality is rising, and at least one-third of children are malnourished. AIDS is a growing threat: officially the infection rate is 3% but is probably much higher.

Two-thirds of the workforce are employed in agriculture. Most are subsistence farmers who use fairly primitive techniques, growing plantains, cassava, sugar cane, and maize. Despite ample fertile land, huge quantities of food need to be imported—a result primarily of government indifference to agriculture, and poor transport facilities, as well as the disruption of warfare. There are also some cash crops such as coffee, palm oil, and cotton, but output has declined steeply in recent years. The country's vast forest potential could also be exploited on a sustainable basis.

In the past much of the country's income has come from its lucrative mineral deposits. Traditionally the major earner has been copper in the Katanga region in the south-east. This, along with the cobalt industry, is the responsibility of the state-owned company, Gécamines. But production has fallen steadily as a consequence of low investment and warfare; copper and cobalt now account for less than 10% of exports.

In 2004 around half of export earnings came from diamonds, most of which come from Kasaï in the south. This is the world's third largest diamond producer, though of industrial rather than gem quality. Diamond production, which is mostly in the hands of the private sector, has been less affected by the recent wars.

The country also has some manufacturing industry, often centred on the processing of minerals. Most foreign investors have stayed away, though Société de Belgique and Unilever have extensive interests.

For most of the period since independence in 1960 the country was under the control of one man, Mobutu Sese Seko. Mobuto came to power in a US-backed military coup in 1965 and established his own political party. In 1991 he changed the

From Congo to Zaire

country's name from the Democratic Republic of the Congo to Zaire. Mobuto was subsequently elected unopposed for three consecutive terms of seven years—a period during which he ruthlessly repressed all opposition and looted the economy.

Mobutu's grip weakened in 1996 following a rebellion in Kivu province

in the east. This was originally centred on a rebellion by an ethnic group he had persecuted, the Banyamulenge, who by then had seized power in neighbouring Rwanda. With Rwandan help they overcame the Zaïran army and under the leadership of Laurent Kabila, as the Alliance des forces démocratique pour la libération du Congo-Zaire (AFDL), they took control in 1997.

Mobutu fled (and died later that year). Kabila declared himself president and changed the country's name back to the Democratic Republic of the Congo.

Hopes for a more democratic and peaceful era were soon dashed. First it became clear that Kabila was going to be just as heavy-handed. He also sufficiently antagonized his former Tutsi allies that in 1998, with backing from Rwanda and Uganda, they launched another rebellion, but this time against him. Zimbabwe, Angola, and Namibia soon sprang to his defence, leading to a complex and dangerous regional conflict.

Early in 2000 there was a ceasefire following a UN-brokered plan for disengagement of all foreign troops. The prospects then brightened, following the assassination of Kabila in January 2001, and his replacement as president by his son Joseph. In December 2002 there was a final settlement, and in July 2003 a transitional administration was formed, headed by Kabila.

In July 2006 the UN helped organize the first multiparty presidential election for 40 years. This was a tense and complex operation supported by EU troops. After the first round Kabila emerged well ahead.

Costa Rica

Long an island of peace in a troubled region, Costa Rica is struggling to maintain its social achievements

Land area: 51,000 sq. km.
Population: 4 million—urban 61%
Capital city: San José, 310,000
People: White or mestizo 94%, black 3%, other 3%
Language: Spanish, English
Religion: Catholic 76%, Protestant 16%, other 8%
Government: Republic
Life expectancy: 78 years
GDP per capita: $PPP 9,606
Currency: Colón
Major exports: Electronics, coffee, bananas, meat

Mountain ranges form a rugged backbone extending down the centre of Costa Rica. On the Pacific side they descend to coastal lowlands and sandy beaches. On the Caribbean side they come down more gradually to heavily forested lowland areas with a rich variety of plant and animal life. Most of the population, however, live in the highlands, in the Meseta Central where the climate is milder and the soil more fertile. Unfortunately, this is also a geologically active zone subject to earthquakes and potentially dangerous volcanoes.

Most people are of European or part-European descent. Few Amerindians remain and the largest minority group consists of black descendants of workers brought from Jamaica at the end of the 19th century to build the railroads and work on plantations. This English-speaking group make up around 30% of the population of the province of Limón on the Caribbean coast.

Costa Rica's development has been relatively equitable and it has created one of the most extensive welfare states in Latin America. Much of this can be traced back to the 1949 constitution which gave the state a strong role in social welfare, established labour rights, and introduced free secondary education. Remarkably, it also abolished the army, to avoid coups and save money—though the country does have a 7,500-strong police force. Although this is one of the most prosperous countries in Central America, with good standards of health and education, around 20% of the population are classified as poor.

Christopher Columbus, who christened Costa Rica (rich coast), had visions of gold and treasure. But the country's wealth has been based on more mundane products— particularly coffee and bananas. Costa Rican coffee is produced largely by 78,000 small farmers in the central highlands—the average grower has ten hectares. Quality and yields are high. In terms of exports, however, coffee has been overtaken by bananas—20% of export value—produced on both the Caribbean and Pacific coasts. Prices have been lower in the past few years, but Costa Rica, which is the world's second largest exporter, stands to benefit from a greater opening of European markets to Central American bananas.

Though still relying significantly on these crops, Costa Rica has nevertheless achieved a very diverse economy. Indeed this soon became primarily an industrial country, initially based on import substitution, in areas like food processing, chemicals, textiles, and metals.

The fabric started to unravel to some extent in the 1970s and 1980s. As public investment increased and revenues fell, the government became deeply embroiled in the Latin American debt crisis in the early 1980s. Since then the economy has gone through several stop-go cycles.

Lately these cycles have been based more on high-tech industry. Since it has the advantage of a sound democracy, a well-educated workforce, and a good legal system, Costa Rica is now interested in capitalizing on these attributes by attracting high-tech investors. It achieved a major success in 1998 when Intel built the first of two chip factories in duty-free zones. This also encouraged at least 15 other high-tech manufacturers, including two major US medical equipment suppliers. Microsoft also carries out some software development here. However there

Home to Intel and Microsoft, Costa Rica is becoming a 'chip republic'

are risks to such a strategy, given the fluctuating fortunes of the computer industry.

Costa Rica is also now a prime tourist destination. More than 1.4 million people come each year to visit the rainforests and volcanoes, though most recent investment has gone into beach resorts. Tourism now brings in $1.3 billion per year.

Politically, Costa Rica's extensive welfare state and the fact that it has had a stable democracy since 1948 gave it an enviable reputation as the 'Switzerland of Central America'. It avoided the 1980s warfare in neighbouring states; indeed Costa Rica was the driving force of the Central American Peace Agreement of 1987, for which then President Oscar Arias Sánchez was later awarded the Nobel Peace Prize.

Over much of this period, Costa Rica has been ruled by the centre-left Partido Liberación Nacional (PLN), or by the more conservative Partido Unidad Social Cristiana (PUSC).

The 1998 parliamentary and presidential elections were won by the PUSC, with Miguel Ángel Rodríguez elected president. And in 2002, the PUSC again took the presidency, this time for Abel Pacheco.

However in parliamentary elections the two main parties had been losing ground, particularly to a PLN offshoot, the Partido de Acción Ciudadana (PAC), led by Otton Solís. In 2004 former presidents of both the PLN and the PUSC were implicated in corruption scandals.

Nevertheless in the 2006 presidential elections the PLN scraped through, with Arias returning as president, just ahead of Solís.

Côte d'Ivoire

Côte d'Ivoire's former political stability has been shattered by a coup and a civil war

Land area: 322,000 sq. km.
Population: 18 million—urban 45%
Capital city: Yamoussoukro, 299,000
People: Akan 42%, Voltaiques or Gur 18%, Northern Mandé 17%, Krous 11%, Southern Mandé 10%, and many immigrants
Language: French, Dioula, and many local
Religion: Muslim 34–40%, indigenous 25–40%, Christian 20–30%
Government: Republic
Life expectancy: 46 years
GDP per capita: $PPP 1,476
Currency: CFA franc
Major exports: Cocoa, palm oil, wood, coffee

Côte d'Ivoire has a narrow coastal strip of land, which includes in the east a series of deep lagoons. Behind this are two forest areas. The one to the west still has dense rainforest, while the one to the east has now largely been cleared to grow plantation crops. Beyond these to the north the land is mostly savannah, used for grazing cattle.

The population is very diverse with four or five main groups and around 60 sub-groups. Thus, the majority group, the Akan, includes the Baoulé who live mostly in the towns and villages of the south-east. Among the Krous are the Bété, living in the forests of the south-west. In the north-east there are a number of Voltaic groups including the Senoufou.

The country has also attracted many people from Burkina Faso, Guinea, and Mali to work on plantations and in domestic service.

And there are also substantial Lebanese, Syrian, and French communities. First- and second-generation immigrants make up around one-third of the population.

This is one of the most urbanized of African countries and Abidjan is one of Africa's more cosmopolitan cities. By West African standards incomes are reasonable. Nevertheless shrinking government budgets have undermined public services, and education and health standards are low: fewer than half the population are literate, and life expectancy has been falling, partly as result of HIV/AIDS which now affects around 7% of the adult population.

Around 80% of the workforce are employed in agriculture. Farmers grow food crops such as cassava, yams, maize, sorghum, and rice. Peasant farmers also cultivate the two main cash crops, cocoa and coffee, which take up around two-thirds of cultivated land. Côte d'Ivoire is the world's largest cocoa producer and the fifth largest coffee producer. For the first 20 years following independence in 1960, these exports contributed to rapid economic growth. Other important agricultural exports are bananas and pineapples along with tropical hardwood and palm oil.

Côte d'Ivoire also has energy and mineral resources—including deposits of oil and gas that could make the country self-sufficient. In addition there is a significant manufacturing sector. This includes garments and textiles along with some local processing of primary commodities like cocoa—though at present only around one-fifth is transformed into higher value products like chocolate.

Export crops helped to develop Côte d'Ivoire but they also increased its vulnerability to international commodity prices. Dramatic price falls in the late 1970s, exacerbated in 1994 by the devaluation of the CFA franc, led to a severe economic crisis and the country was soon plunged deeply into debt. Commodity prices recovered for a few years and growth resumed, but they then fell again and since the onset of the political conflict the economy has suffered a steep decline.

Côte d'Ivoire is vulnerable to changes in cocoa prices

Côte d'Ivoire's previous reputation for political stability rested on the 23-year autocratic reign of the country's first president, Félix Houphouët-Boigny. For many years, his Parti démocratique de Côte d'Ivoire-Rassemblement démocratique africain (PDCI-RDA) was the only legal political party. His liberal economic policies helped to attract investment. But he also promoted some grandiose projects including moving the capital to Yamoussoukro, his home town, where he also built the world's largest Catholic cathedral.

In 1990, Houphouët-Boigny introduced a new multiparty constitution and swiftly held an election in which he won a seventh term, dying in office in 1993.

He was replaced by Henri Konan-Bédié, who was confirmed as president by an election in 1995. This election had been boycotted by the main opposition parties, who came together in 1997 as the Front populaire ivorienne (FPI) led by Laurent Gbagbo. Konan-Bédié then consolidated his grip and forced his presidential rival, Alassane Ouattara of the Rassemblement des républicains (RDR), to flee into exile.

But Konan-Bédié overestimated the support of the army, and in December 1999 they overthrew him in Côte d'Ivoire's first coup, led by General Robert Guéi, who promised a swift return to democracy.

In July 2000 a referendum approved a new constitution and in the election of August 2000, Guéi claimed victory over Gbagbo, the FPI leader. But the election was clearly fraudulent and popular protest forced Guéi to resign allowing Gbagbo to take over.

All seemed to be going well, with politicians resolving their differences, when there was a military uprising that killed Guéi and triggered a civil war. A number of rebel groups emerged and subsequently came together as the 'New Forces' and seized control of the north.

In January 2003 a peace agreement was reached and the New Forces joined a government of national unity led by Gbagbo. But the peace is fragile and the country remains divided in two. Elections planned for 2006 seem unlikely to take place.

Croatia

Croatia's fierce nationalism sparked the Balkan wars. Now the country is more democratic and the economy has revived

Land area: 57,000 sq. km.
Population: 5 million—urban 59%
Capital city: Zagreb, 779,000
People: Croat 90%, Serb 5%, other 5%
Language: Croatian
Religion: Catholic 88%, Orthodox 4%, other 8%
Government: Republic
Life expectancy: 75 years
GDP per capita: $PPP 11,080
Currency: Kuna
Major exports: Transport equipment, chemicals, textiles

Croatia's eccentric shape encloses a variety of landscapes and climates. The north-west includes a part of the Dinaric Alps. To the south the dramatic, broken coastline of Istria and Dalmatia consists of mountains that drop steeply into the sea, creating numerous inlets and bays as well as more than 1,100 islands.

When it was a republic within Yugoslavia, this territory had a clear majority of Croatians, as well as around 12% ethnic Serbs. The differences between these two Slav peoples are cultural rather than physical: the Croats are Roman Catholic and Western in outlook; most Serbs are Eastern-oriented Orthodox Christians. As a result of the wars of the 1990s, more than half of the Serbs fled. In addition there are around 800,000 Croatians in Bosnia.

Croatia was, after Slovenia, the second richest of the republics of former Yugoslavia and has an economy similar to that of Western European countries.

The war caused the GDP to shrink dramatically, but in recent years growth has been strong. Some of this is the result of public investment in infrastructure. But industry also played a part and now accounts for 35% of GDP. The break-up of Yugoslavia cost Croatia many of its markets and forced it to re-orient more towards the west. Initially it re-established some industries such as textiles under contract to foreign companies. But since then textile output has fallen and there has been much stronger growth in other sectors, including publishing and printing as well as in the manufacture of metal and electrical products.

Croatia has been a significant oil producer but its reserves are now exhausted and most crude oil has to be imported. Even so oil refining remains important for both national consumption and exports.

Agriculture is also more significant than in some neighbouring countries. Most agricultural land was already privately owned, though many farms are small and often not very profitable. The best land and most of the larger holdings are to be found in Slavonia in the north-east, which is where most of the capital-intensive production of cereals and other cash crops takes place. Before the war, Croatia was a food exporter but now it is a substantial importer.

As in most industrialized countries, the majority of the workforce—around 60%—are employed in services. Many of these people are employed in the large public sector and most of the rest in trade and transport.

Croatia's spectacular Adriatic coastline had long been a major package-tourist attraction—notably the ancient city of Dubrovnik, which took a severe battering from the Yugoslav army in 1991. Since then tourism has revived, with over nine million visitors in 2004, of whom a high proportion come from Germany and Italy.

Dubrovnik took a battering

Despite Croatia's economic revival, unemployment remains high: in 2004 it was still around 20%. Hopes for economic revival are now placed on accession to the EU for which Croatia is now negotiating.

Croatia's transition from communism was dominated for the first decade by a former general, Franjo Tudjman, who in 1990 created a new party, the Croatian Democratic Union (HDZ). In the first free elections in May 1990 the HDZ won a parliamentary majority and Tudjman became president. Tudjman's hard-line nationalism alarmed Serbs in Krajina and Slavonia who feared breaking with Serbia, and rejected the idea of independence, and in mid-1990 they established 'Serbian Autonomous

Regions' which Tudjman refused to recognize.

When Croatia did declare independence in July 1991, this provoked a full-scale civil war as the Serb paramilitaries united with the Yugoslav army to seize more than one-quarter of Croatia. The UN negotiated a ceasefire in January 1992, but in August 1995 the Croatian forces regained control, causing the Serbs to flee. In December 1995 Tudjman signed the Dayton peace accords.

Croatia is a parliamentary democracy, but in the early years the HDZ had a majority in both houses and Tudjman had a fairly free hand. He was elected president in 1992 and re-elected in 1998, and used his status as a war hero to build a centralized and authoritarian state.

To many people's relief, Tudjman died in December 1999 and the parliamentary elections in 2000 resulted in a surprising turnaround. The main centre-left opposition alliance of Social Democrats (SDP) and Social Liberals (HSLS) won almost half the seats. The SDP's leader, Ivica Racan, became prime minister and a presidential election in February 2000 was won by a political moderate, Stipe Mesic.

But Racan's coalition, riven by policy disagreements, eventually collapsed and further elections in 2003 returned the HDZ which claimed to have abandoned nationalism for a more standard conservative approach. Led by Ivo Sanader, the HDZ formed a minority coalition government with the HSLS and others. In 2005, Mesic, one of the most widely respected politicians, was re-elected president, defeating the HDZ candidate.

Cuba

Cuba now has elements of a market economy. Political change however will have to await the departure of Castro

Around one-quarter of Cuba's territory is mountainous, with three main systems, of which the most extensive is the Sierra Maestra in the south-east. Most of the island, however, consists of plains and gently rolling hills. For a relatively small island, the environment is biologically rich—particularly in its mangrove forests and wetlands.

Cuba's population is racially mixed: most people are mulatto. There is little overt discrimination, but those with darker skins tend to have the worst jobs. Very few members of the Communist Party's leadership are black.

Though they are far from rich, Cubans enjoy some of the world's best levels of social development. Health and education services are free. In the mid-1990s the school system lost many teachers to the tourist industry

Land area: 111,000 sq. km.
Population: 11 million—urban 76%
Capital city: Havana, 2.1 million
People: Mulatto 51%, white 37%, black 11%, other 1%
Language: Spanish
Religion: Roman Catholic, Protestant
Government: Communist
Life expectancy: 77 years
GDP per capita: $PPP 3,500
Currency: Peso
Major exports: Sugar, nickel, tobacco, seafood

but pay rises have stemmed the flow.

Agriculture accounts for only around 7% of GDP and directly employs only 20% of the workforce, but it remains critically important to the economy. The main crop is sugar, which the Soviet Union used to buy at a favourable price. Since then, however, output and income have shrunk dramatically and sugar now accounts for only one-third of exports. In 2002 a decision was taken to restructure the sugar industry which would effectively reduce it by half. Tobacco is also an important source of income but in this case production has increased. Other crops, including fruit and vegetables, have also done better, but Cuba only grows around one-quarter of its staple food, rice, and has to import the rest.

Cuba's land reform created state farms, but also left around 20% of the land with smallholders and co-operatives who have proved more productive—growing about 40% of the food. Since 1993, most state farms have been dissolved and transformed into 'Unidades Básicas de Producción Cooperativa' (UBPCs). This involves the government leasing land rent-free to former workers who are responsible for running financially independent co-operatives. This seems to have raised production, though few of the UBPCs are profitable. Farmers can sell surplus production in 'agropecuarias'—farmers' markets.

Manufacturing has suffered from the loss of the Soviet market and the continuing US boycott. To generate more industrial jobs, the government has now created free-trade zones and has invested over $1 billion in research investment in areas such as biotechnology and pharmaceuticals. Mining is also an important source of foreign exchange. Cuba has large reserves of nickel and cobalt, which are mined in conjunction with the Canadian corporation Sherrit International.

The most vigorous part of the economy has been tourism. Arrivals have increased dramatically to around 2.3 million in 2005, with Canadians, Italians, and Spanish in the lead—chiefly heading for beach resorts on the northern coast. Tourism brings in around $2.6 billion annually but three-quarters of this goes straight out

Tourism to Cuba encourages unofficial enterprise

again to pay for related imports and as profits to the many foreign companies who run the hotels. There is also considerable unofficial income from informal ancillary enterprises, including prostitution.

After 1993, Cubans were able to hold dollars, replaced in 2004 by 'convertible pesos' that they could spend in 'dollar shops'. These are virtually the only source of non-essential consumer goods; prices are around 40 times higher than in the official shops. 'Peso shops' distribute food rations, though Cubans can also buy more expensive produce from farmers' markets.

Despite economic liberalization, political change seems some way off. Cuba remains a 'socialist workers' state' in which the only legal party is the Partido Comunista de Cuba, headed by Fidel Castro, who has been the country's president since the revolution in 1959. The only opposition comes from human rights organizations and small groups linked to Cuban communities in the USA. Several hundred political prisoners are in jail. In March 2003 for example, a further 75 dissidents were arrested and given long prison sentences. Despite the lack of democracy and the economic privations, Castro remains a remarkably popular leader.

Many of Cuba's difficulties are the result of hostility from the USA which, under pressure from a powerful exile lobby in Miami, classifies Cuba as an enemy state and imposes rigid economic sanctions. Relations with the USA are alternately warm and cool. One of the severest freezes resulted from the 1996 US Helms-Burton Act which imposes sanctions on companies that do business with Cuba.

Nevertheless, there have also been some thaws which have enabled flows of emigrants between the two countries and the restoration of some direct flights. In 2000, the US Congress adopted bills that allowed sales of food and medicine.

In August 2006, there were signs of a transition of power when Castro was taken ill and 'temporarily' handed the reins to his brother Raul. But Raul is only five years younger, so the longer-term position is uncertain.

Cyprus

A bitterly divided island that has rejected an opportunity for reunification

Land area: 9,000 sq. km.
Population—urban 69%
Capital city: Nicosia, 296,000
People: Greek 77%, Turkish 18%, other 5%
Language: Greek, Turkish, English
Religion: Greek Orthodox 78%, Muslim 18%, other 4%
Government: Republic
Life expectancy: 79 years
GDP per capita: $PPP 18,776
Currency: Cyprus pound
Major exports: Citrus fruit, potatoes, grapes

The island of Cyprus has two main mountain ranges. The Kyrenian Mountains extend along the northern coast, while the Troodos Mountains are in the south and centre. Between these is the most fertile part of the country, the Mesaorian plain, which includes the capital, Nicosia.

Cyprus is sharply divided between Greek and Turkish Cypriots. Greeks are still in a large majority, virtually all of them living in the south and centre. The Turks are confined to the north; although many have emigrated, others have arrived from Turkey.

The Greek part of the island has shifted decisively away from agriculture and now depends heavily on service industries. The most important is tourism. More than two million tourists arrive each year, half from the UK—accounting for around 15% of GDP. The other main service activity is business and finance, which includes 26 international banks and contributes 5% of GDP. The economy in the north lags far behind: the main activity is agriculture, growing the traditional export crops of citrus fruit and potatoes.

At times, Greek and Turkish Cypriots have coexisted peacefully, and the constitution adopted at independence in 1960 involved explicit power-sharing. But by 1964

there were violent confrontations and the UN sent in a peacekeeping force, which is still there, 1,200 strong. Matters came to a head after a military coup in 1974. Turkey sent in 30,000 troops to protect Turkish Cypriots and established control over the northern one-third. People moved from one part of the island to the other.

In 1983, the Turkish Cypriot leader Rauf Denktash proclaimed the north to be the Turkish Republic of Northern Cyprus, a claim recognized only by Turkey. The rest of the world recognizes the Republic of Cyprus (ROC), even if it controls only 57% of the territory. In 2003 Tassos Papadopoulos of the centre-right Democratic Party won the presidential election for the ROC.

In April 2004, the UN proposed a settlement involving a loose federation. This was put to referendums on both parts of the island. While the Turkish part voted 67% in favour, the Greek part voted 75% against—provoking international criticism. As a result when Cyprus entered the EU in May 2004 the Turkish part was excluded. Hopes for reunification were revived in 2005, however, when a more conciliatory figure, Mehmet Ali Talat, was elected president in the north.

Czech Republic

A formerly communist country that had a 'velvet revolution' and a fairly painless transition to a market economy

Land area: 79,000 sq. km.
Population: 10 million—urban 74%
Capital city: Prague, 1.2 million
People: Czech 90%, Moravian 4%, other 6%
Language: Czech
Religion: Roman Catholic 27%, Protestant 2%, other 11%, unaffiliated 60%
Government: Republic
Life expectancy: 76 years
GDP per capita: $PPP 16,357
Currency: Czech crown
Major exports: Machinery, manufactures

The Czech Republic has two main geographical areas. The largest, to the west, is Bohemia, a broad elevated plateau of plains and low hills centred on Prague and ringed by mountains. Beyond the mountains to the east lies the second main area, the Moravia Valley centred on Brno, whose eastern boundary is formed by the Carpathian Mountains that separate the country from Slovakia.

Ethnically the population is fairly homogenous. But throughout the country there are also small numbers of Slovaks, who moved here when the country was part of Czechoslovakia, and also some Roma, or gypsies, whose lifestyle exposes them to both economic and racial discrimination. Until the 1980s, the Czech Republic had a relatively young population, but recent falls in the birth rate are causing the population to age rapidly.

Levels of human development are fairly high for a formerly communist country. In the past health standards were reasonable, though lifestyles were unhealthy: high fat consumption and smoking reduced life spans. Over the past decade, however, life expectancy has increased.

This is a broadly based industrial economy—though between 1990 and 2004 industry as a proportion of GDP fell from 48% to 35%. Previously it concentrated on heavy industry such as engineering and steel, but more recently some manufacturers have been exporting intermediate goods, particularly to Germany. Lighter industries such as food processing have also been expanding.

Privatization of state-owned assets took place rapidly, mostly through vouchers which small investors rapidly sold to institutions. In 2004 the private sector accounted for more than 85% of GDP. Transferring the ownership in this way did little to transform the management of enterprises and many have since stagnated. But some businesses acquired an injection of expertise from foreign companies: the reputation of Skoda cars, for example, was transformed when it was taken over by Volkswagen—and contributes more than 10% of total exports.

The agricultural sector accounts for only about 3% of GDP and is now entirely in private hands, either as commercial farms or co-operatives. With the country having joined the EU in 2004, farmers will benefit from greater price support.

Meanwhile, many more people are working in services particularly in retailing where many of the large stores are now part of foreign-owned chains. Others work in catering and a range of activities related to tourism which accounts for around one-fifth of total exports. Prague is the main tourist attraction and most new jobs have been created there or in the surrounding regions—heightening the disparities with northern Bohemia and northern Moravia.

From 1918 to 1938, and from 1945 to 1992, what is now the Czech Republic was united with Slovakia. As Czechoslovakia, their joint transition from communism in 1989 was so smooth it was termed the 'velvet revolution'. Within the Czech lands, the democracy movement was called the Civic Forum, one of whose leaders was the writer Václav Havel. Civic Forum won most of the Czech seats in the election of 1990 while a corresponding movement in Slovakia won the seats there. The new parliament elected Havel as president.

From a velvet revolution to a velvet divorce

Civic Forum soon reformed into a number of parties, of whom the most important subsequently have been the right-wing Civic Democratic Party (ODS) and the centre-left Czech Social Democratic Party (CSSD).

The union with Slovakia did not last. The disagreements were partly ideological, since the leading Czech party was the right-wing ODS led by prime minister Václav Klaus, while the leading Slovak party was centre-left. Eventually, and without much public support, the politicians disbanded the federation. On 1 January 1993 they became two countries and Havel was elected president of the new Czech Republic—while Klaus remained prime minister.

The reforms proceeded apace but Klaus lost popularity as the economy slowed and his government became embroiled in corruption. As a result, the 1996 elections reduced the ODS to a minority government which soon collapsed amid economic chaos.

In the ensuing election in 1998 the voters' allegiances swung left. The CSSD became the largest party, the ODS came second, and the unreconstructed Communist Party came third. The new prime minister was Milos Zeman who was obliged to enter into a pact with the ODS.

Prior to the 2002 general election Zeman stepped down as CSSD party leader in favour of Vladimir Spidla who started to distance the party from the ODS. In the election the CSSD strengthened its position and formed a coalition government with two small parties. But divisions within the CSSD forced Spidla to resign in June 2004, to be replaced by Stanislav Gross. In 2003 the parliament had elected Václav Klaus as president.

Parliamentary elections in 2006 led to a dead heat between potential coalitions. Eventually, Klaus asked CSSD leader Mirek Topolanek to form a new government.

Denmark

A bridge from continental Europe to Scandinavia that has had a vacillating relationship with the EU

Land area: 43,000 sq. km.
Population: 5 million—urban 85%
Capital city: Copenhagen, 0.5 million
People: Danish, Inuit, Faroese
Language: Danish, Faroese
Religion: Evangelical Lutheran 95%, other Christian 3%, Muslim 2%
Government: Constitutional monarchy
Life expectancy: 77 years
GDP per capita: $PPP 31,465
Currency: Krone
Major exports: Meat products, furniture, pharmaceuticals

Denmark is one of Europe's more physically fragmented countries. The largest part, with around 70% of the territory of Denmark itself, is the Jutland peninsula. To the east of this lie the two largest of Denmark's 400 or so islands: Funen and Zealand; the capital, Copenhagen, is located on the east of Zealand. In addition, Denmark has two distant dependent states— Greenland and the Faroe Islands. Most of Denmark itself is low-lying and fertile, broken occasionally by hills, particularly in the centre and east of Jutland.

Denmark's population is ethnically fairly homogenous, though flows of immigrant workers from the 1970s, and asylum-seekers during the 1980s and 1990s, added variety. Around 5% of the population are foreign-born, with the largest numbers coming from former Yugoslavia and Turkey.

The standard of living is high and the government gives a high priority to education, on which it spends 8% of GDP, one of the highest proportions in Europe. Danes also enjoy free medical care and extensive welfare benefits. Although public provision is not as generous as in other Nordic countries, social welfare spending is equivalent to more than one-quarter of GDP. Income tax rates touch 60% and there is a flat VAT rate of 25%. Poverty levels are among the lowest in Europe.

Even so, it is doubtful that the Danish welfare state will survive in its current form, given the ageing population. The arrival of immigrants rejuvenated the population somewhat but a falling birth rate is still leading to a rising proportion of older people. In the next 30 years the ratio of working people to those over 60 years old will fall from 3.0 to 2.1.

Agriculture accounts for only around 3% of GDP, but is still vital to the economy, since it feeds into industry and exports. Two-thirds of Denmark is used for crops or pasture. Most activity is concerned with livestock—either growing animal feed or raising cattle and pigs. Farms are small and family-owned but technologically very sophisticated. Since they produce around three times Denmark's own requirements, they export most of their output. Denmark also has a major fishing fleet, though over-fishing and controls by the EU have been constraining output.

Agriculture provides important raw materials for Danish industry—some of the cereal crop finishes up, for example, in cans of Carlsberg beer, one of the world's most venerable brands.

Denmark is the home of Carlsberg and Lego

Other distinctive Danish exports include stylish furniture, the hi-fi equipment of Bang & Olufsen, and the ubiquitous plastic Lego bricks—which have also been used in the construction of Legoland, one of Europe's largest theme parks.

Apart from its soil, Denmark has limited natural resources. It does not have much coal or many minerals, but oil and gas fields in the North Sea supply the bulk of local needs. With a lot of flat and exposed land, Denmark can also generate a significant amount of wind power.

The Danes joined what is now the EU in 1973, but have not been strong federalists. They gave their fellow members a shock in 1992 when they voted against the Maastricht treaty which set the stage for monetary union. After various amendments, they endorsed it the following year. Since then, attitudes towards Europe seem to have become more positive.

Denmark is a constitutional monarchy, with Europe's oldest royal family, currently headed by Queen Margarethe II. Most governments have been coalitions or minority administrations. For decades after the

Second World War, the main party was the Social Democratic Party, which had promoted the welfare state. This pattern was broken decisively in the 1982 election when the reins of government passed to Poul Schluter of the Conservative Party at the head of a centre-right coalition. As a result, Denmark became the first of the Nordic countries to rein in public spending and deregulate its economy. Schluter eventually resigned in 1993, implicated in a scandal around immigration policy.

He was replaced as prime minister by a Social Democrat, Poul Nyrup Rasmussen, at the head of a coalition that included the Radical Liberals, the Centre Democrats, and the Christian People's Party. The Social Democrats continued with many of the neo-liberal economic policies. They were also returned at the head of coalitions in the 1994 and 1998 elections.

Nevertheless, there were growing signs of a shift to the right, particularly over concerns about immigration which led to street violence and contributed to the rise of the right-wing populist Danish People's Party (DPP) led by Pia Kjaersgaard.

In 2001 Poul Nyrup Rasmussen called a snap election which he expected to win. But he miscalculated and was replaced by Anders Fogh Rasmussen at the head of a Liberal–Conservative coalition. This coalition was returned after another snap election in February 2005.

In 2006, Denmark found itself in the international spotlight as a result of newspaper cartoons that mocked Islam and provoked violent demonstrations around the world.

Djibouti

A city-state still suffering from the aftermath of its civil war and regional conflicts

Though there are mountains in the north of this small country, most of Djibouti consists of a bare arid plateau—one of the world's hottest places. Most activity centres on the capital, also called Djibouti, and particularly its port, which is crucial to its now landlocked neighbour, Ethiopia.

The country's brief period since independence in 1977 has been soured by conflict between its two main ethnic groups. More than half the population are Somalis of the Issa and other clans, while around one-third are the Afar, who are of Ethiopian origin. However, the precise ethnic division is uncertain, partly because of the flows of refugees, but also because the ethnic population balance is a sensitive issue. Djibouti's population growth rate is around 2%. Three-quarters of these people live in the capital; most of the remainder are pastoral nomads.

Djibouti is a poor country: 45% of people are below the poverty line and life expectancy is only 53 years. It is also remarkable for its consumption of a mild intoxicant: qat. Every day, Djiboutians together chew their way through a remarkable 11 tons of the drug—which is imported from Ethiopia—devoting a high proportion of household expenditure to the habit. Most of Djibouti's economy depends

Land area: 23,000 sq. km.
Population: 0.8 million—urban 84%
Capital city: Djibouti, 624,000
People: Somali 60%, Afar 35%, other 5%
Language: French, Arabic, Somali, Afar
Religion: Muslim 94%, Christian 6%
Government: Republic
Life expectancy: 53 years
GDP per capita: $PPP 2,086
Currency: Djibouti franc
Major exports: Hides and skins, coffee

on its location as a transit point for neighbouring countries: trade and services make up around 80% of GDP. But this makes Djibouti vulnerable to developments around it, such as wars between Ethiopia and Eritrea.

Road and rail links with Addis Ababa are now in a very poor state—and in need of investment. Most food has to be imported and there is little industry.

Since independence in 1977, Djibouti has been ruled by the Rassemblement populaire pour le progrès (RPP), which from 1981 to 1992 was also the only legal party. But the government could not contain the Somali–Afar rivalry and in 1990 an armed rebellion was started by the Afar front pour la restauration de l'unité et la démocratie (FRUD).

Decision-making is centralized around the president. Until April 1999, Djibouti had had only one president, Hassan Gouled Aptidon, who was then replaced by his close confidant, and former security chief, Ismaël Omar Guelleh. He has continued in the same vein, eliminating political opponents and using military force to crush the rebels. In 2003 in the parliamentary election, the president's supporters won every seat, and in 2005 Guelleh achieved a second six-year term in an uncontested election.

Dominican Republic

The expansion of manufacturing industry and tourism has yet to benefit most of the poor

Land area: 49,000 sq. km.
Population: 9 million—urban 59%
Capital city: Santo Domingo, 2.7 million
People: Mulatto 73%, white 16%, black 11%
Language: Spanish
Religion: Roman Catholic
Government: Republic
Life expectancy: 67 years
GDP per capita: $PPP 6,823
Currency: Dominican peso
Major exports: Manufactured goods, ferronickel, sugar, gold, coffee

The Dominican Republic occupies the eastern two-thirds of Hispaniola, an island that it shares with Haiti. Much of the territory is mountainous—the highest peaks in the central highlands rise above 3,000 metres. The best land for agriculture and the most densely populated area, apart from the capital, is the Cibao Valley, which stretches across the north of the country.

Most people are of mixed race, or mulatto—the result of intermarriage between the predominantly Spanish colonists and the slaves brought to work in mines and sugar plantations. Wealth generally correlates with colour of skin—the blacks being the poorest. Education standards are low: although most children enrol in primary school around one-quarter subsequently drop out. Health standards are also low, though access to services such as clean water and sanitation has been improving.

Poor as the Dominican Republic is, neighbouring Haiti is even poorer, and up to 500,000 Haitians, many coming to work on the sugar harvest, are thought to live in the Dominican Republic. Most have no legal status and live in squalid conditions.

Sugar is still important—and one of the largest sources of export earnings, taking advantage of preferential access to the US market. But shrinking demand and under-investment have reduced production. Other minor export crops, such as cocoa and coffee, have also suffered setbacks due to disease and drought. Exports of tobacco and cigars have done better. Food production too has been slowing. Again this has been the result of low investment—discouraged both by high interest rates and government subsidies for imported food—particularly in election years. By 2004, only 15% of the population worked in agriculture.

Mining is another source of revenue, though it employs few people. The Canadian-owned Bonau nickel mine has been a major source of export earnings but has been hit by falling world prices. A more dynamic part of the economy has been manufacturing which in 2005 accounted for 15% of GDP.

Apart from more traditional industries like food processing, the fastest growth has been in the free-trade zones. More than 400 companies, the majority US-owned, have established factories in 40 or more zones scattered around the country. Around two-thirds of these factories make garments, while others produce footwear, leather goods, and electronics. The zones do provide employment for a predominantly female workforce, but working conditions are poor and wages are desperately low. These factories are also facing increased competition from low-wage countries in Asia.

Another rapidly growing source of income and foreign exchange is tourism. The Dominican Republic welcomes more than three million visitors each year. More than half are from Europe, though arrivals from North America have been rising as US tourists have become concerned about travelling further afield.

Few tourists venture beyond the beaches

Few visitors venture out onto the streets, preferring the self-contained beach resorts. But even here they are vulnerable to the country's poor infrastructure—particularly to pollution from inadequate sanitation. Most of the resorts are owned by multinational, often Spanish, corporations.

The Dominican Republic is still struggling to build a stable democracy. For 30 years after his first election victory in 1966, the government of the Dominican Republic was dominated by Joaquín Balaguer and

his conservative Partido Reformista Social Cristiano (PRSC), whose administrations were corrupt and repressive. These years were marked by instability and violence. There were, however, a couple of interludes of rule by the centre-left Partido Revolucionario Dominicano (PRD).

In 1996, when he was 88 years old and virtually blind, Balaguer's seventh and last term ended prematurely following allegations of electoral fraud. The ensuing election was won by Leonel Fernández Reyna of the Partido de la Liberación Dominicana (PLD), formerly a Marxist split-off from the PRD, though now a party of the centre.

Fernández could not stand again in the May 2000 presidential election. This was won by the PRD candidate Hipólito Mejía who won with just under half the vote. Although he campaigned with promises of reform and investment in public services he failed to deliver on most of these, and his administration was tainted by corruption and the collapse of the country's second largest bank.

In the May 2004 presidential election Fernández and the PLD returned to power with 57% of the vote. He proved quite effective: by 2005 he had managed to reduce inflation from 50% to 7% and boost economic growth to 9%. Fernández had less success in tackling corruption because of dependence in Congress on the PRD. But following the 2006 legislative elections when the PLD gained a majority in both houses he is now in a much stronger position. Other priorities include improving electricity supplies and tackling drug-related crime.

Ecuador

Tensions between mestizos and Amerindian groups make for uneasy government

Land area: 284,000 sq. km.
Population: 13 million—urban 62%
Capital city: Quito, 1.4 million
People: Mestizo 65%, Amerindian 25%, Spanish 7%, black 3%
Language: Spanish, Amerindian languages
Religion: Roman Catholic
Government: Republic
Life expectancy: 74 years
GDP per capita: $PPP 3,641
Currency: US dollar
Major exports: Bananas, coffee, oil

Ecuador has three main geographical areas. To the west, containing around one-quarter of national territory, are generally low-lying coastal plains. These rise abruptly to two Andean mountain chains that run down the centre of the country from north to south and are separated by high valleys. Beyond the Andes lies the 'Oriente'—the Ecuadorian part of the Amazon basin with its tropical rainforests. In addition, Ecuador has sovereignty over the Galapagos Islands 1,000 kilometres to the west in the Pacific Ocean. As its name indicates, Ecuador sits astride the Equator, and outside its mountain regions has a year-round humid climate.

Most Ecuadorians are mestizo—mixtures of Amerindians and white immigrants who came chiefly from Spain and other European countries, though there are also descendants of other immigrants including the Lebanese who arrived in the 1900s. The various Amerindian groups are largely Quechua-speaking and can be found throughout the country but particularly in the highlands. They have not in the past exerted much influence but have recently become more assertive. There are also two main black groups—one in Esmeraldas on the north-east coast and the other in the Chota Valley in the northern highlands. Each year around 100,000 people emigrate, chiefly to the USA or Spain.

This is a very unequal society: in 2004 the richest 10% of people received 42% of national income and around one-third of the population lived below the national poverty line.

Standards of education and health are low, particularly in the highlands. Around one-quarter of the population do not have access to public health facilities—a result of very low health budgets. And people in rural areas are poorly educated: Amerindian groups on average get only two years at school.

Ecuador's distinct geographical areas have tended to impede national integration. People in the mountains, particularly those in the capital, Quito, have generally been more traditional and conservative compared with the more commercially oriented communities on the Pacific coast centred on the port city of Guayaquil.

This divide is evident too in agriculture, which is the country's major employer. In the mountains, production is chiefly for national consumption—typically corn, potatoes, and beans, as well as livestock. But productivity is low and landholdings are small—tempting some farmers to colonize the sparsely populated Oriente, where they face resistance from indigenous groups. On the coast, much of the land is devoted to large plantations, especially of bananas, of which Ecuador is the world's largest exporter. Fishing is also an important source of both food and export revenue—particularly the cultivation of freshwater shrimps.

Ecuador is the world's leading exporter of bananas

Industry and manufacturing in Ecuador consist largely of goods for local consumption. But participation in the Andean Community and the WTO has stimulated exports particularly of paper and wood.

The largest export industry continues to be oil which is extracted from the Oriente both by the state-owned Petroecuador and by foreign oil companies. Production has risen steadily, but because local refining is limited much fuel still has to be imported.

Governance in Ecuador has been fragile. Since the end of military rule and the subsequent democratic elections in 1979, successive governments have switched between centre-right, centre-left and populist—though none has managed to hold the country together sufficiently to follow a coherent development strategy.

One of the stranger fates was that of Abdalá Bucaram of the Partido Roldista Ecuatoriana, who was elected president in 1996 and dismissed by the Congress in 1997 on the grounds of 'mental incapacity'.

He was temporarily replaced by Fabián Alarcón. And then the 1998 presidential election was won by the ex-mayor of Quito, Jamil Mahuad of the centre-right Democrácia Popular. He made an encouraging start in that he ended a long-running border dispute with Peru. But the catastrophic economic situation in 1999 forced Ecuador to become the first government in recent history to default on some of its sovereign debt.

Mahuad's downfall came in January 2000 when he proposed to replace the sucre with the US dollar. Trade unions protested and Amerindian groups marched on Quito. At this point, the army stepped in and removed Mahuad in favour of his vice-president, Gustavo Noboa, who nevertheless pressed on with replacing the sucre by the dollar.

Lucio Gutiérrez, one of the coup leaders, then formed his own party, the Partido Sociedad Patriótica 21 de Enero (PSP), and won the presidential election in January 2003. Despite the left-wing rhetoric of his campaign, Gutiérrez came to terms with the IMF and started to cut public spending. Two years later, however, he was removed from office by the Congress, and was later jailed for sedition. He was replaced by former vice-president, Alfredo Palacio—who also faces public unrest.

Egypt

Egypt, the largest Arab country, has made little progress towards democracy

Land area: 1,001,000 sq. km.
Population: 71 million—urban 42%
Capital city: Cairo, 11.0 million
People: Egyptian 98%; Berber, Nubian, Bedouin, and Beja 1%; other 1%
Language: Arabic, English, French
Religion: Muslim 90%, Coptic 9%, other 1%
Government: Republic
Life expectancy: 70 years
GDP per capita: $PPP 3,950
Currency: Egyptian pound
Major exports: Petroleum products, cotton products

Egypt's vast territory consists of two main desert areas separated by the fertile Nile Valley. The western desert occupies around two-thirds of the country, a rocky sandy plateau with occasional inhabited oases. The eastern desert, with around one-quarter of the territory, is even less hospitable and more sparsely populated. The country's lifeline, the Nile, provides most of the water, and its valley and delta are home to more than 95% of the population.

Ethnically most Egyptians are the result of generations of intermarriage between Arabs and other groups. Most people are Sunni Muslims, but an important minority are members of the Coptic Orthodox Church, whose religion predates the Arab conquest, and who have a strong influence in both commerce and government. The Copts have also at times been the target of Islamic militancy.

Though some Egyptians have prospered, around 40% of the population live on less than $2 per day. One-fifth of the workforce are unemployed or underemployed. Some two million Egyptians also work abroad, chiefly in the Gulf countries, sending home around $3 billion each year in remittances.

Around 30% of the workforce are still engaged in agriculture, mostly in small plots in the Nile Valley. With intensive applications of fertilizer, and manual irrigation systems, these farms are quite productive—though many are affected by waterlogging and salination.

One of the main crops is Egypt's high-quality, long-staple cotton, but output and export income have been in decline. The main food crop is wheat and production is intensive, but output is insufficient for a fast-growing population, and Egypt is one of the world's largest food importers.

To increase production the government has been reclaiming land from the desert; meanwhile these gains have largely been offset by losses to urbanization and industry. The largest scheme is the Southern Valley development project with an estimated cost of $86 billion.

Egypt's most important industrial enterprise is oil production. Though by Middle Eastern standards the reserves are modest, the oil and petroleum products from Egyptian refineries are still the main export earners, and recently the government has made further investment in gas. These industries, along with others like cotton spinning, are in the hands of state enterprises which employ around one-third of the workforce.

Within the service sector, one of the most important sources of income is the Suez Canal. Around 16,000 vessels pass through each year.

16,000 vessels per year through the Suez Canal

Another vital service industry is tourism. Some eight million people arrived in 2004, chiefly visiting the historical sights, spending $5 billion and employing two million people. The industry has, however, been vulnerable to the effects of terrorist attacks and political instability elsewhere in the region, notably the war in Iraq.

Since it became a republic in 1952, Egypt has never had a pluralist democracy. The early years were dominated by Gamel Abdel Nasser, who nationalized many industries, including the Suez Canal. He was replaced on his death in 1970 by Anwar el-Sadat. In 1978, Sadat established the current ruling party, the National Democratic Party (NDP). Sadat was assassinated in 1981 by Islamic terrorists and was replaced by the current president, Hosni Mubarak.

Much recent history has been dominated by the relationship with Israel. Egypt has acted as a peace-broker between other Arab states and Israel for which it has been rewarded with billions of dollars in US aid.

Mubarak and the NDP face very little effective opposition, not least because they maintain tight control. The country is ruled under an almost permanent state of emergency; most kinds of public gathering are prohibited. There are 16 officially recognized, and weak, opposition parties. The most direct political opposition has come from the Muslim Brotherhood.

There have also been a number of Islamic terrorist groups. The most active was the Gamaat Islamiya (Muslim Groups). But in 1999 this announced a ceasefire, and in 2000 the other main group, al-Jihad, also said it would stop operations.

Until the latest election Mubarak had guaranteed his own re-election because the single presidential candidate was nominated by the People's Assembly which he controls. Electors could only say yes or no to him in a referendum.

For 2005, however, Mubarak announced that there would be multi-candidate presidential elections. When these and subsequent parliamentary elections were held, however, they made very little difference. Candidates faced many restrictions and the process was accompanied by the usual violence and fraud. Mubarak was returned with 89% of the vote on a 23% turnout. The electoral changes were probably designed to create a legitimate succession path for his son Gamal. Meanwhile hundreds of opposition figures, including the principal presidential opponent, have been jailed.

El Salvador

El Salvador has moved from a debilitating civil war to an often violent peace

Land area: 21,000 sq. km.
Population: 7 million—urban 59%
Capital city: San Salvador, 1.5 million
People: Mestizo 90%, white 9%, Amerindian 1%
Language: Spanish
Religion: Roman Catholic, Protestant
Government: Republic
Life expectancy: 71 years
GDP per capita: $PPP 4,781
Currency: Colón, US dollar
Major exports: Coffee, manufactures

El Salvador is the smallest and most densely populated country in the continental Americas. It has two highland areas: to the north, a rugged mountain range along the border with Honduras, and to the south a range of 20 major volcanoes separated by a series of basins that make up the central plain. The major lowland areas are the valley of the Lempa River that runs from north to south, and a narrow plain along the Pacific coast.

The volcanoes have dominated El Salvador not just physically but also economically. The rich and porous volcanic soils that cover one-quarter of the land area are ideal for coffee cultivation, and the struggle for control over this and other fertile land has shaped the country's history. Land clearing for cultivation has removed most of the forest cover.

Little of El Salvador's Amerindian culture remains. In the 1930s the army launched a vicious military campaign, La Matanza, the 'killing', which not only cost 30,000 lives but also laid waste to the indigenous culture. The civil war of the 1980s also left its mark. Aside from the human and economic costs there is now a disturbing culture of violence particularly from street gangs.

Levels of human development remain low, though literacy rates and life expectancy have increased. Almost half the population live below the national poverty line.

Agriculture employs one-quarter of the workforce and provides 70% of domestic food requirements. Despite post-war gestures at land reform, the best land is still controlled by a few wealthy people: around 1% of landowners control more than 40% of the arable land. Coffee is still one of the main export earners, grown on the slopes of the mountains. Sugar and cotton are cultivated in the lowlands on the coastal plain. The main food crops are corn, beans, and rice, but much of the food has to be imported.

One of the most pressing issues now is environmental degradation. With most of its forest cover removed, El Salvador suffers from extensive soil erosion. Agriculture and rural development generally are also vulnerable to natural hazards such as hurricanes and earthquakes.

El Salvador's economy is now dominated by industry and services. Manufacturing was originally designed for import substitution, but in recent years the emphasis has been on manufacturing for export in 'maquila' factories engaged in offshore assembly. In 2003 they employed 89,000 people, the majority of whom were women. Most of this involves the production of garments which has overtaken coffee as the leading export earner. However these factories and other exporters are facing stiffer competition from Mexico and other countries in Latin America.

In an effort to increase foreign investment and trade El Salvador has

Adopting the dollar

adopted the US dollar as a legal currency at a fixed exchange rate against the colón and will eventually use only the dollar. Nowadays the largest source of dollars is actually the two million or more Salvadorans living abroad, most in the USA, who send home over $2 billion annually in remittances. Emigration surged during the war but continues at a steady rate, much of it unauthorized.

El Salvador's long history of injustice and bitterness finally erupted into armed revolt in the 1970s. A series of guerrilla groups united in 1980 as the Farabundo Martí Liberation Front (FMLN), and when Archbishop Oscar Romero was assassinated in 1980 his death triggered an armed insurrection that launched the country into a full-scale civil war.

In 1982, the right-wing Alianza Republicana Nacionalista (Arena) party took over the government and was sustained by massive US aid. Death squads roamed the country, killing anyone promoting peaceful reform. The war lasted until 1992 and cost 80,000 lives. Around one million people were displaced—many ending up in the USA.

A UN-supervised ceasefire was finally achieved in January 1992. The FMLN agreed to end the armed struggle, while the government said it would dismantle the death squads, reduce the size of the armed forces, and redistribute some land.

For the presidential elections in 1994 the FMLN had transformed itself into a political party, but lost to the Arena candidate Calderón Sol. Since then the pattern has been polarized in much the same way, with the right-wing Arena holding onto the presidency and the FMLN as a substantial left-wing opposition in the legislative assembly. There is no significant centre party.

In the 1997 legislative elections the FMLN was only one seat behind Arena. Nevertheless in 1999, Arena's Francisco Guillermo Flores Pérez was elected president, albeit on a very low turnout. Then in the 2003 legislative elections the FMLN overtook Arena to become the largest party.

Despite Arena's failure to reduced levels of crime, poverty, or unemployment, the Arena candidate, media baron Antonio Saca, scored a decisive victory in the 2005 presidential elections. The legislative elections in 2006, however, did not give Arena an overall majority so Saca has to negotiate an all-party consensus for his reforms.

Equatorial Guinea

Equatorial Guinea has discovered oil but has yet to find democracy

Land area: 28,000 sq. km.
Population: 0.5 million—urban 48%
Capital city: Malabo, 156,000
People: Fang, Bubi
Language: Spanish, French, pidgin English, Fang, Bubi
Religion: Roman Catholic
Government: Republic
Life expectancy: 43 years
GDP per capita: $PPP 19,780
Currency: CFA franc
Major exports: Oil, timber

The mainland of Equatorial Guinea, called Río Muni, consists of a coastal plain and then forested hills and plateaux that extend east to the border with Gabon. The country also includes several islands, the largest of which is Bioko, 200 kilometres to the north in the Gulf of Guinea and home to the capital, Malabo.

The majority ethnic group are the Fang, though these are divided into many subgroups. Those on Bioko, 15% of the population, are mostly Bubi—many of whom are politically and economically marginalized. This was Spain's only colony in Sub-Saharan Africa, and Spanish is the official language. Health standards are poor, but literacy is higher than in neighbouring countries.

During the Spanish period, the country had thriving cocoa plantations, but most of these withered away during the early years of independence. Around 85% of the labour force still work in agriculture. Most farmers grow subsistence crops like cassava and sweet potatoes, along with some cocoa and coffee. The main agricultural export is timber, mainly okoumé, which is exported to Asia and is used to make plywood.

Equatorial Guinea's prospects were transformed in 1991 by the discovery of oil off the coast of Bioko. By 2005, oil represented 86% of GDP. In recent years growth has been around 9% annually, though much of the benefit has gone to Western oil companies.

Democracy has never had a foothold. At independence the country was ruled by Francisco Macías Nguema, a dictator who dragged Equatorial Guinea into a social and economic abyss. In 1979, after a bloody coup supported by Spain and Morocco, he was overthrown by his nephew, Teodoro Obiang Nguema Mbasogo, who merely delivered a different brand of dictatorship.

In 1991, under pressure from aid donors, Obiang opened the country up to multiparty democracy, but his Partido Democrático de Guinea Ecuatorial (PDGE) remains dominant. In 1996, and in 2002, he was re-elected president, achieving 97% of the vote in elections marked by fraud and intimidation. His repressive reign continues. In 2002, for example, following an alleged coup attempt, hundreds were arrested, many of whom were tortured.

There has been speculation about Obiang's health and he has been grooming his son, Teodorín Nguema Obiang, as his successor, though his brother, Armengol Nguema Mba, also has his eyes on the job.

Eritrea

Eritrea has stopped its pointless war with Ethiopia but relations remain tense

Land area: 125,000 sq. km.
Population: 4 million—urban 20%
Capital city: Asmara, 500,000
People: Ethnic Tigrinya 50%, Tigre and Kunama 40%, Afar 4%, Saho 3%, other 3%
Language: Afar, Amharic, Arabic, Tigre and Kunama, Tigrinya
Religion: Muslim, Coptic Christian, Roman Catholic, Protestant
Government: Republic
Life expectancy: 54 years
GDP per capita: $PPP 849
Currency: Nafka
Major exports: Livestock, textiles, food

Eritrea can be divided into three main areas. The eastern part of the country is a long, arid coastal plain inhabited by nomadic herders or fishing communities. The savannah and shrub-covered western lowlands descend to the border with Sudan. The main part of Eritrea is the central highlands which are part of the Ethiopian Plateau. This is the most densely populated part of the country and has the more fertile land, though many of the hills have suffered severe soil erosion.

Eritrea has a number of ethnic groups of which the largest are the Tigrinya, who speak Tigrinya. They form most of the population of the highlands and are also to be found in the adjacent Ethiopian region of Tigray. Another major group are the similarly named Tigre, who speak Tigre, and are to be found in the north of the plateau, as well as in parts of the lowlands. Though their languages have the same root and script, they are mutually unintelligible. Hundreds of thousands of Eritrean refugees who fled during the wars in the early 1990s still live in Sudan and Ethiopia, though many were expelled from Ethiopia in 1999 during a border war.

Decades of war have kept Eritreans desperately poor. More than 40% of children are malnourished and only 43% of rural people have access to improved water sources. Education levels are also low. During the Ethiopian period, schools were forced to teach in Amharic, though most schooling was confined to urban areas. Only 45% of children are enrolled in primary school.

More than two-thirds of the population depend on agriculture or fishing. Farmers grow subsistence cereal crops and raise a small number of animals, but output is low. They have little irrigation and have to rely on erratic rainfall. And in the densely populated highlands their land has become increasingly eroded. Given Eritrea's long coastline, the fishing industry has considerable potential for local consumption and export.

Before its incorporation into

Ethiopia in the early 1950s, Eritrea had started to develop a number of industries. But Ethiopian priorities steadily undermined Eritrean enterprises and subsequent conflicts destroyed many of the rest. Even so, there are still light industries processing food and producing textiles and garments, mostly in state-run enterprises.

When it achieved independence from Ethiopia in 1993, Eritrea gained the entire coastline, so in principle Eritrea's ports should be a useful source of income—though this does depend on an abiding peace with Ethiopia. Another important source of income is remittances from Eritreans abroad—worth around $400 million in 1999.

Eritrea got Ethiopia's coastline

The government prefers to manage without external aid, but has accepted funds for post-war reconstruction as well as food aid.

Formerly an Italian colony, Eritrea was federated into Ethiopia in 1952, and in 1962 was formally, and illegally, annexed by Ethiopia. By then an independence struggle had already started but it was weakened by a long conflict between two groups. The Eritrean Liberation Front (ELF), which emerged in 1958, was organized along ethnic and regional lines. But in 1973 a socialist splinter group appeared—what is now the Eritrean People's Liberation Front (EPLF). The two groups fought each other as well as the Ethiopians.

By 1991, the EPLF was dominant and had seized control of the territory. Meanwhile the Ethiopian government in Addis Ababa had been overthrown by the Tigrayan People's Liberation

Front, another Marxist-inspired guerilla group which had good relations with the EPLF. In 1993 this led to an amicable parting, and independence for Eritrea.

Early in 1994, the EPLF disbanded and re-emerged as the People's Front for Democracy and Justice (PFDJ), which was to incorporate all political parties, including some former ELF members, with Isaias Afwerki as president.

Eritrea achieved a new and supposedly pluralist constitution in 1997, though in practice opposition parties are banned. The PFDJ imposes this partly through moral force derived from long years of struggle, but also through repression.

Eritrea soon entered into conflicts with all its neighbours. The first was with Sudan: each country accusing the other of harbouring its dissidents. Then there were arguments with Yemen and Djibouti.

The most damaging and surprising conflict, however, was with Ethiopia. In May 1998, a border skirmish erupted into a full-scale war that by mid-1999 had cost around 50,000 lives. Ostensibly, the war was fought over a few barren hectares of loosely defined border which Eritrea claims that Ethiopia had been encroaching on, and taxing its people.

In December 2000 a peace agreement was signed and in April 2001 the UN started to monitor a Temporary Security Zone along the disputed border. Although it looked as though a border agreement had been reached, by mid-2004 the two sides were again quarrelling and plans to demarcate the border have been postponed indefinitely.

Estonia

Estonia is one of the more successful ex-Soviet republics and has turned decisively towards Europe

Land area: *45,000 sq. km.*
Population: *1 million—urban 70%*
Capital city: *Tallinn, 398,000*
People: *Estonian 68%, Russian 26%, Ukrainian 2%, Byelorussian 1%, other 3%*
Language: *Estonian, Russian, Ukrainian*
Religion: *Evangelical Lutheran, Russian Orthodox, Estonian Orthodox*
Government: *Republic*
Life expectancy: *71 years*
GDP per capita: *$PPP 13,539*
Currency: *Kroon*
Major exports: *Mechanical equipment, food, textiles, wood products*

Estonia's mainland consists largely of a plain that includes many lakes. Around 40% of the territory is covered with forests. In addition, it has more than one thousand islands and islets in the Baltic Sea.

Estonians were disturbed when many Russians moved into the country following the Soviet invasion in 1940. Once Estonians recovered their independence in 1991 they were therefore determined to re-exert their national identity and demanded that anyone who wished to become a citizen, or to work in public administration, should speak Estonian—not an easy language to learn.

For the Russians, who made up one-third of the population, this was a significant obstacle. Some have been emigrating but most have stayed. Since then there has been some relaxation: for parliamentary candidates, for example.

Russia has frequently complained about discrimination against Russians, and Estonia's relations with its giant neighbour remain tense. Estonia has, however, been strengthening links with Finland: Estonian is related to Finnish, which helps to sustain cultural and commercial ties.

Education standards are quite high and in response to the increasing demand for educated labour more students are entering tertiary education. Immediately following independence, health standards and life expectancy fell but have since largely recovered.

Of the former Soviet republics, Estonia made one of the most rapid transformations towards a market economy. The government started the privatization programme with land and housing by distributing vouchers to all households, and later extended the scheme to large companies. By 2001, more than 85% of the economy was run by the private sector.

After independence, Estonia's economy contracted suddenly as ties with Russia were cut. But from around 1995 the economy revived, thanks in part to foreign investors impressed by Estonia's enthusiastic liberalization, open trading regime, and low wage rates. This meant that basic industries such as textiles and timber products, along with chemicals, plastics, and metals, revived fairly quickly, finding new markets in Europe. Estonia joined the EU in May 2004.

Finland has been one of the main investors. The electronics company Elcoteq, for example, attracted by wages one-sixth of those in Finland, assembles mobile phones and other items for Nokia and other Scandinavian companies, and is now Estonia's largest exporter.

Estonians assemble mobile phones for Nokia

Estonia has some of its own energy supplies, based on extracting oil from shale, but this is an expensive and polluting process, and the country still relies on Russia for additional, and more expensive, supplies of oil.

Agriculture, which is largely based on livestock, has not performed so well. Between 1989 and 2005, its share of GDP fell from 20% to 4%. The government rapidly dismantled the former collective farms to create more private farms, many of which have now consolidated into larger enterprises. While food output has been declining, forestry has been buoyant as a result of increasing demand for pulp and paper. Fishing has suffered from reductions in EU quotas.

Estonia's market economy has also opened up the potential for a larger and more vigorous service sector that had been stifled during the Soviet years. Tourism has certainly benefited. Estonia abolished visa requirements for visitors from Nordic countries and now welcomes around three million tourists a year—the majority from Finland arriving by ferry and often to take advantage of low prices for liquor.

Estonia, which had first gained independence in 1918 after centuries of rule by Sweden or Russia, regained its independence in 1991 following the demise of the Soviet Union and adopted a new constitution in 1992. This provides for a single-chamber parliament, the Riigikogu, which elects a president whose role is largely ceremonial.

Since independence Estonia has had a large number of political parties, and a series of coalition governments. The 1992 elections produced a nationalist coalition which pushed through rapid reforms, but its popularity was undermined by economic contraction and political scandals. The 1995 elections resulted in a swing to the left—though the government led by Mart Siimann carried on many economic reforms.

The 1999 elections were contested by more than 30 parties. The outcome was a government formed by a centre-right coalition with Mart Laar as prime minister. However the coalition was fractious and Laar resigned in 2002 with the Reform Party leader Siim Kallas taking over.

The parliamentary elections in March 2003, which were contested by a modest 12 parties, resulted in a centre-right coalition government led by Juhan Parts, of Res Publica, a right-wing party. He resigned in 2005 to be replaced by Andrus Ansip, of the Reform Party, heading Estonia's 12th government since independence.

Ethiopia

A former favourite country of Western aid donors has become increasingly repressive

| Land area: 1,221,900 sq. km. |
| Population: 74 million—urban 16% |
| Capital city: Addis Ababa, 2.4 million |
| People: Oromo 40%, Amhara and Tigrean 32%, Sidamo 9%, other 19% |
| Language: Amharic, Tigrinya, Orominga, Guaraginga, Somali, Arabic, English |
| Religion: Muslim 45%, Ethiopian Orthodox 35%, animist 12%, other 8% |
| Government: Republic |
| Life expectancy: 48 Years |
| GDP per capita: $PPP 711 |
| Currency: Birr |
| Major exports: Coffee, hide and skins, qat |

More than half of Ethiopia consists of two high tablelands—the western and eastern highlands which are split, south-west to north-east, by the East African Rift Valley. To the east, much of the territory consists of the lower arid plains of the Danakil depression and the Ogaden.

Ethiopia's population is very diverse, the result of many centuries of linguistic and racial mixing. Officially, there are 64 recognized ethnic groups, but there are probably 200 or more languages. Following a new constitution in 1995, the country became a federal republic and now has nine regions or 'nationalities', though each of these actually contains a mixture of ethnic groups. The largest group are the Oromo, who are to be found in the centre and south, followed by the Amhara and the Tigrayans, who occupy the north and the north-west. Many are Christians, belonging to the previously dominant Ethiopian Orthodox Union Church, but nowadays more are Muslims.

Ethiopia remains one of the world's poorest countries. Half the population are malnourished, infant mortality is high, and life expectancy short. Fewer than 20% of Ethiopians have access to modern healthcare services and their health status is among the worst in the world. Malaria has resurged and AIDS is a major threat: more than 4% of Ethiopians were infected with HIV in 2004. Education standards are also very low: only 40% of people are literate. The population grows by around 2.5% per year and by 2020 it could exceed 100 million.

Some 85% of the population live in the rural areas, and survive mostly through farming—growing a range of subsistence crops, including teff, which is a cereal grass. But agricultural output has been extremely variable. The high population density in the highlands and ensuing deforestation have led to extensive soil erosion. And since almost all agriculture is rain-fed it is very vulnerable to the unreliable climate. Per capita food production has declined sharply. In a good year

Ethiopia is just about self-sufficient, but when the rains fail it has to rely on emergency shipments of food aid. Ethiopia's strategies to improve food security include, on the one hand, stronger early warning systems, and on the other, efforts to increase output through greater use of fertilizer.

But some of the problems are political. In the past farmers had greater freedom to migrate temporarily to help harvest cash crops and supplement their incomes. Now they need permission to travel. In 2002/3, Ethiopia was again in the grip of a famine that affected an estimated 13 million people.

The main cash crop is coffee, which is grown mainly in the south by smallholders and which has provided around two-thirds of export earnings—though less recently because of low world prices. Another important agricultural export is qat, a bush whose leaves can be chewed and are a mild stimulant popular in Somalia and Djibouti.

Cultivating qat for its neighbours

Ethiopia has little industry. Most is controlled by the state and is concentrated in food processing. However, some industries, particularly textiles and garments, are being privatized, and there is probably potential for more private-sector activity in the processing of hides and skins, which are also major exports.

In the past three decades, Ethiopia has been through a series of political traumas and civil wars. Following a famine in 1972–74, which cost an estimated 200,000 lives, political unrest led to the overthrow of the monarchy of Haile Selassie and in 1975 to the establishment of a socialist republic. But civilian groups failed to cohere into viable political parties, leaving the way clear for the establishment of a 17-year military regime, the Provisional Military Administrative Council, known in Amharic as the 'Derg'.

From 1977, the Derg was controlled by Colonel Mengistu Haile Mariam, whose repressive regime tried to transform Ethiopia into a Marxist-Leninist state. It survived until 1991 when, having lost the support of the Soviet Union, and following uprisings in Eritrea and Tigray, it was defeated by the Ethiopian People's Revolutionary Democratic Front (EPRDF). This led in 1993 to independence for Eritrea—which left Ethiopia landlocked.

Meles Zenawi, the Tigrayan leader who was head of the transitional administration has remained in charge. The EPRDF, a group which is still dominated by the Tigray People's Liberation Front, regularly wins most of the seats for the legislative assembly, most recently in 2005, and chooses Zenawi as prime minister.

The legislature also elects the largely ceremonial president, currently Girma Wolde-Giorgis.

Strangely, the former allies Ethiopia and Eritrea declared war in 1998 over a border dispute. The fighting stopped in 2000 but the border dispute is unresolved.

Western governments had previously been attracted by what appeared to be one of Africa's more progressive and least corrupt regimes, but they have had their illusions dashed as Zenawi has become increasingly repressive, locking up political opponents and journalists.

Fiji

Indigenous Fijians and Indians struggle for power

Land area: 18,000 sq. km.
Population: 0.8 million—urban 52%
Capital city: Suva, 176,000
People: Fijian 51%, Indian 44%, other 5%
Language: English, Fijian, Hindi
Religion: Christian 52%, Hindu 38%, Muslim 8%, other 2%
Government: Republic
Life expectancy: 68 years
GDP per capita: $PPP 5,880
Currency: Fiji dollar
Major exports: Sugar, garments, gold

Fiji comprises hundreds of islands and islets in the South Pacific. The two largest islands, Vitu Levu and Vanua Levu, account for more than 80% of the territory.

The population is finely balanced between Fijians, who are of mixed Melanesian and Polynesian origin, and Indians, who are descendants of 19th-century immigrant sugar workers. Indians were narrowly in the majority until 1987, when many fled after the first coup; now the balance again favours Fijians.

Sugar remains the backbone of agriculture, but drought and worries about land tenure have affected production. Sugar processing also used to dominate industrial production, but manufacturing industry is now more diverse. Garments now account for one-quarter of export income, though will suffer from a loss of preferential access to Australia and the USA. Earnings from the Vatukoula gold mine are also important. In addition, Fiji depends heavily on tourism: in 2004 more than 500,000 tourists arrived.

Politics in Fiji has long been afflicted by conflicts between ethnic Fijians and those of Indian origin. A major issue is the ownership of land which has largely been monopolized by indigenous Fijians who lease it to Indians—who produce virtually all the sugar. Many leases are now becoming due, and a proposal to offer new leases sparked the latest conflict.

From independence in 1970, indigenous Fijians had controlled the government. The 1987 election, however, was won by two parties that were largely supported by Indians. The Fijians responded with a military coup and a constitutional change to bar Indians from power. This led to international isolation that only ended in 1994 when a new constitution gave the Indians a fairer deal. Eventually, in 1999, with the Fijian vote split among five parties, the Indian-dominated Fiji Labour Party won a majority in parliament and Mahendra Chaudhry became the first Indian prime minister.

But his promise of 30-year land leases to Indians incensed the Fijians. In May 2000 businessman George Speight, along with a section of the army, staged a coup. After a period of marshal law the military installed an interim civilian government led by Laisenia Qarase. He then formed a new indigenous party, the People's Unity Party, which won a majority in the 2001 election and again in 2006. As required by the constitution he has to offer government posts to the FLP, who came second. But he now faces opposition from the military.

Finland

In a short time, thanks to Nokia, Finland has become a leader in telecommunications

Land area: *338,000 sq. km.*
Population: *5 million—urban 61%*
Capital city: *Helsinki, 560,000*
People: *Finn 93%, Swede 6%, other 1%*
Language: *Finnish, Swedish*
Religion: *Lutheran National Church 84%, other 2%, none 14%*
Government: *Republic*
Life expectancy: *79 years*
GDP per capita: *$PPP 27,619*
Currency: *Euro*
Major exports: *Metals and machinery, forestry products*

Most of Finland is fairly low-lying and the country as a whole is heavily forested. One of its most striking features is the complex of thousands of shallow lakes in the south. Here the climate is fairly temperate but even some of the ports on the Baltic Sea are icebound during the long winter. Most of the higher ground is in the northern one-third of the country, and since this lies within the Arctic Circle the climate is severe.

Finland's population, which is concentrated in the south, is ethnically fairly homogenous. Nevertheless, there are also significant numbers of Swedes, mainly in the southern coastal areas and on the Åland Islands, which lie between the two countries. Swedish is also an official language. Another minority, though now very small in number, are the Sami, or Lapps, who herd their reindeer in the far north. The break-

up of the Soviet Union also caused some immigration of ethnic Finns from Russia and Estonia, but the proportion of foreign-born is less than 2% and immigration has often been exceeded by emigration. By European standards, Finland is also fairly rural—more than one-third of people live in the countryside.

Even so, agriculture is not a major source of income, employing only 5% of the workforce. The fierce climate and poor soils mean that less than one-tenth of the land is suitable for farming—though the country is largely self-sufficient in basic foods.

The most important rural activity is forestry. Finland's vast forests of pine, spruce, and birch support a number of major industries. The extensive pulp and paper sector, which is dominated by three main companies, is responsible for around 12% of world exports of paper and paperboard. In addition, Finland is an important producer of building materials and furniture, and has used its forestry experience to specialize in the production of forestry machinery.

Finland's other industries have also been expanding fast. Remarkably, most of this development has come from one company, Nokia, which is the world's largest manufacturer of mobile phones and produces other telecommunications equipment. As a result, industry still contributes more than one-quarter of GDP, a higher proportion than in most developed countries.

Finland had made steady economic progress for decades but was rocked by a deep recession in the early 1990s and by the collapse of trade with the Soviet Union, which had taken around one-fifth of exports. The government made determined efforts to tackle the problems. It made deep cuts in public expenditure and this, combined with contributions from the forestry and electronics industry, helped Finland to emerge from the worst of the crisis. Over the period 1995–2000 economic growth averaged 4% per year. Unemployment has also been cut and by 2004 was down to 9%.

However this progress has been quite narrowly focused on telecommunications equipment and is over-reliant on the fortunes of Nokia. Although Oulu in the north is also a technology centre, the industry is increasingly concentrated in the south of the country in the region bounded by Helsinki, Turku, and Tampere, to which many people have been migrating.

Tourists head for the land of the midnight sun

In addition, Finland has been developing its tourist industry, based on skiing and on visitors wanting to experience Lapland's midnight sun.

Finland joined the EU in 1995 and is an enthusiastic member, being the only Nordic country to switch to the euro. This is due as much to security as to economic concerns. Finland

had a mutual security treaty with the Soviet Union but since 1990 it has tried to follow a neutral course. This means that Finland is reluctant to join NATO but Finns hope that the EU will offer similar security and has been a strong supporter of the EU's expansion to the east.

For the past hundred years or so the government of Finland has been in the hands of coalitions or minority administrations, typically led either by the Social Democratic Party or the rural-based Centre Party.

Since the 1995 election, however, the mosaic has taken on a new pattern. In that year the votes were more evenly shared. The Social Democrats, led by Paavo Lipponen, had the most, though only 25%, and formed a 'rainbow', five-party coalition that included all shades of the political spectrum from the Conservative Party to the Left-Wing Alliance.

The coalition reformed after the 1999 election still with Lipponen as prime minister. In 2000, the Finns elected to the largely ceremonial role as president Tarja Halonen, the first woman to hold the office.

In the 2003 general election the rainbow's colours changed. This time the Centre Party came out slightly ahead and formed a coalition with the Social Democrats and the Swedish People's Party. Anneli Jaatteenmaki, the Centre Party leader, became Finland's first woman prime minister. However she quickly resigned following allegations that she had leaked official documents concerning Iraq. The Centre Party's new leader Matti Vanhanen took over. In January 2006 he tried to become president but lost to Tarja Halonen.

France

One of Europe's most centralized states, with a distinctive and influential culture

Land area: 552,000 sq. km.
Population: 60 million—urban 76%
Capital city: Paris, 11.1 million
People: French, including ethnic minorities from former colonies
Language: French
Religion: Roman Catholic 85%, Muslim 7%, Protestant 2%, Jewish 1%, other 5%
Government: Republic
Life expectancy: 80 years
GDP per capita: $PPP 27,677
Currency: Euro
Major exports: Machinery, cars, food and agricultural products, chemicals

France has the largest territory in Western Europe. Around two-thirds is lowlands, chiefly to the north and west, including the Paris basin and the Loire Valley. But there are also dramatic mountain ranges. To the south-west the Pyrenees mark the border with Spain. To the south, occupying one-sixth of the territory is the Massif Central, consisting mostly of plateaux at between 600 and 900 metres. And to the south-east, forming a barrier with Italy and Switzerland, are the French Alps, beyond which to the south is the Mediterranean coastline—the Côte d'Azur.

Though France has an increasingly diverse population the government has been determined to sustain a national identity. Thus for more than 250 years, the Académie Française has protected and promoted the French language. The French government has also made strenuous efforts to resist the encroachment of American culture. Nevertheless France does have strong regional identities, and even regional languages such as Breton and Catalan.

The French population continues to be stimulated by immigration. Around 6% of the population have foreign nationalities, the majority coming from Algeria and Morocco. But millions more immigrants have become French citizens. There may also be up to half a million unauthorized immigrants. Rather than encouraging multiculturalism, France has been determined to ensure assimilation, a policy which has led to conflict over girls who wear Muslim headscarves to school. The education system has also promoted a centralized style of management: most of the country's leadership has been processed through the 'grandes écoles', so they tend to share the same values.

Most French workers are now employed in services and often work for the government. The French state continues to play an important part in the economy—spending around half of GDP and employing more than one in four workers. Many service workers are involved in tourism. France is by far the world's leading tourist destination, with over 160 million visitors per year.

Agriculture nowadays employs only 4% of the labour force—chiefly on smaller farms. Even so, output has increased and France is the EU's leading food producer. The country is largely self-sufficient in food and is a leading exporter of wheat, beef, and other foodstuffs—as well as the world's leading exporter of high quality wine. France is also one of the main supporters of the EU's expensive system of agricultural protection.

France is the world's fourth largest industrial power and has many globally important companies. Danone, for example, is the world's largest dairy products firm, and Peugeot-Citroën is Europe's second largest car maker. One distinctive manufactured export has been the high-speed train, the TGV. France has also been a leading arms exporter. With no oil, France has invested heavily in nuclear electricity, some of which it exports.

Exporting high-speed trains

Like many other European countries, France has been afflicted by high unemployment—around 10% overall, but often above 25% for young people. Many employers blame this on over-regulation of the labour market and a high minimum wage. To reduce unemployment and share work more widely, the government in 2000 introduced an official 35-hour week.

Government in France frequently requires 'cohabitation' between a president of one party and a prime minister of an opposing one. The current constitution, which dates from 1958, strengthened the position of the president, who is directly elected for a seven-year term. The president appoints the prime minister and chairs the weekly Council of Ministers.

François Mitterand of the Parti socialiste (PS) won the presidential election in 1981 but the centre-right party, the Rassemblement pour la république (RPR), subsequently made gains in the National Assembly and obliged Mitterand to appoint a prime minister from its ranks.

Then, after a series of elections, the situation was reversed. The RPR had won the 1993 elections to the National Assembly and in 1995, their candidate, Jacques Chirac, also won the presidential election, defeating Lionel Jospin of the PS. Chirac called a snap National Assembly election in 1997. This was a miscalculation. The PS won, obliging Chirac to appoint Jospin as prime minister.

Chirac won the presidential election in March 2002. Then the mainstream right-wing parties, united as the Union pour la majorité présidentielle (UMP), won the legislative elections, with Jean-Pierre Raffarin as prime minister. Raffarin proved fairly ineffective, however, and following France's rejection of the proposed EU constitution was replaced by Dominique de Villepin. He in turn was undermined in late 2005 by rioting in the poverty-stricken suburbs and in 2006 when attempts to free up the labour market provoked fierce opposition from students.

This has cleared the way for minister of the interior Nicholas Sarkozy to be the UMP's presidential candidate in 2007, probably facing Ségolène Royal of the PS.

French Guiana

One of the richest, but most impenetrable, countries of South America

French Guiana's hot and humid territory is largely flat. There are low mountains in the south along the border with Brazil, but most of the country is a vast plateau covered with dense tropical rainforest. The majority of people live along the swampy coastal plain or in the capital, Cayenne.

France used the territory as a penal colony until 1953—most notoriously Devil's Island. Even after they had served their sentences, it was difficult for prisoners to leave. Most of those who stayed succumbed to tropical diseases.

Today, most people are creole or black, but there is also a significant French population. The others include the original Amerindians in the jungles of the south; the Marons, who are descendants of prisoners who escaped to the interior; and immigrants from other French territories. There are even some people from the Hmong ethnic group from Laos.

The country benefits from generous aid from France, particularly since the establishment in 1968 at Kourou, 60 kilometres west of the capital, of the rocket-launching base now used by the French and European space agencies, as well as by a commercial company. There are around a dozen launches each year.

Land area: 90,000 sq. km.	
Population: 0.2 million	
Capital city: Cayenne, 55,000	
People: Black or mulatto 66%, white 12%, Amerindian 12%, other 10%	
Language: French	
Religion: Roman Catholic	
Government: Overseas collectivity of France	
Life expectancy: 77 years	
GDP per capita: $PPP 8,300	
Currency: Euro	
Major exports: Shellfish, gold, timber	

This and other government services employ not just French people but also many local workers. As a result, the average per capita income is one of the highest in South America, though this is distributed somewhat unequally between rocket scientists and Amerindian hunter-gatherers.

The other main economic activities are agriculture, fishing, and forestry. Most farms are small, growing subsistence crops, but there are also a few large commercial fruit plantations that export their produce to France. In addition, fishing for shrimp provides an important source of export income. Because of the European tastes of many inhabitants, the country remains highly dependent on imports from France of food and other goods.

As an overseas territory of France, the country elects two representatives to the French National Assembly and one to the Senate. Locally, there are also consultative general and regional councils.

In the 2004 elections to the General Council, the Socialist Party of Guiana took the most seats. Local people are in favour of greater autonomy but also want more aid; few want independence from France, which for many people could result in a catastrophic drop in income.

French Polynesia

Tahiti and the other islands have become increasingly reliant on tourism

Land area: 4,000 sq. km.
Population: 0.2 million
Capital city: Papeete
People: Polynesian 78%, Chinese 12%, French 10%
Language: French, Tahitian
Religion: Protestant 54%, Roman Catholic 30%, other 16%
Government: Overseas collectivity of France
Life expectancy: 76 years
GDP per capita: $PPP 17,500
Currency: CFP franc
Major exports: Pearls, coconut products

French Polynesia consists of around 130 islands in five groups in the Pacific, of which the largest are Tahiti and nearby Moorea, which have three-quarters of the population. These two islands are volcanic and mountainous, while many others are little more than coral atolls with scant vegetation.

The majority of people are Polynesian, but there is also an influential French minority. Since the islands are a French 'overseas country', EU citizens are free to travel and work there, and many have chosen to do so. But there have been concerns that they may be taking the better jobs and altering the social balance.

The islanders' relatively high standard of living reflects substantial French investment, particularly during the period of nuclear testing that ended in 1996. To compensate for the end of testing France gives an economic restructuring payment of around $170 million per year.

In recent years the economy has shifted decisively towards tourism, which accounts for one-quarter of GDP. Many people also work in government. As a result, more than two-thirds of the population are now employed in services. But many people still work in various forms of agriculture and in fisheries, notably oyster farming and the production of cultured black pearls, which are the main source of export revenue. The islands continue to enjoy other forms of aid from France, though unemployment is increasing.

Amendments to the French Constitution in 2003 paved the way for greater autonomy. After passing an 'organic law' agreed between the assemblies of France and French Polynesia, the latter in 2004 became an 'overseas country' within the French republic and can pass its own laws.

As citizens of a French overseas country, the islanders elect members to the French parliament, but they also have their own territorial assembly. For 20 years the most seats went to Tahoeraa Huiraatira-Rassemblement pour la république led by Gaston Flosse who served as president. But the elections of 2004 produced a surprisingly strong showing by opposition coalition, the Union for Democracy, led by Oscar Temaru of the pro-independence Tavini Huiraatira party who was chosen as president. He says he still wants independence but in ten to twenty years when the "political, economic, and social conditions are ripe".

Gabon

Oil-rich Gabon is one of Africa's most prosperous countries, but the oil is running out

Land area: *268,000 sq. km.*
Population: *1 million—urban 84%*
Capital city: *Libreville, 578,000*
People: *Four major Bantu groupings: Fang, Eshira, Bapounou, Bateke. Many immigrants*
Language: *French, Fang, Myene, Bateke, Bapounou/Eschira, Bandjabi*
Religion: *Christian, animist, Muslim*
Government: *Republic*
Life expectancy: *55 years*
GDP per capita: *$PPP 6,397*
Currency: *CFA franc*
Major exports: *Oil, timber, manganese, uranium*

Gabon consists almost entirely of the basin of the Ogooué River and its tributaries, broken by mountains in the centre and the north, with a narrow strip of coastal lowlands. Three-quarters of the country is covered in dense tropical forest—a rainforest second in area only to the Amazon Basin. This has more than 250 species of tree and provides living space for more than four-fifths of the world's chimpanzees and gorillas.

Gabon's rapid development as an oil producer has made it one of the most urbanized countries in Sub-Saharan Africa—more than half the population are now thought to live in urban areas. This has also left its forest resources relatively unscathed by development. Though there has been some deforestation, particularly along the coast and the river banks, Gabon still has thousands of plant species and thriving wildlife.

The population is very diverse, with more than 40 ethnic groups. But there is also a strong French influence and most of the population is Roman Catholic. The urban population is largely to be found in the coastal strip—in Libreville, the capital, one of the world's most expensive cities for visitors, or in Port Gentil, the centre of oil production.

The interior, on the other hand, is sparsely settled, largely along river banks. Communications generally are poor. Even Port Gentil has no road link with the rest of the country; it must be reached by boat or by air.

Oil has given Gabon one of Africa's highest per capita GDPs but the wealth is unequally distributed. In the rural areas services for health and education are in a poor state. And even in the urban areas 20% of the population live below the poverty line. There are also many foreign workers—around one-fifth of the workforce—who come from neighbouring countries to perform the more menial tasks.

Agriculture is largely for subsistence, but only 0.6% of the total land area has been cultivated and more than half the country's food has to be imported. Most of this, from vegetables to cheese, comes from France, at considerable cost.

The most important rural product is timber. Around three-quarters of the country is forested and before the discovery of oil, timber was the country's leading source of wealth. Timber is still a major export and is thought to employ around half the labour force. Gabon is the world's leading producer of okoumé and ozigo, softwoods that are used for making plywood. Other traditional cash crops include rubber, coffee, and cocoa. There are also efforts to expand output of palm oil and sugar.

Source of timber for making plywood

Since the late 1960s, Gabon's development has been fuelled by oil, most of which is offshore. Oil accounts for around half of GDP and government income. Some 15 international companies are involved in exploration or production. The most important oilfields are Gamba, which is operated by Shell, and the extensive Rabi Kounga field and its satellites, which are operated by Shell and Elf. Future prospects are uncertain. Most of Gabon's fields are maturing, output is falling, and reserves could be exhausted within the next few decades.

Another major mineral export is manganese. Here the prospects are better. Gabon is the world's third largest producer and its open-cast mines in the Moanda region in the south-east have around one-quarter of global reserves. Mining is in the hands of the Comilog company, which is largely foreign owned, though there is some state and Gabonese participation. The other important mineral is uranium, but the reserves are now almost exhausted.

Gabon embarked on an economic structural adjustment programme in 1995, liberalizing the economy and privatizing a number of state enterprises. But life without oil will be difficult; apart from the possibilities of eco-tourism there do not seem to be many new options on the horizon.

For most of the period since independence in 1960, Gabon has been ruled by President Omar Bongo Ondimba. He came to power in 1967, and in 1968 established the Parti démocratique gabonaise (PDG) as the only legitimate political party. Subsequent oil wealth allowed Bongo and his supporters to entrench their position and distribute patronage. Protests were met with violent repression. The regime maintained close relations with France and even today Gabon has a contingent of French soldiers.

A fall in the oil price in the late 1980s, combined with general extravagance and corruption, produced a financial crisis. The resulting austerity measures led to violent protests and in 1991 eventually forced Bongo to amend the constitution and introduce a multiparty system. He nevertheless continues to be re-elected, most recently in 2005 with 79% of the vote.

In 2001 his PDG won 86 of the 120 seats in the National Assembly and, thanks to Bongo's system of patronage, can also rely on support from independents and others. Opposition parties are very ineffective having been split by factional infighting and by co-option by Bongo.

The Gambia

A holiday destination with a repressive government

Land area: 11,000 sq. km.
Population: 1.4 million—urban 26%
Capital city: Banjul, 34,000
People: Mandingo 42%, Fula 18%, Wolof 16%, Jola 10%, Serahuli 9%, other 5%
Language: English, Mandinka, Wolof
Religion: Muslim 90%, Christian 9%, other 1%
Government: Republic
Life expectancy: 56 years
GDP per capita: $PPP 1,859
Currency: Dalasi
Major exports: Groundnuts

This long, thin country extends within Senegal for around 300 kilometres inland from the coast, along both banks of the lower Gambia River. The river is fringed with swamps beyond which are flats that give way to low hills and a sandy plateau. The better land is up-river and most people have settled between the river flats and the uplands.

The Gambia has a large number of ethnic groups, but the majority are the Malinke (also called the Mandingo), who are largely to be found higher up the river. Only one-third of the population is urban, but the capital, Banjul, is more than usually crowded since it effectively occupies an island at the mouth of a river and has few opportunities for expansion. There are also a large number of middle-class refugees from Sierra Leone.

The Gambia is one of the world's poorest countries, and poverty has been increasing. According to official estimates, more than half the population is poor and more than one-third are extremely poor. Three-quarters of people make their living from agriculture. They use the lower-lying land that benefits from annual flooding to grow rice, which is the staple food. But yields are low and most of the country's rice now has to be imported. Other subsistence crops include millet, sorghum, maize, and vegetables.

Most farmers get their cash income by growing groundnuts, which are ideally suited to the sandy soil on the higher ground. Groundnuts occupy around half the country's cultivable area and most of the crop is sold through Senegal. The government has made some attempts to diversify by promoting the cultivation of flowers and fruit for air-freighting to Europe, but so far has had little success.

In fact, more than 80% of the country's exports are 're-exports'—manufactured goods that have been imported into the port of Banjul but are en route to neighbouring countries such as Senegal, Mali, Guinea, and Guinea-Bissau.

The other major export earner is tourism. The Gambia's coastline may be short but it has a number of sandy beaches that have attracted package tourism from Europe, particularly from the United Kingdom which provides more than half the 90,000 visitors. Tourism creates hotel jobs for local workers but the benefits do not extend very far beyond the resorts. The military coup in 1994 deterred visitors for a few years, but they have now returned and there has been more investment in new hotels.

Politically and geographically, the Gambia seems an anomalous country that one might expect to be part of Senegal. Its status derives from its history as a colony of the British, whose main preoccupation was to deprive the French of an important navigable waterway. The Gambia's relations with Senegal have sometimes been fraught.

For many years following independence in 1965, the Gambia was held up as a model, albeit imperfect, of African democracy. The first prime minister was the leader of the People's Progressive Party (PPP), Sir Dawda Jawara, who was the first

Gambia gets aid despite human rights abuses
president in 1970 when the country became a republic. Jawara survived a Libyan-backed coup

attempt in 1981 with the help of troops from Senegal. This intervention also led in 1982 to the formation of a federation—'Senegambia'—that seemed destined to unite the two countries. Jawara resisted this, however, and the federation was disbanded in 1987.

In 1992 Jawara was re-elected as president but in 1994 was overthrown in a military coup led by Lieutenant Yahyah Jammeh. The coup, following

which all political parties were banned, was widely condemned internationally. The Gambia found itself isolated and had most of its international aid withdrawn.

International pressure did eventually result in a democratic opening. In 1996, Jammeh formed a new party, the Alliance for Patriotic Reorientation and Construction. Just before the election, however, he took the precaution of banning the three largest opposition parties—and anyone who had held ministerial office for the previous 30 years. This left the United Democratic Party as the main opposition. Jammeh won the presidential election with more than half the vote.

The government has shown scant regard for democratic freedoms. It continues to harass the United Democratic Party, whose leaders it arrests and detains. There have also been regular raids on radio stations and newspapers.

Throughout his regime, Jammeh has been subject to real or supposed conspiracies. In 1998 three lieutenants were executed for plotting against him. And early in 2000 Jammeh claimed to have uncovered another plot hatched by his palace guards, several of whom died while 'resisting arrest'. In the same year troops fired on student demonstrators.

Despite these abuses, aid donors have largely returned, reassured by the façade of democracy, and give around $50 million per year.

Jammeh easily won the 2001 presidential election and will win again in 2006. The 2003 legislative election was boycotted by the UDP because of irregularities.

Georgia

A 'rose revolution' has brightened the prospects for a country blighted by war and economic crisis

Land area: 70,000 sq. km.
Population: 5 million—urban 52%
Capital city: Tbilisi, 1.3 million
People: Georgian 83%, Azeri 7%, Armenian 6%, Russian 2%, other 2%
Language: Georgian 71%, Armenian 7%, Azeri 6%, Russian 9%, other 7%
Religion: Orthodox Christian 84%, Muslim 10%, Armenian-Gregorian 4%, Catholic 1%, other 1%
Government: Republic
Life expectancy: 71 years
GDP per capita: $PPP 2,588
Currency: Lari
Major exports: Food, chemicals, machinery, tea, wine

Georgia is dominated by the Caucasus mountains. Its northern border is formed by the Greater Caucasus and the southern border by the Lesser Caucasus; more than one-third of the mountains are forested. Between these are the major rivers and their valleys—the Kura which flows east into Azerbaijan and the Rion and others flowing west into the Black Sea. The latter have deposited masses of silt to form the Kolkhidskaya lowlands which make up most of the coastal land. Formerly a sub-tropical swamp, this has been drained to create valuable agricultural land.

Most people live in the lower parts of the country. While ethnic Georgians—who call themselves Kartveli—make up around 80% of the population, they are divided into many regionally based subgroups. Georgians have a distinctive language, alphabet, and culture. Most are Christian, though an important subgroup in the south-west are Muslim. Two of the non-Georgian ethnic groups have been trying to secede: the South Ossetians on the northern border with Russia; and the Abkhazis, who live in the north-west.

Levels of human development have been high, with good standards of education and health: some Georgians are reputedly very long-lived. But war and economic crisis have devastated the health and education systems and standards are falling. Over half of Georgia's people now live below the poverty line and the population has been declining as a result of emigration and a falling birth rate.

Agriculture has been the most important employer, but has shrunk to only 16% of GDP. The country's geographical diversity permits a wide variety of crops, including tea, citrus fruits, grapes, and tobacco. Around 60% of state-owned land has been privatized; the rest is leased out. Food production is low.

The Soviet Union had promoted heavy industry, but much of this is now obsolete. Georgia has some oil and gas of its own but chiefly relies on imports from neighbouring former-Soviet republics which have been trying to charge higher prices. Georgia does have some important mineral resources, including manganese, iron ore, and gold, but high energy costs and falling demand have affected production.

Georgia has attractive scenery and beach resorts so could have a tourist industry, but political instability and poor infrastructure keep people away.

Many important businesses are still run by the state, but the government has sold majority stakes in some others to foreign companies. The Borjomi mineral water plant, for example, is run by French and Dutch investors. After a catastrophic economic decline following independence, economic growth has shown a strong revival.

Georgia's has struggled to build a coherent democratic state. There have been a number of secessionist conflicts. The first has been in South Ossetia which in 1990 hoped to unite with Alania (North Ossetia) in the Russian Federation. Fighting here ended with a ceasefire in 1992 but the issue is far from resolved.

Struggles for secession

The second secessionist conflict, in Abkhazia, has proved more intractable. The Abkhazis declared independence in 1992, and subsequently expelled more than 250,000 Georgians. Various ceasefires have been negotiated, with the involvement of 1,500 Russian peacekeeping troops. But fighting has persisted between the Abkhazis and the Georgian guerrillas who have been supporting the refugees. The Abkhazis also want associate status with Russia, which would threaten Georgia's potential Black Sea outlet for oil pipelines from the Caspian.

The first president, following independence in 1991, was Zviad Gamsakhurdia, who was driven from office in 1992 following accusations of corruption and of human rights violations—sparking a civil war that ended only with his death at the end of 1993.

He was replaced by Eduard Shevardnadze, who had been Soviet foreign minister and was elected president in 1995. His new party, the Union of Georgian Citizens, also took the most seats in parliament. The party was returned to power in the 1999 parliamentary elections, and Shevardnadze was re-elected president in April 2000.

Somewhat surprisingly, since Shevardnadze was in little danger of losing, the election was marred by ballot stuffing and other irregularities. After similar activities in the November 2003 parliamentary elections, the opposition took to the streets and finally invaded the parliament building and forced Shevardnadze to resign in what was called the 'rose revolution'.

A presidential election in January 2004 was won overwhelmingly by Mikhail Saakashvili, a 36-year-old US-educated lawyer and leader of the National Movement. He has energetically set about tackling the corruption that helped undermine the previous regime and has promised to resolve the secessionist conflicts, but faces an uphill task.

Germany

Politically powerful in Europe, Germany also has a large and successful economy, though it is becoming less competitive

Land area: 357,000 sq. km.
Population: 83 million—urban 88%
Capital city: Berlin , 3.4 million
People: German 92%, Turkish 2%, other 6%
Language: German
Religion: Protestant 34%, Roman Catholic 34%, Muslim 4%, unaffiliated or other 28%
Government: Republic
Life expectancy: 79 years
GDP per capita: $PPP 27,756
Currency: Euro
Major exports: Machinery, cars, chemicals, manufactured goods

Germany has three main geographical regions. From the North Sea and Baltic coasts southwards, covering roughly one-third of the country, are the lowlands of the North German Plain. This leads to a belt of central uplands running west to east, cut through by major rivers, including the Rhine and the Weser. Further south still are the South German Highlands culminating in the Alps at the borders with Austria and Switzerland.

The 1990 reunification of Germany created by far Europe's largest national population. Today there are 15 million in what was East Germany and 67 million in what was West Germany. Germany also has the largest number of immigrants in Europe—around 9% of the population are foreign nationals, the largest proportion from Turkey. Many of these are former 'guest workers', who

arrived in the 1970s and have chosen to settle. Others are refugees and asylum seekers who came at the end of the 1980s, including more than two million people of German ancestry.

From the 1950s, West Germany developed into Europe's most dynamic industrial nation. East Germany lagged far behind and the merger of the two economies proved difficult and expensive. Germans still have to pay a 'solidarity tax' which adds around 5% to the normal tax bill. This pays for extending social benefits to the East, raising wages, and repairing extensive environmental damage. Even so, wide gaps remain. In 2005, unemployment in the West was 8% but in the East was 19%.

Nevertheless, this remains a powerful industrial economy—the world's third largest after the USA and Japan. Its strength has been highly efficient manufacturing industry, which accounts for around one-fifth of GDP and two-thirds of exports of goods and services. The most important sectors are machinery, cars, and chemicals. Built around this is a large service sector. Agriculture employs few people but the country is still 70% self-sufficient in food.

The economy has been organized

in a fairly coherent way—built around the idea of the 'social market'. Rather than relying on the short-term vagaries of stock markets, investment has been based on long-term partnerships between banks and companies. Another feature has been co-operation between companies and the powerful trade unions. Though membership has been falling, one-third of workers still belong to trade unions and wages are settled annually in national rounds. In addition, workers are entitled to participate in management through works councils and have seats on the boards of the largest companies.

Germany also has one of the most generous welfare systems with high unemployment benefits and pensions. Employers argue that these costs make the country uncompetitive. Hourly wage costs are one-third higher than in the USA or the UK.

The German model has, as a result, come under increasing pressure. Many years of low growth, combined with the costs of reunification, have been eroding Germany's economic position,

Germany's economic model under fire

and have taken the budget deficit beyond the levels permitted by the EU's growth and stability pact.

In response, early in 2004, the government introduced Agenda 2010, a major reform of the welfare state, which included making it more difficult to get unemployment benefits.

Another notable feature of the German economy and society is a concern for the environment. The country has intensive systems of recycling—and spends around 2% of GDP on environmental protection.

Germany is a federal state with 15 states, or länder (ten from the West, five from the East), plus the federal capital Berlin. Each land has its own parliament and government responsible for such issues as policing and education. The state governments are also represented in the federal parliament's upper house, the Bundesrat. But most power rests in the 662-member lower house, the Bundestag. In 2004, the parliament elected Horst Köhler to the largely ceremonial function of president.

After 1949, most West German governments were coalitions and the tradition continues in the new Germany. From 1982 the head of government, the chancellor, was Helmut Kohl, leader of the centre-right Christian Democratic Union (CDU) who was the architect of German reunification.

This era ended with the 1998 election which resulted in a coalition of the Social Democratic Party (SPD) with the Green Party. The SPD is further left than the CDU, but in many respects their policies are similar.

Gerhard Schröder of the SPD became chancellor and the post of foreign minister and deputy chancellor went to Green politician, Joschka Fischer, and they narrowly won the 2002 election. In 2005 Schröder decided to call an early election but did not gain enough ground to form a new coalition.

Instead the government was to be a grand coalition of the CDU/CSU and SPD with Angela Merkel of the CDU as chancellor. She has had some success abroad, repairing relations with the USA that were damaged by opposition to the Iraq war, but could struggle with reforms at home.

Ghana

Ghana has achieved one of Africa's few transfers of power through the ballot box

Land area: *239,000 sq. km.*
Population: *21 million—urban 45%*
Capital city: *Accra, 1.6 million*
People: *Akan 44%, Moshi-Dagomba 16%, Ewe 13%, Ga 8%, Gurma 3%, other 16%*
Language: *English, Twi, and other African languages*
Religion: *Christian 63%, indigenous beliefs 21%, Muslim 16%*
Government: *Republic*
Life expectancy: *57 years*
GDP per capita: *$PPP 2,238*
Currency: *Cedi*
Major exports: *Gold, cocoa, timber*

Apart from scattered hills and plateaux along the eastern and western borders, Ghana consists mostly of lowlands. These include a coastal plain that extends up to 80 kilometres inland and incorporates many lagoons and marshes. Almost two-thirds of the country, however, consists of the basin of the Volta River, which now includes Lake Volta, which was formed by the Akosombo dam—the largest artificial lake in the world. There is a sharp climatic division between the dry north and the wetter and lusher south.

Ghana's population comprises dozens of ethnic groups of whom the largest are the Akan. The population has been growing rapidly—at around 3% per year, though this rate is expected to drop.

By the standards of Sub-Saharan Africa, Ghana is relatively well off. Even so, 40% of the population live below the poverty line—and poverty is generally more severe in the north. Less than half of rural Ghanaians have access to health services, though in 2003 the government introduced a National Health Insurance Scheme which everyone must join. Education standards are also low: only 60% of children enrol in primary school; but literacy rates are starting to rise.

Around 50% of people still make their living from agriculture, which accounts for around 35% of GDP. The main food crops are maize, yams, and cassava, while the main cash crop, and one of the leading export earners, is cocoa. Cocoa is produced by more than 1.6 million peasant farmers, mostly on plots of under three hectares, and although production collapsed in the mid-1980s it has now bounced back—partly because the government has raised producer prices and farmers have been replacing their trees.

Even so, overall agricultural productivity has been falling, a consequence of low investment and the removal of subsidies on inputs such as fertilizers, which many farmers can no longer afford. Some are now reverting to subsistence production. Another important agricultural export is timber, though there are concerns about deforestation.

Ghana, whose colonial name was the 'Gold Coast', once again has gold as its leading export. The main producer is the Ashanti Goldfields Corporation. Other important mining activities involve manganese and diamonds. Ghana also has a diverse collection of manufacturing industries, a legacy of earlier efforts at self-sufficiency.

In 1983, Ghana started to follow the market-led structural adjustment recipes of the IMF and the World Bank, selling off public enterprises and reducing protection for local manufacturers. Donors rewarded Ghana with considerable aid. This stimulated entrepreneurial activity, particularly in financial services and trade, and imports flooded in, making Accra a much more stylish city. Tourists also started to arrive to visit Ghana's ecological reserves as well as historical sites related to slavery.

Ghana has important historical slavery sites

But development did not reach far into the rural areas, and the loans steadily piled up. By 1999, Ghana's total debt was equivalent to four years of export earnings and the economy was in steep decline. In the past few years, however, the economy has been growing at around 5% annually.

Most of this activity was presided over by Jerry Rawlings. He first entered the political scene in 1979 as 28-year-old Flight-Lieutenant Rawlings, leading a military coup that resulted in the murder of a number of senior figures. He briefly handed power over to a civilian administration but returned in a second coup in 1981. His Provisional National Defence Council (PNDC) government opted for 'party-less' government to be run by technocrats. Initially this had a strong socialist orientation but, following a crippling drought and economic recession, Rawlings turned sharply to the right into the embrace of the IMF, and from 1983 embarked on the sustained programme of structural adjustment that aid donors held up as a model of success.

Donors were less happy with Rawlings' dictatorial, if popular, rule. In 1992 he responded with a new constitution that allowed multiparty elections and he refashioned the PNDC as a political party—the National Democratic Congress (NDC). Rawlings won the 1992 presidential election, overcoming the candidate of the main opposition party, the business-oriented New Patriotic Party (NPP). In 1996 the NDC again proved victorious, though the NPP did gain one-third of the seats. Rawlings was re-elected with 57% of the vote.

Rawlings' intended successor was his vice-president, John Atta Mills, but he lost the 2000 presidential elections in a run-off with the NPP candidate John Agyekum Kuofor and for the first time Ghana achieved a transfer of power through the ballot box.

Kuofor turned his attention to the previous regime and prosecuted some ex-ministers for corruption. This proved divisive but, with the economy growing, the government remained fairly popular and he won a second term in the 2004 election. Kuofor has also become an international figure as one of Africa's leading statesmen.

Greece

Greece has seen considerable benefits from EU membership and is now mending fences with its old adversary, Turkey

Land area: 132,000 sq. km.
Population: 11 million—urban 61%
Capital city: Athens, 3.8 million
People: Greek 98%, other 2%
Language: Greek
Religion: Greek Orthodox
Government: Republic
Life expectancy: 78 years
GDP per capita: $PPP 19,954
Currency: Euro
Major exports: Manufactures, food

Most of Greece is mountainous. The mainland is dominated by the rugged Pindus Mountains, which extend from the northern border to the hand-shaped peninsula of the Peloponnese. The main lowland areas are the plains in the north-east along the Aegean coast. Greece's coastline is highly indented so the country is penetrated from all directions by the sea. Around one-fifth of the territory consists of two thousand or more islands, the largest of which is Crete, to the south.

The country is subject to regular earthquakes, one of which in 1999 killed 150 people. A more beneficial natural phenomenon is sunshine which allows Greece to use Europe's greatest surface area of solar panels.

Officially, the population is almost entirely ethnic Greek, though it can also be considered a mixture of the many different groups that have

arrived from neighbouring countries, including Turkey, Macedonia, Albania, and Bulgaria. Greeks may not be rich but they have been relatively long-lived—thanks to a healthy diet.

Greece continues to be a country of emigration and immigration. Emigration was still continuing to richer countries of the EU in the 1990s, notably to Germany. But many people are also arriving. The break-up of the Soviet Union encouraged more than 60,000 ethnic Greeks to move to Greece. There are also around 500,000 unauthorized immigrants, mostly from Albania. By 2004, 9% of the population were foreign born.

One-fifth of the population still work in agriculture, but only one-quarter of the land is cultivable, and since the soil is poor and there is not much rain, yields are often low. The coastal plains are used for the cultivation of major cereal crops and sugar beet as well as cash crops such as cotton and tobacco. Greece is also a major producer of olives, tomatoes, grapes, and other fruits.

Although the fishing industry declined in the 1990s because of an ageing fleet and over-fishing, the sea catch now seems to be growing again, as is fish farming in which Greece is Europe's largest producer.

For a European country, industry

is underdeveloped. Greece has few minerals, apart from lignite, and its manufacturing tends to be small-scale. Even so, wages have risen in recent decades and some of the simpler industries such as garments and textiles have migrated to poorer neighbouring countries.

Many of the larger industrial activities are controlled by the state, either owned directly or through state-owned banks—though this is as much a legacy of military centralization as of socialist government. So far the government has been slow to deal with the inefficient state-owned enterprises.

Historically, one of the most successful industries has been shipping. Greece owns around one-sixth of the global fleet, 3,300 vessels, typically bulk carriers, though most of these sail under other flags using foreign—and cheaper—crews.

Another important source of foreign exchange has been tourism—an industry that employs 18% of the labour force and contributes a similar proportion to GDP.

More tourists than its population

With many historical sites, and spectacular scenery and beaches on its many islands, Greece often attracts more tourists than its total population. While these are welcome for the income they bring, the sheer numbers have been causing environmental damage.

Greece was the poorest member of the EU when it joined in 1981. Subsequent economic development, however, often driven by substantial EU subsidies, contributed to rapid growth and stability that allowed Greece to adopt the euro in 2001.

After a coup in 1967 Greece had a military government. But the military were ousted in 1974 following their backing for a failed coup in Cyprus that resulted in a Turkish invasion. Since then Greece has been ruled by one of two parties: the conservative New Democracy Party (NDP), or the Panhellenic Socialist Movement (Pasok).

The period 1989–90 saw a series of hung parliaments until the NDP, led by Constantine Mitsotakis, took over and tried to balance the books. But the sacrifices proved unpalatable and early elections in 1993 were won by Pasok, led by Andreas Papandreou. When the latter fell ill he was replaced in 1996 as prime minister by Costas Simitis. Pasok steadily drifted towards the political centre and won a third consecutive term in 2000, though only narrowly defeating the NDP, then led by Constantine Karamanlis.

The elections in March 2004 marked something of a turning point. Both parties promised a fresh start, with a more open and merit-based form of administration. And they had both moved on to a new generation of leaders, albeit from familiar political families: Costas Karamanlis for the NDP and George Papandreou for Pasok. This time the NDP won a clear overall majority.

At home the priorities are to improve health care and education. Internationally, the most important issue is the relationship with Turkey, for which Greece is officially supporting accession to the EU, though against the grain of public opinion. Another is the reunification of Cyprus for which Greece now seems to offer less practical support.

Guadeloupe

An overseas department of France that occasionally agitates for independence

Land area: 2,000 sq. km.
Population: 0.5 million
Capital city: Basse-Terre
People: Black or mixed race 90%, other 10%
Language: French
Religion: Roman Catholic
Government: Department of France
Life expectancy: 78 years
GDP per capita: $PPP $7,900
Currency: Euro
Major exports: Sugar, bananas

Guadeloupe comprises two volcanic islands in the Caribbean, Basse-Terre and Grande-Terre, that together form a butterfly shape, along with several smaller islands. The western wing, Basse-Terre, is mountainous, with dense rainforests as well as mangrove swamps. It also has a smoking volcano, La Soufrière, though this has not erupted for 30 years, along with some spectacular waterfalls. Much of the island has been designated a national park. Grande-Terre has lower hills and plains and is more suitable for agriculture.

The population is mostly black or of mixed race and speaks French, though many also use a creole dialect. The majority of the workforce are employed in the service sector, particularly in tourism. Most tourists come from the USA, many on brief visits from cruise ships, to enjoy the French colonial atmosphere of the largest town, Basse-Terre, or to head for the quiet beaches, the rainforests, or the outer islands.

Agriculture is less significant nowadays and much of the country's food has to be imported from France. Historically, sugar was the major crop but efforts have been made to diversify into bananas and other fruits as well as aubergines and flowers.

There is also some light industry, including rum manufacture.

Guadeloupe enjoys substantial subsidies from France—which help to offset high levels of unemployment. Since Guadeloupians are French citizens many young people take the opportunity to emigrate to France.

As an overseas department, Guadeloupe sends representatives to the French parliament—two senators, and four representatives in the National Assembly. But it also has its own elected general council and regional council.

During the 1960s and 1970s, there was some agitation for independence. The pro-independence movement organized strikes and demonstrations, but never gained control of the local assemblies. Since then enthusiasm for independence seems to have waned.

The local assemblies have usually been dominated by centre and right-wing parties, though the left gained power during the 1980s. In the 1998 elections the majority of seats in the regional council went to the centre-right Rassemblement pour la république, and Lucette Michaut Chevry was elected as its president. But following a severe drought in 2002 and a series of strikes, she lost in the 2004 elections to Victorin Lurel's Socialist Party.

Guam

As a US military outpost, Guam is sensitive to changes in defence priorities

Land area: 1,000 sq. km.
Population: 0.2 million—urban 38%
Capital city: Hagåtña/Agaña
People: Chamorro 37%, Filipino 26%, Japanese and other Asian 27%, white 10%
Language: English, Chamorro, Filipino
Religion: Roman Catholic
Government: Unincorporated territory of the USA
Life expectancy: 79 years
GDP per capita: $PPP $21,000
Currency: US dollar
Major exports: Transhipments of oil products, construction materials, fish

Guam is one of the Mariana Islands in the Pacific Ocean between Hawaii and the Philippines. The northern half of the island is a limestone coral plateau, much of which has been levelled to form airfields. The southern half has a range of volcanic hills. Around one-third of the territory is owned by the US military—and the country has suffered from much indiscriminate dumping of waste, either on the land or in the ocean.

The oldest inhabitants are the Chamorro people who still make up more than one-third of the population. They have their own distinctive language and culture. But years of colonization, by Spain and then by the USA, have taken their toll. Guam has, for example, the world's highest per capita consumption of Spam.

Many other people have arrived. In addition to more than 20,000 US military and other federal employees and their dependants, the bases have drawn immigrants from elsewhere in the Pacific, particularly the Philippines. In a few decades it seems likely that Filipinos could outnumber Chamorros. More recently, Guam has also been attracting unauthorized immigrants.

Guam's military bases, mostly those of the navy, with around $10 billion-worth of infrastructure, are one of the mainstays of the economy. Military cutbacks in the 1990s saw some defence-related jobs disappear. One of the first to close was a ship-repair operation. However, the proposed transfer of some 8,000 marines and their dependants from the US base at Okinawa in Japan to be completed by 2014 should be a major boost for the island.

The second main source of income is tourism. Around one million visitors arrive each year, mostly from Asia and the Pacific, and particularly Japan. At present most simply head for the beaches. Guam, however, wants to attract higher value tourists so is building more luxury spas. Other suggestions for increasing visitor numbers range from promoting Chamorro culture to introducing casino gambling.

Guamanians are US citizens and send a delegate to the US House of Representatives with limited voting rights. In 2004, Democrat Madelein Bordallo was re-elected as the delegate. The 2002 elections for the local governor were won by a Republican, Felix Camacho. The Republicans also won a majority in the legislative assembly.

Guatemala

Guatemala endured 36 years of civil war. But the injustices that led to the war remain unresolved

Land area: 109,000 sq. km.
Population: 12 million—urban 46%
Capital city: Guatemala, 3.0 million
People: Ladino and European 59%, Amerindian 40%, other 1%
Language: Spanish, Amerindian
Religion: Roman Catholic, Protestant
Government: Republic
Life expectancy: 67 years
GDP per capita: $PPP 4,148
Currency: Quetzal
Major exports: Coffee, sugar, bananas

Guatemala is one of the most picturesque countries in Central America. The northern, sparsely inhabited part juts out between Mexico and Belize. But around 60% of the population live either on the narrow Pacific coastal plain or in the mountainous areas of the south and west that are studded with spectacular lakes and volcanoes.

Guatemala's population distribution still reflects the legacy of the Spanish conquest. The invaders seized the most fertile soil on the plains and lower slopes, driving the Indian population up to the steeper, less valuable land—where their descendants largely remain.

Today more than half of the population would consider themselves 'ladinos'—though the term is more social and cultural than racial: apart from Spanish descendants it also includes Amerindians who have adopted Spanish language and culture.

Most of the rest of the population consist of 23 Mayan indigenous groups, whose women are particularly noticeable because of their unique and colourful form of dress. The majority are Roman Catholics but Protestant evangelical groups have been making increasing inroads.

More than one-third of the population make their living from agriculture. But this covers vast disparities. Guatemala has probably the most unequal land distribution in Latin America. The most recent data refer to 1979, when 3% of farms had 65% of the land, and 88% of farms had only 16%, and since then the situation has probably deteriorated.

The larger holdings mostly grow export crops such as sugar and bananas on the coastal plantations, while on the lower mountain slopes they grow coffee and cardamom. The poorest subsistence farmers concentrate on corn, rice, and beans, but as many as 650,000 have to make the annual pilgrimage to work on the plantations.

Industry accounts for only 13% of employment, mostly concentrated around Guatemala City. Some efforts have been made to expand simple assembly work for export to the USA, with Korean-owned plants making T-shirts and underwear. But competition from Mexico and elsewhere has been stiff and many factories have closed.

Agriculture also accounts for around one-third of export earnings, led by coffee, sugar, and bananas. But poor prices in the 1990s encouraged diversification into such crops as mange-tout, fruit, and flowers. The production of staple crops, like corn and beans, has also fallen as a result of, among other things, cheap imports and the high cost of inputs.

The unequal distribution of resources and overt discrimination have kept the Mayan population

Discrimination against Mayan Indians poor. Compared with the Ladinos, the Mayans' life expectancy is 17 years less and their literacy rates are only half. The governing élite have made few efforts to redistribute income: only 3,000 of the nation's estimated 100,000 professionals pay income tax. Tax revenues account for less than 10% of GDP.

This inequality has been sustained by a long history of repression. For 39 years after a US-inspired coup in 1954, the country was dominated by the army and its right-wing supporters. Many of those on the left joined guerrilla groups that in 1982 came together as the Unidad Revolucionária Nacional Guatemalteca (URNG). From the early 1980s in particular, during the presidency of Efraín Ríos Montt, the government tried to eliminate this opposition with an onslaught on the countryside that wiped out 400 villages. This 'dirty war' is thought to have cost up to

100,000 lives and a further 40,000 'disappearances'.

Formal democracy was restored only in 1985, and in 1990 the government started negotiations with the URNG. International support for the process included awarding the 1992 Nobel Peace Prize to the Mayan woman activist, Rigoberta Menchú.

The peace agreement was signed in December 1996. This required constitutional amendments to rein in the military, to outlaw discrimination, and to allow the use of Mayan languages in schools. But most of the programmes established in the accords are behind schedule, under-funded, and beset by political bickering.

Nor has there been much progress in curbing the human rights abuses. In April 1998 a Roman Catholic bishop, Juan Geradi, was murdered after issuing a critical report on the army, and since then the country has become increasingly lawless, beset by violent crime and drug smuggling.

From 1999 Guatemala was ruled by Alfonso Portillo, of the right-wing Frente Republicano Guatemalteco (FRG). His administration had little economic impact and faced a number of corruption scandals.

After a presidential election he was replaced in January 2004 by Oscar Berger of the conservative Gran Alianza Nacional which consists of the Partido Patriótico, the Movimiento Reformador, and Solidaridad Nacional. Since Berger does not have a majority in the congress, he has signed a two-year 'governability pact' with two of the other parties. He has had some success in reducing the influence of the army but not of the many criminal gangs.

Guinea

Despite rich agricultural and mineral potential, this West African state has yet to make much economic progress

Land area: 246,000 sq. km.
Population: 9 million—urban 35%
Capital city: Conakry, 1.1 million
People: Peuhl 40%, Malinke 30%, Soussou 20%, other 10%
Language: French, Sousou, and others
Religion: Muslim 85%, Christian 8%, indigenous beliefs 7%
Government: Republic
Life expectancy: 54 years
GDP per capita: $PPP 2,097
Currency: Guinean franc
Major exports: Bauxite, alumina, gold, coffee

Moving in from the coast, Guinea has four main geographical regions. The coastal plain reaches around 50 kilometres inland, leading to the Fouta-Djallon Highlands which rise to more than 1,500 metres. Then there are the savannah plains of Upper Guinea which slope towards the Sahel in the north-east, and finally the forested highlands of the far south-east. Guinea is also the 'reservoir' of West Africa. The region's major rivers, the Senegal, the Gambia, and the Niger, all have their origins in Guinea.

Ethnic divisions correspond to the four regions. The largest group are the Peuhl (also called the Fulani), who are mostly to be found in the Fouta-Djallon Highlands. Then there are the Malinke in Upper Guinea, the Soussou who occupy the coastal plain, and a number of smaller groups, many

of whom can be found in the forested highlands. The local population was also swollen in the late 1990s by refugees from fighting in neighbouring Liberia and Sierra Leone. In the 1980s, up to two million Guineans also fled the repressive regime in their own country and do not seem to have returned in significant numbers.

Their lack of enthusiasm reflects the country's persistently low achievements in human development. Around 40% of the population live below the poverty line. Life expectancy is short, health services are poor, and the literacy rate is only 41%. Ironically, for a country that has the sources of many major rivers, two-thirds of the population do not have access to clean drinking water.

Two-thirds of the population depend for survival on agriculture—the main food crops being rice, cassava, maize, plantains, and vegetables. Despite having much fertile agricultural land, only around 20% of the available land is cultivated. Most production is for subsistence, leaving a wide food gap that must be met with imports of rice.

With the help of international donors, the government has been struggling to increase food production and to diversify into cash crops such as coffee, cotton, and fruits. This has required extensive investment in infrastructure, particularly in roads.

Nevertheless, Guinea's fortunes still depend critically on bauxite. Guinea has half of global reserves—which are to be found both in the coastal region and in Upper Guinea. Guinea is the world's second largest producer. In the past, bauxite has been responsible for 90% of the country's

The world's second largest bauxite producer

exports and 70% of government revenue. These proportions have been falling in recent years as a result of falling world prices and inefficient production at the major smelter that converts part of the output to alumina (aluminium oxide). Much of the industry has been controlled by state-owned enterprises—though there has been greater foreign investment in the past few years, with the participation of companies from Russia, Ukraine, Iran, and Australia.

Guinea also has significant exports of gold and diamonds. But mining has mostly been in the hands of smaller enterprises and the trade is largely clandestine: only around 15% of diamond production is recorded. In addition, Guinea has extensive reserves of other minerals, particularly iron ore (6% of world reserves), that remain largely unexploited. And it has enormous hydroelectric potential.

Guinea's economic weaknesses and its over-centralized and corrupt bureaucracy are a legacy of the 26-year dictatorship of Sékou Touré. He led the country to independence in 1958 and his Marxist-tinged and

ultimately repressive and isolated regime only ended with his death in 1984. At this point the army took over, led by Colonel Lansana Conté who embarked on limited economic reforms putting more emphasis on the free market.

Political reform had to wait until 1991 and a new 'fundamental law' that offered the potential for greater democracy. In 1993, Conté, as leader of his newly constituted Parti de l'unité et du progrès (PUP), won the presidential election, and in 1995 the PUP won the majority of seats in the National Assembly. The PUP's power base is mostly in the coastal areas—while opposition parties, grouped as the Co-ordinated Democratic Opposition, are more regionally confined.

In 1996, Conté withstood a coup attempt by officers protesting against poor working conditions. The coup's failure seemed to help sustain the regime. Then in 1998, Conté beat four other candidates to win a second presidential term. After the election he cracked down and started arresting opposition leaders—saying that security needed to be tightened because of the chaos in neighbouring Sierra Leone.

In June 2002 the PUP duly won a large majority in the legislative elections and in December 2003 Conté was re-elected president with 96% of the vote—after the leading opposition party the Front républicain pour l'alternance démocratique boycotted what they reasonably expected to be an unfair poll.

Conté is ill and seems unlikely to complete his seven-year term, but has no obvious successor.

Guinea-Bissau

Guinea-Bissau in West Africa is probably the world's most indebted country, and has recently suffered a civil war

Land area: *36,000 sq. km.*
Population: *2 million—urban 34%*
Capital city: *Bissau, 397,000*
People: *Balanta 30%, Fulani 20%, Manjaca 14%, Mandinga 13%, Papel 7%, other 16%*
Language: *Portuguese, Criolo, African languages*
Religion: *Indigenous beliefs 50%, Muslim 45%, Christian 5%*
Government: *Republic*
Life expectancy: *45 years*
GDP per capita: *$PPP 711*
Currency: *CFA franc*
Major exports: *Cashew nuts*

Guinea-Bissau is low-lying and monotonously flat. Much of the territory comprises coastal swamps, and daily tidal flows through the mangrove forests stretch up to 100 kilometres inland. Beyond the swamps and plains are dense forests leading in the north to broad savannah.

The country has a large number of ethnic groups, though around half the population are either Balanta or Fulani. Many follow traditional animist religions, but Islam has been making steady inroads, particularly among the economic and political leadership.

Standards of human development are low. Half the population live below the poverty line and do not have access to safe water or sanitation. Life expectancy at 45 years is one of the lowest in the world. The country's relative isolation may initially have

protected it from the worst ravages of HIV/AIDS but the infection rate is now around 4%. Education standards are also very poor. Years of economic crisis and structural adjustment have taken their toll on the school system: only around half of children enrol. Girls in particular lose out: only one-quarter of women are literate.

Most economic life centres on agriculture, which accounts for almost 60% of GDP, more than 80% of employment, and more than 90% of exports. The staple food crop is rice, which is grown on one-third of arable land, along with millet, sorghum, cassava, and other root crops. But the country still has to import around 40% of its rice. In the interior, the Fulani and other groups raise cattle.

The main cash crop is cashew nuts, which provide most export income. Cashew nut trees can survive even on very poor land and require little or no cultivation. Production has been rising but Guinea-Bissau has earned less than it could from this. It has few facilities for processing, so most of the nuts are exported to India and then make their way to Europe where they are eaten as a cocktail snack. There are other cash crops such as groundnuts, and mangoes are another with export potential.

One of the obstacles to better agricultural performance is land tenure or, more precisely, deciding who owns what. Most food is produced by the 90,000 small villages, the tabancas. But there are also 1,200 larger farms, the poteiros, that are operated under state concessions. Unfortunately, many of these two types of holding overlap and it has proved difficult to resolve ownership.

Confusing patterns of land-holding

Guinea-Bissau also has considerable fishing potential. Fishing employs only 10% of the workforce but provides additional revenues as a result of licences sold to foreign fleets. The agreement with the EU involves funds to develop local capacity but also allows EU fleets to overexploit local resources.

Until around the mid-1980s, Guinea-Bissau's government attempted to develop along state-socialist lines. But recent decades have seen a decisive shift towards a market economy. This made the country more acceptable to international donors, particularly the World Bank and the IMF, though they have at times suspended their programmes because of political uncertainty.

The country has also received many other forms of aid, particularly from Portugal, the ex-colonial power. But not all aid has been well used and a combination of corruption and economic crisis has made Guinea-Bissau one of the world's most heavily indebted countries.

In an effort to stabilize the economy, the government in 1996 took the unusual step of abandoning its currency, the peso, and entering the currency zone of the former French West African countries.

Guinea-Bissau became independent in 1974, at which point it was federated with Cape Verde. Between 1974 and 1999 the country was ruled by one party—the Partido Africano da Independência de Guiné e Cabo Verde (PAIGC). In 1980, the party leadership was seized by General João Bernardo Vieira in a bloodless coup. He broke the union with Cape Verde and from 1985 he steered the PAIGC away from state socialism, and from the early 1990s opened the political system to other parties.

Faced with a divided opposition he won subsequent presidential elections in 1991, and the PAIGC won the 1994 parliamentary election. By mid-1998, however, a combination of corruption and economic mismanagement had let to an army mutiny and a civil war, though it was not until May 1999 that Vieira was finally overthrown. In November 1999, presidential and legislative elections were won by the Partido para a Renovaçao Social (PRS), with Kumba Yala as president.

Yala proved erratic and incompetent and lost the confidence of most people. In September 2003, with the country in chaos, the military intervened. After a bloodless coup that was supported even by the PRS and an interim civilian government, elections in March 2004 returned the PAIGC to power. But presidential elections in 2005 then brought back Mr Vieira as an independent. He appointed a multiparty cabinet, so the PAIGC is now in opposition.

Guyana

Guyana in South America remains split between its Indian and black communities

Guyana has a narrow coastal plain that soon leads into the dense tropical rainforests that cover more than 80% of the territory—though there are mountains in the west along the borders with Venezuela and Brazil.

The country's ethnic mix reflects its colonial history. Most of today's Guyanese, concentrated along the coast, are descendants of slaves from Africa and subsequent indentured labourers from India who came to work on coastal sugar cane plantations. The surviving Amerindians are now in scattered communities along the riverbanks.

The Guyanese have made significant progress in health and education standards, but 3% of the population are now HIV-positive.

The country has strong agricultural potential. Sugar is the main crop and investment has reinvigorated the industry. Rice too has seen record production for export. As yet, the country's forests, with 1,000 varieties of tree, have been relatively under-exploited but the arrival of Malaysian logging companies should change that—and raise fears of deforestation.

Guyana also has important mineral deposits. The opening of the Canadian-owned Omai gold mine led to a rapid increase in production. Bauxite from two government-owned

Land area: 215,000 sq. km.
Population: 0.7 million—urban 38%
Capital city: Georgetown, 137,000
People: East Indian 50%, black 36%, mixed 7%, Amerindian 7%
Language: English, Amerindian languages
Religion: Christian 50%, Hindu 35%, Muslim 10%, other 5%
Government: Republic
Life expectancy: 63 years
GDP per capita: $PPP 4,230
Currency: Guyanese dollar
Major exports: Sugar, gold, bauxite, rice, shrimps, molasses

enterprises used to be a leading export but high production costs have reduced their competitiveness.

Politics in Guyana has usually been sharply divided along racial lines. The two main parties are the People's Progressive Party (PPP), which represents the Indo-Guyanese, and the People's National Congress (PNC), which represents the Afro-Guyanese. Elections have typically been bitterly contested affairs. In 1997, for example, the majority of seats went to a coalition of the PPP and Civic, a small party formed by professionals. The PNC refused to accept the result, leading to street protests and rioting.

In 1998 an intervention by the Caribbean Community (Caricom) produced the 'Herdmanston Accord', which included constitutional reform.

The elections in 2001 again produced a victory for the PPP-C coalition, led by Bharrat Jagdeo and, although Caricom observers declared the elections fair, again there were PNC protests followed by rioting. The lead-up to the postponed 2006 elections seemed to show an improvement—until the assassination of the agriculture minister. The PPP won a narrow majority.

Haiti

The poorest country in the Americas—decades of violence have pushed Haitians even deeper into poverty

Land area: 28,000 sq. km.
Population: 8 million—urban 38%
Capital city: Port-au-Prince, 1.4 million
People: Black 95%, mulatto plus white 5%
Language: French, creole
Religion: Roman Catholic 80%, Protestant 16%, other 4% (including voodoo)
Government: Republic
Life expectancy: 54 years
GDP per capita: $PPP 1,742
Currency: Gourde
Major exports: Coffee

Haiti occupies the western one-third of the island of Hispaniola, the rest of which is the Dominican Republic. The territory consists of mountain ranges interspersed with fertile lowlands. Around half takes the form of two peninsulas to the north and south of the Golfe de la Gonâve, within which there are a number of offshore islands.

Decades of plunder by the dictatorial Duvalier family, followed by years of violent political turmoil, have reduced an already poor population to destitution and even starvation. This is the only state in the Western Hemisphere classified by the UN as a Least Developed Country. Average income is now one-third of that in the Dominican Republic. Around half the population are undernourished, half lack access to safe water, and half are illiterate. There is scarcely any tax revenue

and social services are weak or non-existent. Only 10% of schools are government run; most of the rest are in the hands of the Catholic church or non-governmental organizations.

Unsurprisingly, many Haitians emigrate—either across the border to work on the sugar plantations of the Dominican Republic, where up to 800,000 Haitians live, or to the USA; around 300,000 live in New York. Migrant remittances are worth around $900 million annually—one of the country's main sources of income.

Two-thirds of the population struggle to make a living from agriculture. Much of this is subsistence farming on tiny plots growing maize, sorghum, millet, and beans. The better, irrigated land is also used to grow rice and beans for sale in the urban areas. The main cash crop is coffee but export earnings have slumped as a result of reduced output and low world prices. Sugar used to be an important export earner but falling prices and foreign competition have discouraged production.

Agriculture in Haiti has also suffered from environmental degradation. Over-exploitation of the soil has been compounded by erosion resulting from deforestation. Most people rely on wood for fuel and have steadily stripped the hillsides of trees—only around 2% of the land is now forested.

In the early 1980s, there was some prospect of industrial development in two free-trade zones near Port-au-Prince, but the fighting and the trade embargo drove most companies away. That leaves the majority out of work: 60% are unemployed or underemployed and many rely on feeding programmes and other support from international aid agencies.

The country's current troubles are the persistent legacy of the years of dictatorship. From 1957 to his death in 1971, 'Papa Doc' Duvalier ruled through fear and repression, whilst draining the country of its meagre resources. His son Jean Claude, 'Baby Doc', continued in the same fashion until national and international pressure ousted him in 1986. Several years of upheaval ensued until 1990, when a fiery Catholic priest, Jean-Bertrand Aristide, heading a grass-roots movement called Lavalas ('avalanche'), became president in February 1991. However, his radical policies favouring the poor so upset the army and the business community that he was ousted that September in a military coup by General Raul Cédras.

The arrival of Aristide

The coup in turn outraged the international community. Eventually Cédras withdrew—and 20,000 US troops arrived to supervise the restoration of democracy and Aristide's return.

Constitutionally, Aristide was not eligible to stand in the 1991 elections. In his place the Lavalas candidate, duly elected, was René Préval. The government was obliged to impose IMF-style spending cuts and privatization programmes.

Aristide carefully distanced himself from these measures and formed a breakaway party called Fanmi ('family') Lavalas. Struggles between the two Lavalas parties led to disputes over Senate elections.

After several postponements an election in May 2000 gave Aristide's Lavalas a resounding victory. But the election was marred by irregularities, and 15 opposition groups united as Convergence Démocratique to challenge the legitimacy of the government. Meanwhile, in an election boycotted by the opposition, Aristide was re-elected president in November 2000.

Aristide unfortunately had developed the autocratic characteristics of previous Haitian leaders and sought absolute control, using street gangs where necessary. In 2004 the opposition and students stepped up demonstrations calling for his resignation. In February 2004 an improvised rebel army ousted Aristide who was flown out in a US plane.

The Supreme Court Chief Justice Boniface Alexandre took over as interim president supported by a UN-authorized peacekeeping force.

Elections were finally held in May 2006 and, after some confusion, won again by René Préval, now estranged from Aristide, at the head of multiparty coalition Lespwa (Hope). As with Aristide most of his support comes from the rural areas and the slums. He is opposed by the business community and does not have a legislative majority. But he does have the backing of the international community.

Honduras

Honduras remains a deeply divided country, now suffering from violent street gangs

| 0 | Miles | 200 |
| 0 | Km | 320 |

Caribbean Sea

La Ceiba
San Pedro Sula

Tegucigalpa

NICARAGUA

PACIFIC OCEAN

Land area: 112,000 sq. km.
Population: 7 million—urban 46%
Capital city: Tegucigalpa, 858,000
People: Mestizo 90%, Amerindian 7%, black 2%, white 1%
Language: Spanish, Amerindian dialects
Religion: Roman Catholic 97%, Protestant 3%
Government: Republic
Life expectancy: 68 years
GDP per capita: $PPP 2,665
Currency: Lempira
Major exports: Coffee, bananas

Honduras occupies the centre of the Central American isthmus. Most of the land is mountainous—around three-quarters has slopes of 20 degrees or more. Lowland areas are confined to the narrow coastal strips and to the river valleys.

The majority of Hondurans are of mixed Indian-European extraction. There are some Indians who live mostly along the Guatemalan border and on the Caribbean coast and there are groups of Garifuna, who are descendants of the Carib Indians and black Africans.

Around half of Hondurans live below the national poverty line and the income distribution is very skewed: the top 10% of the population get 42% of the income, while the bottom 10% get only 0.9%. The rural population in particular have poor standards of education and health and there are increasing problems of

TB and malaria. Around one-fifth of the population are undernourished. The northern commercial capital, San Pedro Sula, has also become an epicentre for the AIDS epidemic in Latin America. Nationally, around 2% of adults are HIV-positive.

Agriculture employs around one-third of the workforce though only 14% of the total land area is suitable for arable farming. Half the population are in the rural areas struggling to make a living, cultivating corn and beans on the steep slopes. Agriculture is also responsible for more than half of export earnings. Nowadays the leading export crop is coffee, which is produced by around 105,000 independent producers—though the quality can be low.

The other export crops are grown in the lowland areas which have the most fertile land. The Caribbean coast in particular has one of the world's best climates for bananas.

Since the beginning of this century, the Honduran banana industry has been dominated by two corporations which are subsidiaries of US multinationals: Standard Fruit (selling under the Dole brand name) and the Tela Railroad Company (Chiquita). These companies have had a powerful influence on commercial and political life and employ around 15% of the workforce. Hurricane Mitch in 1998 destroyed about 70% of the banana plantations. Since then there has been replanting but output has not fully recovered.

There have also been efforts to diversify into other crops such as melons, pineapples, and mangoes.

Recent developments in shrimp farming have offered new export markets along with lobsters, most of which come from 5,000 Miskito Indian divers who risk their lives for the catch.

Perilous diving for lobsters

Land distribution remains a critical issue. Pressure from peasant farmers in the 1960s and 1970s resulted in some land reform. In 1992, however, an Agrarian Modernization Law removed most of the grounds for automatic confiscation.

In the urban areas, one of the most important new sources of employment has been the 'maquila' factories which assemble goods for export. Around 93,000 people work in a number of export processing zones in the north, chiefly in garment manufacture using imported US cloth and taking advantage of well-developed port facilities around San Pedro Sula. Maquila output has grown strongly, and is the leading source of foreign exchange. It could also expand further as a result of diversification and a new regional trade agreement with the USA.

Tourism is also becoming an important source of foreign exchange. More than one million people come to the beaches, for example, or to the coral reefs or to visit historical Mayan sites. Another importance foreign exchange contribution comes from Hondurans abroad whose remittances in 2005 were $1.8 billion.

Honduras has many of the same contrasts between rich and poor as its neighbours, but somehow it managed to escape the civil wars that devastated other countries in Central America. Nevertheless, it was still deeply involved and heavily militarized, since the country offered military bases both for the 'contra' fighters from Nicaragua and also for the US forces—for which it was rewarded generously with military aid.

The military have also had a major influence on politics, though this has diminished in recent years. A more worrying security threat comes from the drug cartels who use Honduras as a transit point, and from the violent street gangs, the 'maras', who are responsible for most street crime.

Politics in Honduras has long been a two-party struggle between liberals and conservatives—strongly influenced by US commercial and political interests. The centre-right Partido Liberal (PL), headed by Carlos Flores Facussé, won the 1997 presidential elections. But the 2001 elections saw a victory for the more conservative Partido Nacional (PN) with Ricardo Maduro Joest emerging as president. Maduro promised to crack down on crime though with little success, and at the cost of human rights abuses by the security forces.

The presidential election at the end of 2005 resulted in a victory for the PL candidate, Manuel Zelaya Rosales. The PL does not have a majority in Congress, so has to rely on support from smaller parties.

Hungary

Hungary is one of the more successful of the former communist countries that have now joined the EU

Land area: *93,000 sq. km.*
Population: *10 million—urban 65%*
Capital city: *Budapest, 1.7 million*
People: *Hungarian 92%, Roma 2%, other 6%*
Language: *Hungarian*
Religion: *Roman Catholic 51%, Calvinist 16%, Lutheran 3%, other 30%*
Government: *Republic*
Life expectancy: *73 years*
GDP per capita: *$PPP 14,548*
Currency: *Forint*
Major exports: *Machinery, manufactured goods*

Hungary is divided into roughly two halves by the Danube, which cuts through the country from north to south. The area to the west is called Transdanubia. The land is predominantly flat but has two higher areas: the uplands of Transdanubia in the south-west and the loftier Northern Mountains along the border with Slovakia. Hungary's plains also come in two main parts. In the north-west of Transdanubia is the Little Hungarian Plain. To the east of the Danube the Great Hungarian Plain accounts for more than half the country.

Most of Hungary's people are related to Finns and Estonians and are ethnically distinct from their Slavic neighbours. The most significant minority population are 190,000 Roma, or gypsies, who are among the country's poorest people. Most Hungarians suffered a steep drop in living standards after 1990. Since then there has been a steady if slow recovery. Education standards are high, but health is worse than might be expected: middle-aged men tend to have particularly poor health, due to bad diet, smoking, and high alcohol consumption. Hungary also has one of the world's highest suicide rates.

Around one-quarter of ethnic Hungarians live outside Hungary. As a result of the historical redrawing of boundaries there are around 1.6 million in Romania, half a million in Slovakia, and many others in Serbia and Ukraine. This has frequently been a source of friction: Hungary has protested to Slovakia in particular about discrimination against Hungarians, though Hungary tends to have better relations with Romania.

Industry employs around one-third of the workforce and their output of machinery, other equipment, and chemicals accounts for most exports. Production slumped when markets in Eastern Europe collapsed, but thanks to extensive foreign investment these have now been replaced by sales to Germany and Austria—mostly of component parts, particularly for the car industry. Hungary produces around one million of Audi's engines annually, for example. Many of the companies involved come from the EU, the US, and Japan, but one-quarter of sales are by the local firm, Raba. Other important industries include pharmaceuticals and chemicals and more recently electronic goods and computer software. Around 70% of all Hungary's exports go to other countries in the EU.

Even in the communist era, Hungary had been creating elements of a market economy. And the government has since carried out a fairly effective privatization programme. The private sector is now responsible for 80% of GDP.

With fertile soils and a helpful climate, Hungary can grow a wide range of crops. And though agriculture accounts for only 4% of GDP and employs 6% of the workforce, the country is generally self-sufficient in food. The government has converted collective farms to co-operatives.

Workers leaving agriculture and industry have mostly been absorbed into a mushrooming service sector which generates around two-thirds of GDP. Tourism plays an increasingly important part, generating up to 10% of GDP; expenditure per visitor is increasing. Unemployment has come down steadily, to around 7%.

Hungary's political orientation has been steadily westward. In 1999 it joined NATO and, after a favourable referendum in 2003, Hungary joined the European Union in May 2004 and hopes to adopt the euro in a few years.

Over the last decade the political scene has seen regular swings between left and right. By 1989, the Communist Party had itself become a leading advocate of reform. Renamed the Hungarian Socialist Party (MSZP), it had brought together the more liberal intellectuals, social democrats, and trade unions. Even so, it lost the 1990 election to the right-wing Hungarian Democratic Forum, which formed a coalition with two other conservative parties.

In the 1994 election, the MSZP came out ahead. Led by prime minister Gyula Horn, they entered into an alliance with the country's principal liberal party, the Alliance of Free Democrats (SZDSZ). But faced with an economic crisis they were forced to make cuts in welfare spending. And although foreign investment was creating new wealth the benefits were spread unevenly—concentrated around Budapest, leaving much of the east of the country behind.

Wealth is concentrated in Budapest

In the 1998 elections, the voters rejected the socialists. But this time they chose a different alternative—a coalition headed by Fidesz, which started life in 1988 as a liberal student organization but now has a populist conservative stance.

The elections in 2002 produced yet another switch, bringing back the MSZP and the SZDSZ, though with only a small majority. The prime minister was Peter Medgyessy, though he was replaced in late 2004 by Ferenc Gyurcsany of the MSZP, who has been modelling his policies on those of Tony Blair in the UK. The coalition won a second term in 2006, and strengthened its position. However, in September there were street protests when Gyurcsany admitted he had lied "morning, noon, and night" to get re-elected.

Iceland

One of the world's richest countries, thanks to fish

Reykjavik

ATLANTIC OCEAN

Land area: 103,000 sq. km.
Population: 0.3 million—urban 93%
Capital city: Reykjavik, 113,000
People: Icelandic
Language: Icelandic
Religion: Lutheran 86%, other 14%
Government: Republic
Life expectancy: 81 years
GDP per capita: $PPP 31,243
Currency: Krona
Major exports: Fish, aluminium

Despite its proximity to the Arctic Circle, the climate in most of Iceland is relatively mild. The landscape varies from Mount Hvannadals at 2,119 metres, which is on the edge of a vast glacier, through grassy lowlands to a complex coastline indented with fjords. Iceland is one of the most geologically active places on earth, and has regular minor earthquakes, numerous active volcanoes, and bubbling hot springs.

The country's isolated position has sustained a very homogenous population, but they are far from insular. In addition to Icelandic, most speak English, and education standards are high: Iceland publishes more books per person than any other country. Its people are enthusiastic users of the internet and more than 90% of adults have mobile phones. They are also among the richest people in the world and have a good system of welfare.

Much of the wealth has been based on fish. The fishing industry still employs around 6% of the workforce and accounts for 60% of exports. The main catches are of cod and capelin, though over-fishing has required the introduction of quotas. In 1975 Iceland imposed a 200-mile fishing zone—provoking the 'Cod War' with the UK—and since it also takes 15% of its catch from distant waters it frequently has disputes with other fishing nations.

Iceland's geological activity is also a major resource, providing geothermal energy that heats more than 85% of homes. The fractured landscape also has numerous steeply falling rivers that have huge hydroelectric potential. This is being exploited for smelting aluminium and, with a cable to the UK, Iceland could even export electricity.

A major ongoing issue is that of EU membership. At times there has been a majority of the population in favour, but there are worries about the effects on the fishing industry.

Iceland has one of the world's oldest parliaments, the Althing. Since independence from Denmark in 1944, the system of proportional representation has never given one party an absolute majority and politics has usually been consensual.

The 1999 Althing elections gave the most seats to the centre-right Independence Party, and returned David Oddson as prime minister in coalition with the agrarian-based Progressive Party. The coalition won a third term in the elections of 2003. Oddson stepped down in September 2004 and was replaced by the Progressive Party leader, Halldor Asgrimsson, who is more in favour of EU membership.

India

The world's largest democracy has a rich and diverse culture. Now, at last, it also seems to be achieving more rapid economic growth

Land area: 3,288,000 sq. km.
Population: 1.1 billion—urban 28%
Capital city: New Delhi, 12.8 million
People: Indo-Aryan 72%, Dravidian 25%, Mongoloid and other 3%
Language: Hindi 30%, plus 14 other official languages and English
Religion: Hindu 81%, Muslim 12%, Christian 2%, Sikh 2%, other 3%
Government: Federal republic
Life expectancy: 63 years
GDP per capita: $PPP 2,670
Currency: Rupee
Major exports: Garments, gems and jewellery, cotton textiles, cereals

India's vast territory can be divided into three main regions from north to south. The far north and north-east of India cover part of the Himalayas including Kanchenjunga, the world's third-highest mountain. To the south, these mountains descend to the second region, a broad northern plain formed by the deposits of a number of major river systems, the most important of which is the Ganges, which emerges from the southern slopes of the Himalayas, flowing slowly south-east across the plain before spreading out into a broad delta in the Bay of Bengal.

To the west, the plain becomes a desert that extends into Pakistan. To the east, beyond the Ganges delta, India's territory circles round that of Bangladesh. The third main region, to the south of the plain beyond the Narmada River, is the Deccan plateau, a triangular region of low hills and extensive valleys, bounded along the coasts by two mountain systems: the eastern and western Ghats.

India now has over one billion people and on present trends will be the world's largest nation by 2030. The population can be subdivided in many different ways. The broadest distinction is in terms of religion. India's constitution declares it to be a secular state, but religious intolerance, known here as 'communalism', is never far below the surface. Hindus now account for over 80% of the population and Muslims 12%, and there has always been tension between the two. In 2002, for example, communal riots in Gujarat killed 900 people. The other main source of friction has been with the Sikh community concentrated in the northern state of Punjab. Christians too have come under attack.

Beyond religious divisions, India is also a source of vast cultural and linguistic diversity. The national language, Hindi, is widely spoken in the north but elsewhere it is displaced by dozens of other languages, from

Malayalam to Tamil to Bengali, many of which use different scripts.

Another way of dividing India is by caste—a hereditary social system that has Brahmins at the top and the 'untouchables', who perform the most menial social tasks, at the bottom. The latter, who are now referred to as 'scheduled castes', have specific allocations or 'reservations' of government jobs and parliamentary seats. Similar positive discrimination is exercised in favour of 'scheduled tribes'— a smaller disadvantaged group of tribal communities.

In terms of average income, India is still very poor. But the population is so large that the top 10% of the population would on its own constitute one of the world's largest industrial nations. Another significant group are the 16 million or so Indians who now live abroad, whether as permanent residents or contract workers in the Gulf and elsewhere.

At home, around one-third of the population live below the poverty line and around 40% are illiterate.

India has striking regional contrasts But there are striking regional contrasts. Thus, one of the wealthier states, Punjab, has a per capita income twice that of West Bengal. And the state of Kerala, with many enlightened social policies, has a 90% literacy rate.

In India it is also important to consider gender distinctions. In most states women fare worse than men. Around half of women are illiterate, and women's and girls' health and nutrition are frequently neglected. As a result of this, and selective abortion, there are only 933 women per 1,000 men.

Two-thirds of India's people are still to be found in the rural areas and rely on agriculture for survival. Land is scarce. By the mid-1980s, the average farm size was down to 1.7 hectares and today one-third of rural households have no land at all. Despite the shortage of land, India's food production, particularly on the larger landholdings, has, thanks to the 'Green Revolution', largely kept pace with population growth. Nevertheless, millions still go hungry because they cannot afford to buy the food; one-fifth of the population are undernourished.

India is one of the world's major industrial powers. For the first 40 years after independence, industrial policy was based on state-owned heavy industry, combined with private production of consumer goods which was protected from foreign competition by high tariffs. At the same time, entrepreneurs had to fight their way through thickets of regulations and licences and negotiate with often corrupt officials.

All this began to change from 1991 when the then finance minister, Manmohan Singh, took the first steps by cutting tariffs, liberalizing imports, and encouraging foreign investment. Liberalization has a long way to go. The state still employs around 70% of formal-sector workers who are deeply resistant to change. But India is now much more open to the outside world. And economic growth, which between 1960 and 1990 had averaged only 4% per year (known as the 'Hindu rate of growth'), has at times touched 8%.

All industries, including textiles, steel, and petrochemicals, have been affected by liberalization, but some of the most visible changes have been in areas such as cars and other consumer goods. There has also been a boom in consumer electronics and computers. The latter can take advantage of

India is a nursery for software talent
India's large pool of computer programmers, especially in Bangalore, India's 'cybercity'. Software now accounts for around one-fifth of exports. India's English-speaking graduates also staff many call centres for companies in the USA and Europe and perform many other 'outsourced' functions.

These developments have been transforming the face of India's cities. TV programming has now been opened up to a multitude of satellite channels. Life in the rural areas is slower to change. Villagers may have also have satellite TV but little prospect of buying the wares on offer.

India can rightly be proud of having maintained a democratic government for almost the entire period since independence. This is based on a federal constitution with 29 states and 6 union territories. Each state has its own chief minister and regional assembly. Historically, the states have had relatively little power. Nevertheless, Kerala and West Bengal have had communist governments that have pursued distinctive policies of income redistribution.

India's federal government consists of a lower house, the Lok Sabha, which elects the prime minister, and the upper house, the Rajya Sabha, whose members are elected by the regional assemblies.

For most of the first 50 years the government was the domain of the Congress Party, which was usually led by the Nehru dynasty; first by the leader of the independence struggle, Jawaharlal Nehru, later by his daughter Indira Gandhi (assassinated in 1984 by Sikh extremists), and then by her son Rajiv (assassinated in 1991 by a Tamil extremist).

Congress's almost uninterrupted rule ended with the election in 1996 that gave the majority of seats to the right-wing Hindu nationalist Bharatiya Janata Party (BJP) led by Atal Behari Vajpayee who headed a series of coalition governments until 2004.

While India's economy seemed to be healthy under the BJP, most progress was in the cities, leaving the rural areas behind. The Congress Party, now led by Rajiv's Italian-born widow, Sonia, capitalized on this discontent and in the elections of May 2004 took the largest share of the vote. With only 145 of the 545 seats in the Lok Sabha, however, Congress had to form a coalition with a number of smaller parties, and also had to rely on support from the communist-dominated Left Front. Sonia Gandhi refused to become prime minister, instead nominating elder statesman Manmohan Singh.

Domestically the most violent opposition comes from the Maoist 'Naxalites' in more remote areas.

The most important foreign policy issue concerns relations with the old enemy, Pakistan. The main problem is Kashmir, a territory whose status the two countries have disputed since the partition of India in 1947. Up to 100,000 people have died as a result of cross-border fighting. The enmity has at times escalated alarmingly. However, the two countries are moving towards reconciliation.

Indonesia

Indonesia's vast archipelago has managed to avoid political break-up and is making steady economic progress

| 0 | Miles | 750 |
| 0 | Km | 1200 |

Land area: 1,905,000 sq. km.
Population: 217 million—urban 46%
Capital city: Jakarta, 8.4 million
People: Javanese 45%, Sundanese 14%, Madurese 8%, coastal Malays 7%, other 26%
Language: Bahasa Indonesia, English, local languages
Religion: Muslim 88%, Protestant 5%, Roman Catholic 3%, Hindu 2%, Buddhist 1%, other 1%
Government: Republic
Life expectancy: 67 years
GDP per capita: $PPP 3,361
Currency: Rupiah
Major exports: Oil, gas, plywood, garments, rubber

Indonesia is one of the world's largest and geographically most dispersed countries, consisting of more than 17,000 islands. Of the southern chain of islands, the largest are, from west to east: Sumatra, Java, Bali, and Lombok.

To the north of this chain is the island of Borneo, the southern part of which consists of the Indonesian provinces of Kalimantan. To the east of Borneo is the island of Sulawesi, and furthest east are the Molukas and New Guinea, the western half of which is the Indonesian province of Papua. Most of these islands have volcanic mountains covered in dense tropical rainforests that descend to often swampy coastal plains.

Indonesia's people are as diverse as its topography. By some counts, there are more than 300 ethnic groups with almost as many languages. Many of these groups are related and most are of Malay origin. The largest is the Javanese, who live on the centre and east of Java—the western portion of which is occupied by the Sundanese.

Other large groups are the Madurese and the coastal Malays. The many smaller groups include tribal peoples, such as the Dayak, who inhabit the interior of Kalimantan, and people of Melanesian descent in Papua. In addition, Indonesia also has eight million people of Chinese origin who live mostly in the cities and allegedly control half the economy. Ethnic and religious differences have often led to violence particularly in the Malukus.

Indonesia has also suffered from natural disasters. In recent years these have included a tsunami that struck Sumatra in 2004, killing 233,000 people and levelled the capital of Aceh, and in 2006 an earthquake in Java that killed 6,300 people.

Despite the large land area, Indonesia's population is fairly concentrated. Half the islands are uninhabited and around two-thirds of the population live on Java, Madura, and Bali. The majority of people profess to be Muslim, though religion has not so far been a dominant force in political life. The official language is Bahasa Indonesia, which is based on Malay. Only around 10% of people use this as their first language; most have it as their second.

Faced with a fast-growing population, the government in the 1970s and 1980s instituted a programme of family planning and the annual growth rate has slowed to around 1.5%. A more controversial demographic initiative was the 'transmigration' programme which moved people from densely settled areas like Java—which has around 60% of the population—to other islands. There has also been extensive labour migration—particularly to Malaysia, where over one million Indonesians work on plantations.

Around one-quarter of the population live below the national poverty line. And although standards of education and health have improved, many schools and health centres, particularly in the rural areas, are now in a poor condition as a result of low public spending.

More than half the population still make their living from agriculture. Only about one-tenth of the country is suitable for agriculture, but the rich volcanic soil on the major islands is very fertile. Farmers here, most of whom are smallholders, primarily grow rice in the lower areas, and fruit, vegetables, tobacco, or coffee on the higher slopes. By the mid-1980s, with heavy use of 'green revolution' techniques, they had made the country self-sufficient in rice, though imports have now resumed. The land on the outer islands is less fertile and is primarily used for tree crops such as oil-palm, rubber, and coconuts, as well as cocoa and coffee.

Forests cover around 60% of the land area and the timber industry has grown rapidly, mostly based on hardwoods such as teak and ebony. With a general breakdown in law and

Choking forest fires

order there has been an increase in illegal logging which has added to deforestation. In addition there are forest fires—a result of slash-and-burn cultivation as well as of companies clearing land for plantations. These fires can send a vast pall of smog across neighbouring countries.

Indonesia's rich mineral deposits have been a major source of income. State companies exploit these through production-sharing agreements with foreign investors. Among the major minerals are tin, bauxite, copper, nickel, gold, and silver.

But the most important mineral is oil, and Indonesia is also the world's largest producer of liquefied natural gas. Oil and gas account for around 20% of exports. The industry is controlled by the state company Pertamina which has production-sharing agreements with transnational oil companies. Nevertheless, oil output is low and falling, and Indonesia could become a net importer.

Indonesia has also developed a highly diversified manufacturing sector. This includes iron and steel, aluminium smelting, oil refining, cement, and, more recently, pulp and paper production. Manufacturing for export has concentrated more on labour-intensive items such as garments and footwear, though this is now facing low-wage competition from countries like China.

Indonesia's political system has in the past been authoritarian with a strong military influence. Between 1950 and 1967, following independence from the Dutch, the country was controlled by President Sukarno. During the mid-1960s his army chief, Mohamed Suharto, suppressed an allegedly communist coup—and followed up with a slaughter that cost up to one million lives. In 1968 Suharto himself was elected president and backed by his political group, Golkar, was subsequently re-elected unopposed for six further terms.

Suharto's notorious nepotism

Suharto's 'New Order' administration reinvigorated the economy but seized much of the wealth for itself—a period characterized by 'corruption, collusion and nepotism'. Meanwhile the army brutally suppressed most opposition.

Suharto's reign ended in 1998 as a result of economic collapse. In Indonesia, as elsewhere in Asia, companies had over-borrowed, banks became insolvent, and thousands of companies crashed. Faced with rises in food prices many people rioted, and scapegoated Chinese businesses.

In May 1998, Suharto abruptly resigned, handing over power to his vice-president, B. J. Habibie. He made some crucial decisions. One was to offer the people of East Timor their independence. Another was to decentralize significant political authority to Indonesia's hundreds of districts—a move designed to ward off further secessionist threats, not just from historically dissident provinces like Aceh and Papua but also from richer provinces like Riau which complained that income from oil was being siphoned off to Jakarta.

Habibie was replaced when the June 1999 elections for the People's Consultative Assembly gave most seats to the Democratic Party of Struggle (PDI-P), led by Megawati Sukarnoputri—Sukarno's daughter. Golkar came second, followed by the Islamicist National Awakening Party (PKB). The assembly surprisingly chose as president Abdurrahman Wahid of the PKB—whose chaotic administration ended in 2001 with impeachment for incompetence.

Megawati took his place, but her administration achieved very little other than stability and she was punished in the assembly elections in April 2004 when Golkar emerged as the largest party.

Over this period Indonesia suffered from Islamic terrorism, notably a 2002 bomb attack on a nightclub in Bali that killed 202 people and another on the Jakarta Marriott Hotel in 2003.

The 2004 presidential election, the first to be based on a direct popular ballot, was won by a former general, Susilo Bambang Yudhoyono. He has only a small party of his own and rules with the tacit support of Golkar. He has made significant progress, including reducing expensive fuel subsidies while introducing a national programme of welfare payment for the poorest 25% of families.

His most notable political achievement has been to end a 29-year armed rebellion in Aceh by offering autonomy. Rebels in Papua, however, continue to demand independence.

Iran

Political reform in Iran has now ground to a halt as Islamic conservative forces strengthen their grip on the state

Land area: 1,648,000 sq. km.
Population: 68 million—urban 67%
Capital city: Tehran, 6.8 million
People: Persian 51%, Azeri 24%, Gilaki and Mazandarani 8%, Kurd 7%, Arab 3%, Lur 2%, Baloch 2%, other 3%
Language: Persian and dialects 58%, Turkic and dialects 26%, Kurdish 9%, other 7%
Religion: Shia Muslim 89%, Sunni Muslim 10%, other 1%
Government: Islamic republic
Life expectancy: 70 years
GDP per capita: $PPP 6,995
Currency: Rial
Major exports: Oil and gas, carpets, fruit

Much of Iran consists of a high central plateau ringed by mountains, the highest of which are in the volcanic Elburz range in the north. Iran is not very fertile—more than half the territory is barren wasteland, most of which is salt desert and uninhabited. Only around 10% is suitable for arable farming, much of this in the fertile northern plateau. Another 20% can be used for grazing. Iran is vulnerable to severe earthquakes, the most recent of which in 2003 in the city of Bam killed 43,000 people.

Ethnically, Iran is quite diverse. Just over half the people are ethnic Persians, who are to be found throughout the country. But one-quarter are Azeris, who live in the north-west. The least assimilated minority are the Kurds, who make up 7% of the population and have a largely nomadic existence in the

mountains of the west. This diversity is also reflected in language: Persian (Farsi) is the official language but one-quarter of people use Turkic languages. Iran has also received millions of refugees, chiefly from Afghanistan and Iraq. On the other hand, around 200,000 Iranians leave each year, seeking a better life abroad.

Most people are Shia Muslims, but around one-tenth, including the Kurds, belong to the Sunni sect. In theory there is religious freedom, but in practice those who are not Shia Muslims are second-class citizens. Islam also places restrictions on women, particularly on their dress. But by the standards of other Islamic countries, Iran is relatively liberal— women have good access to education and can enter most professions.

Despite Iran's hostile terrain, agriculture accounts for about one-tenth of GDP and one-quarter of the workforce. The main crops are wheat, barley, and potatoes. But only one-quarter of the potential area is being cultivated and productivity is low—hampered by poor management and political uncertainties. Hopes of reaching food self-sufficiency were dashed by a severe drought from 1998 to 2001. Around one-third of agriculture is based on livestock, chiefly sheep and goats.

At the heart of Iran's industrial economy is oil. Iran has enough reserves for 75 years of extraction at current rates. Over the period 1994–2004, oil accounted for over 80% of export earnings. Reserves are in the south-west, and also offshore in the Persian Gulf. Iran also has the world's second largest reserves of natural gas but these remain under-exploited.

Iranian industry was nationalized after the 1979 revolution. Most is inefficient and runs far below capacity. Foreign investors have been discouraged by official coolness to foreigners, and also by a US ban on trade and foreign investment imposed since 1995. The most promising manufacturing area is petrochemicals, though steel and cement also offer prospects for diversification from oil.

Most Iranians struggle to survive. Unemployment is probably around 20%. Many professional people are on such low salaries that they need to take on two or three extra jobs. Corruption is widespread.

Political life in Iran was transformed by the 1979 revolution that swept aside the Pahlavi dynasty of the Shah and by 1981 had established a strict Islamic Republic which was to be run by 'religious jurists'. At the head of these was the formerly exiled Ayatollah Khomeini, supported by another cleric, Hashemi Rafsanjani, who in 1980 became the speaker of the Majlis, the 207-member parliament.

An Islamic republic run by religious jurists

The early years of the revolution were chaotic and turbulent and included an eight-year war with Iraq, 1980–88, and a persistent enmity with the 'Great Satan', the USA. Khomeini's death in 1989 left no obvious successor. Instead, political and religious functions were re-allocated. The role of rahbar, or religious leader, was taken by Sayed Ali Khamenei, while most political functions were vested in the president. Rafsanjani was elected president.

This set up a tension between the conservative religious leadership and the nominally secular presidency. Rafsanjani tried to open up the economy, with limited success. He was succeeded in 1997 by an even more progressive cleric, Muhammad Khatami. The parliamentary elections in February 2000 altered the picture again, giving a resounding majority to the reformers, but they made little progress against the conservatives.

Since 2004, the conservatives have been firmly in power. Prior to the 2004 parliamentary election the hardline Council of Guardians barred 2,000 reformist candidates, giving the conservative forces an overwhelming majority.

They repeated this success in 2005 with the election as president of Mahmoud Ahmadinejad. He is much more of a populist figure, addressing himself to the poor at home and taking a very aggressive line abroad, particularly towards Israel.

But the most contentious issue has been Iran's pursuit of a uranium enrichment programme that would give it the capacity to build a nuclear bomb—leading to a confrontation with the EU and the USA.

Iraq

Post-war Iraq has descended into chaos, verging on civil war between Sunni and Shia

Land area: 438,000 sq. km.	
Population: 27 million—urban 67%	
Capital city: Baghdad, 6.3 million	
People: Arab 75%–80%, Kurdish 15%–20%	
Language: Arabic, Kurdish	
Religion: Muslim 97%, other 3%	
Government: Transitional democracy	
Life expectancy: 59 years	
GDP per capita: $PPP 3,400	
Currency: Iraqi dinar	
Major exports: Oil	

The heart of Iraq, forming around one-third of the territory, is the alluvial and often marshy basin of the Tigris and Euphrates. These rivers flow the length of the country before uniting as the Shatt al-Arab River, which flows into the Persian Gulf. To the north of this basin are uplands, and in the north-east is the mountainous region of Kurdistan. The rest of the country, to the west and south, around two-fifths of the territory, is largely desert.

There are three main population groups. The largest, accounting for around half the population, are Arab Shia Muslims, who live mostly in the centre and south. Next are the Arab Sunni Muslims who are around one-third of the population and are concentrated in the east around Baghdad. Then there are the Kurds, also Sunni Muslims, who make up one-fifth of the population and live

in the north and north-east. There are also around three million exiled Iraqis.

The lives of most Iraqis have been devastated both by the former oppressive political regime, by 13 years of UN trade sanctions, and by the 2003 war and its aftermath. Until the 1990s health standards were good, but then fell steeply, chiefly as a result of a lack of access to clean water and sanitation. Child malnutrition also increased and remains high. Schools were also in a poor condition though most have now been rebuilt.

Iraq's economy has long depended on the export of oil. Iraq has around 10% of world reserves, third after Saudi Arabia and Iran, along with the world's tenth largest reserves of natural gas. But production has been low. Prior to the war, Iraq was unable to export even its UN-sanctioned quotas and now the industry faces widespread sabotage that deters investment.

Manufacturing remains limited. Before the war efforts had been made to diversify industry away from oil but there has been little progress. Many people are still on the payroll of defunct state-owned enterprises.

Iraq's farmers should have been less affected by the fighting. The rich alluvial soil with ample supplies of water ought to make the country very productive. In 1992, 14% of the labour force were engaged in farming. In practice, many peasant farmers reverted to subsistence production. Production has revived somewhat; even so around half of the country's food needs now have to be imported and one-quarter of the population rely on the public food distribution system.

For 25 years political life in Iraq was dominated by Saddam Hussein who came to power in a coup in 1979. He summarily executed his rivals and subsequently dealt ruthlessly with any opposition. Saddam was politically cunning but he also made disastrous mistakes—notably the war with Iran over the period 1980–88, and the 1990 invasion of Kuwait that provoked intervention and defeat by a US-led coalition in the Gulf War.

Saddam was cunning but made disastrous mistakes

Following that war there were a number of attempts to topple Saddam but uprisings by Kurds in the north and the Shia in the south were easily crushed. The international community made efforts to monitor his weapons production using the UN inspection force UNSCOM. But Saddam denied access to most sites and, after the team was withdrawn in 1998, the USA and the UK responded with bombing raids.

In 2002 US President George Bush referred to Iraq as part of a global 'axis of evil' and made it clear he wanted to get rid of Saddam. Later that year the UNSCOM team briefly returned, but they were not given much time to achieve anything. Attempts to get UN backing for military action failed so in March 2003 the USA started to bomb Baghdad following which US and British forces invaded from the south. Within three weeks the regime had collapsed.

The US established a Coalition Provisional Authority subsequently led by US proconsul Paul Bremer, and appointed a 25-member Governing Council. It also had more military successes, notably the killing of Saddam's sons and the capture of Saddam himself in December 2003.

In June 2004, the CPA handed over power to a transitional administration. Despite continuing violence, a general election took place in January 2005. This resulted in a Shia-dominated transitional government headed by prime minister Ibrahim Jaafari.

The first regular parliamentary elections were held in December 2005, again producing a Shia majority. After six months of political wrangling a government of national unity finally emerged under a new prime minister, Nuri al-Maliki of the main Shia block, the United Iraqi Alliance. But jobs have also gone to the largest Sunni Arab grouping, the Iraqi Consensus Front, as well as to Kurdish parties.

However, the government's authority is weak and the security position remains grim. An insurgency that initially targeted the occupying forces has now moved closer to civil war between disparate Sunni and Shia groups who are slaughtering civilians with little compunction.

Unless the new government can exert some authority over the insurgents there seems little prospect of a US withdrawal. Unfortunately, Maliki seems unwilling to act.

Ireland

Ireland has prospered as a member of the EU and has become a centre for high-tech manufacturing

Land area: 70,000 sq. km.
Population: 4 million—urban 60%
Capital city: Dublin, 495,000
People: Irish
Language: Gaelic, English
Religion: Roman Catholic 88%, other 12%
Government: Republic
Life expectancy: 78 years
GDP per capita: $PPP 37,738
Currency: Euro
Major exports: Machinery, chemicals, food

The Republic of Ireland occupies 80% of the island of Ireland. The remainder, Northern Ireland, is a part of the UK. The interior is largely lowlands interspersed with numerous lakes and bogs. Ringing the lowlands, around most of the coast, are low mountain ranges—the highest peak is only 1,040 metres above sea level.

Ireland's people are largely of Celtic descent and Roman Catholic, though there are also small numbers of Anglo-Irish Protestants, a legacy of the period of English rule. Irish, or Gaelic, is the first official language but only a tiny minority use it; English is spoken almost universally.

Until recently, Ireland was a country of emigration, chiefly to the UK and the USA. More than one million Irish-born people live abroad—around 40 million Americans claim Irish ancestry. But recent years

of rapid development have encouraged some of the diaspora to return—and attracted many other nationalities. In 2005, 54,000 more people arrived than left—a record.

Agriculture remains a significant, though declining, part of the economy. Much of this is based on the rearing of livestock for milk and beef which, along with the export-oriented food processing industry, still employs around one-fifth of the workforce. Ireland's agriculture benefited greatly from the EU's Common Agricultural Policy: in 2005, for example, around 40% of farm income came from EU subsidies.

Irish workers now have plenty of alternatives to farming. In the past decade manufacturing and service industries have been booming. Tourism is also a major earner. The six million visitors each year employ 6% of the workforce.

The Irish government can take credit for the industrial transformation. It offered generous grants and tax concessions while restraining wage claims through 'social contracts' with the trade unions. It has also invested heavily in higher education.

Attracted by financial incentives and an inexpensive, well-educated, English-speaking population within the European Union, foreign companies understandably flocked. Ireland now has affiliates of more than 1,000 foreign enterprises, including major computer companies such as IBM, Intel, Fujitsu, and Dell. Computer equipment alone made up around one-fifth of exports. Per capita, Ireland is now one of the world's leading exporters.

This stimulated rapid economic growth—turning Ireland into an 'emerald tiger' with one of Europe's most rapid growth rates during the 1990s. Unemployment, which was 16% in 1993, is now steady at around 5%. But the wealth has not been evenly spread: property prices in Dublin have been rocketing while the west of the country trails behind.

Ireland has also been going through rapid social and political changes. Under the influence of the Roman Catholic Church, Ireland

The Catholic Church's influence has waned

has been very conservative in such matters as abortion, homosexuality, and divorce. But the Church's influence has waned and divorce finally became legal in 1997.

At the same time, the political landscape is being transformed by the prospect of settlement of the long-running conflict in Northern Ireland. The island was partitioned by the 1921 Anglo-Irish Treaty and the south became independent in 1922. Since then, Irish politics has been shaped by attitudes towards the north. The largest party, the nationalist Fianna Fail, was formed in opposition to the treaty, insisting on Irish jurisdiction over the whole island—the name means 'militia of Fál', a stone monument. The second largest is the

centre-right Fine Gael ('Irish nation'), which represented landowners and the business community who accepted the terms of the treaty. Irish politics did not therefore split along right–left lines, but according to attitudes to partition. There is also a Labour Party, but it has always been much smaller.

In recent years, electoral outcomes have usually required coalitions of one of the major parties with one of the smaller ones. After the 1997 election for the parliament, the Dáil, a Fine Gael–Labour coalition gave way to one between Fianna Fail and the Progressive Democratic Party, which was a 1985 right-wing breakaway from Fianna Fail. Bertie Ahern became Taoiseach—prime minister. Mary McAleese took the largely ceremonial post of president.

In 1998 Ahern signed the 'Good Friday' Northern Ireland peace agreement with the UK. This required Ireland to remove from its constitution a claim on Northern Ireland— subsequently confirmed with 94% approval in a referendum. In 2005 the Provisional Irish Republican Army decommissioned its weapons.

The same coalition won the election in May 2002 with Ahern again as Taoiseach, the first time a government has been returned in over three decades. Nevertheless some of the smaller parties, including Sinn Fein, the political wing of the Irish Republican Army, and the Green Party, also increased their representation. Subsequently an indecisive government has seen its popularity fall.

Ireland has been a keen EU member. It joined the euro zone and held the presidency in 2004.

Israel

After half a century of war and hostility, the peaceful creation of a Palestinian state is still a long way off

Land area: 21,000 sq. km.
Population: 7 million—urban 92%
Capital city: Jerusalem, 693,000
People: Jewish 80%, others, mostly Arab, 20%
Language: Hebrew, Arabic, English
Religion: Judaism 77%, Islam 16%, Christian 2%, other 5%
Government: Republic
Life expectancy: 80 years
GDP per capita: $PPP 20,033
Currency: Shekel
Major exports: Industrial goods, cut diamonds

Israel can be considered to have four main geographical regions. To the north is a hilly region that includes the hills of Galilee and extends down through the Israeli-occupied West Bank. Along the Mediterranean coast there is a narrow plain that is home to most of the country's commerce and population. The border in the north-east is formed by the valley of the river Jordan which flows south into the Dead Sea—the lowest point on earth—and the same fissure continues south to the Gulf of Aqaba. Most of the south is the Negev Desert.

The state of Israel was established in what was formerly Palestine in 1948 as a Jewish homeland and has since attracted immigrants from almost every country. By the end of the 20th century the population was 80% Jewish, of whom more than half were native-born and the remainder

immigrants. Immigration accelerated after 1989 with the arrival of more than 800,000 Jews from the former Soviet Union. Around one-fifth are descendants of the original Arab population—who are typically the poorest people. In addition there are hundreds of thousands of temporary migrant workers from Eastern Europe and Asia to replace excluded Palestinian workers.

The Jewish population is usually divided into two groups. First there are those of European and North American origin—the Ashkenazim—who tend to be richer. Second are those from North Africa and the Middle East—the Sephardim or Misrahi—who form most of the working class. But there are many other distinctions, principally between secular and religious Jews.

Israelis have generally good standards of health and are also highly educated. About 22% of immigrants in the 1990s were professionals. Israel has the highest proportion of scientists and engineers per capita in the world.

Over its lifetime, Israel has shifted from a simple agricultural society to an export-driven high-tech industrial economy—particularly during the 1990s, with the influx of new professionals and the establishment of communications technology and software enterprises concentrated around Tel Aviv.

Israel also has high-tech agriculture which has to make sophisticated use of limited water resources. As well as being self-sufficient in food, Israel is a major exporter of citrus and other

Israel has high-tech farms that work with little water fruits. Most farming, reflecting the country's socialist origins, is still organized by kibbutzim (collectives) or moshavim (co-operatives). Although fairly efficient, their survival still depends on cheap supplies of water and subsidies and many are now heavily in debt.

Israel's Zionist and socialist origins have also influenced its politics, which have largely been dominated by the Labour Party—though now more a liberal party. Labour was for decades the majority ruling party, but was dislodged in 1977 by the conservative Likud. Since the mid-1980s, neither has had a majority and each has had to govern in coalition with a plethora of smaller religious parties.

Throughout its existence, Israel's governments have been embroiled in conflicts with their Arab neighbours, most notably in the six-day war of 1967 as a result of which Israel seized the Gaza Strip from Egypt, the West Bank from Jordan, and the Golan Heights from Syria. The status of these territories has been disputed ever since. Generally, Likud and the orthodox religious parties have wanted to keep them—and encouraged Jewish settlements there—while Labour has been more equivocal.

In the longer term it seems likely that the West Bank and Gaza will constitute an independent Palestine. Major steps towards this were taken in 1993 when, following secret negotiations in Oslo, Labour prime minister Yitzhak Rabin signed a peace agreement with the Palestine Liberation Organization (PLO) that allowed for Palestinian self-rule. Rabin was assassinated in 1995 by an Israeli right-wing nationalist and was replaced by Shimon Peres, who continued negotiations.

The peace process faltered after 1996 when Likud returned to power, though it seemed to come back on track following the 1999 direct election for prime minister, which was won by the moderate Ehud Barak of the One Israel party. Barak met with the PLO at Camp David in July 2000 but the talks stalled and September 2000 saw the start of a second 'intifada'—uprising—by the Palestinians. Hopes of peace faded further with the election in February 2001 of hard-liner Ariel Sharon of Likud as prime minister in coalition with Labour, who started to build a high security fence along the border.

In January 2003 a general election allowed Likud to make many gains. Sharon then secured the withdrawal of settlers from Gaza, and founded a new party, Kadima, but he suffered a massive stroke in January 2006.

He was replaced by Ehud Olmert who led Kadima for the election in April 2006, emerging at the head of a new broad coalition. He has been even more aggressive towards the new Hamas govermment of Palestine and launched an unnsuccessful attack on Hizbullah in southern Lebanon.

Italy

Despite deep regional disparities and unstable governments, Italy has had rapid economic growth

Land area: 301,000 sq. km.
Population: 58 million—urban 67%
Capital city: Rome, 2.7 million
People: Italians
Language: Italian and some regional
Religion: Roman Catholic 90%, other 10%
Government: Republic
Life expectancy: 80 years
GDP per capita: $PPP 27,119
Currency: Euro
Major exports: Machinery, textiles, clothing, cars

Italy is dominated by mountains. The Alps loom over the north of the country and merge with the Apennine Mountains, which run down the centre. Mountains also make up much of the Italian islands of Sicily and Sardinia. Overall, more than one-third of Italy's territory is covered by ranges higher than 700 metres.

The difficult terrain is one reason why the country has been regionally divided. Italy was united as one country only in 1871, at which point only 3% of the population spoke Italian. Since then there has been a steady process of integration, but strong regional identities persist.

For much of the 20th century, Italy's population grew rapidly, and with limited prospects of work at home, millions of Italians emigrated: even as late as 1968–70 net emigration was over 250,000. Now

the position is reversed. The fertility rate, at 1.3 children per woman, is among the lowest in the world.

Italy has become a country of net immigration: currently it has about two million immigrants, most of whom have come from outside the EU, chiefly from Morocco, former Yugoslavia, Tunisia, and Albania.

Thin soils and steep terrain have always made agriculture difficult, and as a result Italy has been a net importer of most kinds of food, except for fruit and vegetables. It is also poorly endowed with other natural resources: it has no coal and has to import 75% of its energy needs.

In economic terms, the country's great strength has been in manufacturing which accounts for around one-quarter of GDP. It does have some large companies, notably Fiat, Pirelli, and also Fininvest which is controlled by former prime minister Silvio Berlusconi. But manufacturing is dominated by networks of thousands of small firms, chiefly in clothing, furniture, kitchen equipment, and 'white goods' such as refrigerators and cookers. These often cluster together regionally: wool textiles in Prato, for example, shoes in Verona, spectacles in Belluno.

These small and medium-sized

enterprises have been the engine for economic growth. There have certainly been periods of recession, particularly after the 1974 oil shock, but overall the country has made striking progress. Much of this activity has, however, been in the underground economy which is thought to control one-quarter of GDP.

Economic activity is concentrated in the north. Here manufacturing accounts for 30% of value-added, compared with 13% in the south. Indeed, the GDP of the south is only 70% of the Italian average.

There are also striking regional differences in unemployment—4% in the north but 15% or more in the south. The impact of unemployment is also cushioned by the extended family which remains stronger in Italy than in most other industrial countries. Around two-thirds of unmarried men under 30 still live with their mothers. And when they marry they do not move very far.

Many Italian men still live with their mothers

Tourism also makes an important economic contribution, with 40 million visitors per year.

Italy's economic progress has not been matched by its system of governance. Until 2001 the post-war period saw a sequence of weak and transient administrations: between 1947 and mid-2000 it had no fewer than 58 governments. This weakness was to some extent deliberate. In an effort to avoid repeating the experience of the fascist governments of the 1930s and 1940s, the constitution adopted in 1948 resulted in reduced powers for the prime ministers—who were often beholden to party leaders who were not themselves in government. Governance was also undermined by widespread corruption and the power of the Mafia.

Between 1947 and 1992, all coalition governments were dominated by the Christian Democrats. After 1993, the political scene shifted dramatically when Italians moved from a system of proportional representation to one in which most seats in both houses of parliament would be determined by straight majorities. This produced a different set of alignments. The 1996 election was won by the 'Olive Tree' coalition: chiefly centrist and left parties. Romano Prodi took over as prime minister. Prodi's government fell in 1998 following a dispute between the coalition partners, and Massimo D'Alema, a former communist, took over at the head of yet another coalition and headed the next government in 1999, but was replaced by Giuliano Amato who put together yet another frail coalition.

The 2001 election saw a victory for Silvio Berlusconi of the right-wing Forza Italia who formed a centre-right coalition, Casa delle Libertá. Berlusconi is a multimillionaire who owns, among other things, three TV channels. He was one of Italy's most determined leaders and survived various corruption cases launched against him and his business interests.

In the general election of May 2005, in a tight finish that he disputed to the end, Berlusconi lost to the centre-left Unione coalition, led by a returning Romano Prodi. But with only a small majority and a fractious coalition Prodi may not last long.

Jamaica

Culturally vigorous and a tourist attraction—but prone to gang violence

Land area: 11,000 sq. km.
Population: 3 million—urban 52%
Capital city: Kingston, 658,000
People: Black 91%, mixed 7%, East Indian 1%, other 1%
Language: English, Creole
Religion: Christian 65%, other 35%
Government: Parliamentary, with the British monarchy
Life expectancy: 71 years
GDP per capita: $PPP 4,104
Currency: Jamaican dollar
Major exports: Alumina, bauxite, sugar

Jamaica's mountainous, and often thickly wooded, territory rises to 2,200 metres at the peak of the Blue Mountain in the east—a landscape criss-crossed by rivers and attractive waterfalls. In the centre of the island, and accounting for around half the country, is a limestone plateau that includes extensive systems of underground caves. The most densely cultivated parts of the country, however, are the coastal lowlands of the south and west.

Jamaica's people are predominantly descendants of slaves brought to work on the sugar plantations, along with smaller communities of Asian and European origin. While officially speaking English and mostly professing Christianity, Jamaicans have developed their own characteristic culture, notably their distinctive dialect, the Rastafarian religion, and reggae music.

On the whole, Jamaicans are well educated and have a good health service. But limited economic opportunities have driven successive generations of young Jamaicans to emigrate—in the 1960s and 1970s to the UK, but more recently to the USA and Canada.

Around 20% of the population still make their living from the land. Apart from food crops such as plantains, yams, and corn, Jamaica also has major agricultural exports, notably sugar on plantations in the coastal plain, bananas grown on small farms, and coffee in the highlands. But agriculture in general has been stagnating—hit by droughts, floods, and hurricanes. One of the more consistent crops has been ganja, or marijuana, which provides solace to about half the male population.

Jamaica's most dependable export earner has been bauxite—of which it is the world's fourth largest producer. Most deposits are in the centre of the island and because they are not very deep they are relatively easy to extract. Some is exported as raw ore, but most is now first converted to alumina (aluminium oxide). With around half of export earnings from this source, Jamaica is vulnerable to world prices, particularly because it is a relatively high-cost producer,

Jamaica has also been trying to develop exports of manufactured goods. The most significant success in the 1980s was the growth of a garment industry though this has shrunk dramatically as a result of competition from elsewhere.

The other main source of foreign exchange and employment is tourism. Jamaica attracts more than 2.5 million visitors each year, most of whom come from the USA and head for the beautiful beaches of the north and west. Directly and indirectly, tourism probably employs around a quarter of a million Jamaicans. The industry grew rapidly in the 1980s and early 1990s, but stalled in the late 1990s, perhaps due to the relatively high rate of exchange—as well as the country's reputation for street violence. But it seems to be reviving with the construction of new Spanish-owned hotels.

Tourism employs 250,000 Jamaicans

Another source of export income is drugs. Some of this is marijuana, but the greatest profits come from transhipping Colombian cocaine. The trade is though to be the equivalent of 7% of GDP.

Politics in Jamaica has frequently been a dangerous business, involving armed confrontations between the two main parties. The violence has its roots in the mid-1970s. Between 1972 and 1982, the prime minister was former union organizer Michael Manley who led the People's National Party (PNP). Manley tried to pursue socialist policies and gain greater control over the foreign-dominated

bauxite industry, and became a leading international figure. Opposing him was the pro-business, anti-communist, Edward Seaga, who led the Jamaican Labour Party (JLP). But their left–right confrontations spread far beyond the parliament to become a kind of tribal warfare with pitched battles between gangs in the streets of Kingston.

Manley could not deliver the economic prosperity he had promised and Seaga won convincing victories in the 1980 and 1983 elections, pursuing more market-friendly policies and opening the country up to foreign investment. But with low prices for bauxite, economic development was slow and Seaga's popularity waned.

In 1989, the PNP and Manley returned to power, though continuing the same liberalization policies as the JLP. Following Manley's retirement in 1992, Percival Patterson took over as prime minister and won convincing majorities in 1993 and 1997. In October 2002 he led the PNP to a fourth successive victory, but by a much narrower margin.

The JLP was still led by Seaga, though he stepped down in 2005 to be replaced by Bruce Godling. A small third party, the National Democratic Movement, was created in 1995 to challenge the two main parties but has had little impact.

When Patterson retired in 2006 he was replaced by one of his ministers, Portia Simpson-Miller, who seems to have wide popular support.

Politics may be less ideologically charged nowadays, but the gangs live on, thriving on income from drugs and extortion. Jamaica still has one of the world's highest murder rates, with

Japan

Japan had an enviable record for growth and prosperity, but its economic and political frailties have been exposed

Land area: *378,000 sq. km.*
Population: *128 million—urban 66%*
Capital city: *Tokyo, 8.5 million*
People: *Japanese 99%, other 1%*
Language: *Japanese*
Religion: *Shinto and Buddhist 84%, other 16%*
Government: *Constitutional monarchy*
Life expectancy: *82 years*
GDP per capita: *$PPP 27,967*
Currency: *Yen*
Major exports: *Cars, semiconductors, office machinery, chemicals*

Though the Japanese archipelago has more than 1,000 islands, most of the territory comprises four main islands. From north to south these are: Hokkaido, Honshu, Shikoku, and Kyushu, of which the largest, with more than half Japan's land area, is Honshu. These islands have mountain chains running north to south divided by steep valleys and interspersed with numerous small plains. The mountains include both extinct and active volcanoes, of which the highest, at 3,700 metres, is the distinctively symmetrical and snow-capped Mount Fuji on Honshu.

This is one of the world's most unstable geological zones and Japan experiences more than 1,000 tremors per year. The most recent severe earthquake hit the city of Kobe in 1995, killing more than 6,000 people. Japan's climate runs from the sub-arctic to the sub-tropical but most of the country enjoys fairly mild temperatures with plentiful rain that supports lush vegetation.

Japan has one of the world's most homogenous populations. This is partly due to isolationism, since for centuries Japan effectively cut itself off from the rest of the world. There is one small indigenous ethnic group, around 24,000 Ainu people on Hokkaido, but the largest group of non-Japanese are the 690,000 Koreans, many of whom are descendants of people brought to Japan during the Second World War to work as forced labourers. During the post-war boom years, however, Japan was the only developed country not to rely extensively on foreign workers to meet labour shortages. Although immigrants arrived from elsewhere in Asia, they often did so unauthorized.

The only group actively welcomed were the 'nikkeijin', the descendants of a previous generation of Japanese emigrants. From the early 1900s, many Japanese had emigrated to the Americas, and in the 1990s thousands of their descendants came back—around 230,000 from Brazil.

Japan's rapid economic development has been accompanied

by striking social changes. With a healthy diet and a good health service average life expectancy, at 82 years, is among the highest in the world. Education standards are high—though learning by rote tends to stifle creativity. Japanese houses are well equipped but cramped, typically with less than half the floor area found in other rich countries. Many workers commute long distances.

Japan's population is ageing rapidly. There are now only 1.5 children born per woman of childbearing age—well below the replacement rate. One-fifth of the population is now over sixty-five. With a steady breakdown of the extended family, this has serious implications for the country's pensions system. It has also contributed to a steady rise in inequality in what was formerly a very egalitarian society.

Social security has already come

Salarymen replaced by freeters
under pressure as economic crisis has forced many large companies to abandon their former commitment to loyal employees, the 'salarymen', whom they had offered jobs for life. As a result of corporate downsizing and the recession, unemployment in 2005 was still over 4%, alarmingly high by Japanese standards. In addition there are now more than four million 'freeters', young people who only work part time.

Japan's development surged after the Second World War—boosted by demand from the Korean War and by high levels of savings that could be channelled into heavy industries such as steel and chemicals, and later into the provision of consumer goods

and services. This process was based on 'keiretsus'—tightly knit groups of manufacturers, suppliers, and distributors that work closely with the banks. The Ministry of Trade and Industry (MITI) also had a pivotal influence. This, combined with protectionism, helped to trigger two major economic booms: 1965–70 and 1986–91.

Manufacturing still accounts for around one-fifth of GDP, and cars and semiconductors are the leading exports. But rising labour costs, an appreciating currency, and the need to operate closer to overseas markets have long obliged Japanese companies to invest in other countries. By the mid-1990s around 10% of manufacturing was located overseas.

Japan's manufacturing success has been all the more striking since it has few mineral resources and has to import more than 80% of its energy needs. This has meant exploiting nuclear energy—Japan has more than 50 nuclear power plants.

Agriculture employs relatively few people—less than 5% of the labour force. But unlike in many other rich countries, they work predominantly on small family farms, chiefly growing rice. This is the legacy of an extensive post-war land reform imposed by the USA. Since then, farmers have remained a politically powerful lobby and have benefited from protection and high subsidies. In recent years the government has deregulated the system somewhat, but Japan still has highly protected food markets.

Japan's economic expansion began to falter after 1991 when the escalating property prices and booming stock market could not be sustained. This exposed the weaknesses of the keiretsu system. Many banks had continued to lend to unsound companies. Companies were reluctant to restructure to reduce massive over-capacity and the government was slow to force banks into liquidation. By 1998, the economy had slipped into recession. Since 2003, however, growth has revived and was almost 3% in 2005.

Japan has found it difficult to take tough decisions because its political system is neither as robust nor transparent as might be expected in one of the world's most advanced countries. For most of the period since 1955 Japan has been ruled by the conservative Liberal Democratic Party (LDP). The LDP has always had close links with both the business community and the farmers—each of whom allied themselves with different LDP factions. For a long time the main opposition party was the centre-left party now called the Social Democratic Party (SDP).

The LDP continued in office despite a series of scandals that included the prosecution of one

A weak and scandal-hit government

prime minister for taking bribes in 1976 from the Lockheed corporation, and the Recruit scandal in 1989, when almost all the LDP factions were found to have received bribes from that Japanese conglomerate.

Throughout this period, it had seemed that the Japanese electorate would continue to forgive a party that had delivered such a long period of prosperity. But the LDP's rule came to an end in 1993. A series of defections caused it to lose a no-confidence

motion in the Diet—the lower house of parliament. It was replaced by a coalition that included the SDP and others. But this too collapsed in 1994 and the LDP returned—this time in a coalition that included the SDP.

For the rest of the 1990s Japanese politics was in a considerable state of flux—delivering 10 prime ministers between 1993 and 2001. The LDP remained the largest party, though ruling in coalition with a succession of smaller parties, currently with New Komeito.

Meanwhile the opposition parties have gone through numerous upheavals, changing names and policies with bewildering rapidity. The main opposition party, following the elections of 2003, is the Democratic Party of Japan, first formed in 1996 and which in 2003 merged with the Liberal Party. But this encompasses a wide range of views and has many internal groupings.

Japan's next LDP leader, who became prime minister in 2001, was Junichiro Koizumi. He arrived on a wave of public acclaim as a relatively youthful figure who would finally embark on the vigorous reforms that Japan needed so badly. He had a number of successful reform initiatives and supported the US-led coalition in Iraq. In 2006, following a rebellion within the LDP over the privatization of Japan Post, Koizumi called an election in which he scored a resounding victory.

Koizumi had said, however, that he wanted to leave office and in September 2006 was replaced by finance minister, Shinzo Abe. Abe is a more conservative figure and also probably more of a nationalist.

Jordan

Jordan's King Abdullah has started to establish a more stable democracy

Land area: 89,000 sq. km.
Population: 5 million—urban 79%
Capital city: Amman, 2.1 million
People: Arab
Language: Arabic 98%, other 2%
Religion: Sunni Muslim 92%, Christian 6%, other 2%
Government: Constitutional monarchy
Life expectancy: 71 years
GDP per capita: $PPP 4,320
Currency: Dinar
Major exports: Phosphates, potash

Jordan has three main zones. First, there is the Jordan Valley region, which lies to the east of the Jordan river and finishes in the south at the Red Sea, which is the world's lowest body of water. Second, further to the east and running parallel with the valley, are the eastern uplands, whose elevation is mostly 600 to 900 metres. Thirdly, beyond the uplands, and occupying around 80% of the country to the east and north, is the desert.

Most people live in the upland areas and almost everyone is Arab. But there are numerous social divisions based on politics and history. One group consists of the 'East Bankers'—descendants of the people who lived on the East Bank of the Jordan prior to 1948. The second, and now making up around 60% of the population, are descendants of the Palestinian refugees who crossed the river from the West Bank following

the formation of the state of Israel in 1948—who were followed by another surge in 1967 when Jordan lost control of the West Bank to Israel. A third group are the 180,000 or more immigrant workers—the majority from Egypt, with some from India and elsewhere in Asia.

The population is young and growing fast—at around 3% per year. This has put severe pressure on public services. Nevertheless the government does seem determined to invest in health, and especially in the education system on which it spends 20% of the budget.

With relatively little by way of industry or natural resources, Jordan has relied heavily on the service industries that account for more than 70% of GDP. Many of these, including transport and communications, derive from Jordan's close links with its neighbours.

Agriculture is divided between fairly primitive rain-fed cereal production in the uplands and the higher-tech irrigated farms of the Jordan valley that are largely given over to fruit and vegetables, some of which are exported, and rely significantly on immigrant labour, generally from Egypt. Water is a perennial problem. Groundwater is rapidly being exhausted. Planned projects include conveying water from Disi in the south to Amman.

Industry is limited. The most significant enterprises are the state-owned mining operations that extract potash and phosphates. Jordan is the world's second largest exporter of raw phosphates. However the government has also been creating special duty-free economic zones, such as the one at Aqaba, to encourage investors.

In recent years industrial output has been boosted by manufacturing, particularly of garments, in 'qualifying industrial zones' which have duty-free access to the US market.

One of the most reliable sources of foreign exchange, however, is migrant remittances. Though Jordan is home to many immigrants, it also has around 900,000 Jordanians overseas, most of whom work in the Gulf states, and who send home more than $2 billion in remittances each year.

$2 billion in migrant remittances

Since the late 1980s, Jordan has been burdened with a large external debt and has had to turn to the World Bank which required a far-reaching programme of structural adjustment. In addition, however, good relations with many donor countries have enabled Jordan to benefit from substantial flows of aid. Jordan is the fourth biggest recipient of US aid.

In principle, Jordan is a constitutional monarchy with executive power vested in its 80-member parliament and its prime minister. In practice, the parliament has always been directed by the king. For most of the second half of the 20th century, this was King Hussein,

who reigned in an often autocratic fashion—bolstered by his links with tribal leaders and with the army. Somehow, he managed to cultivate close ties with Israel and keep the USA as an ally while also placating his vociferous Palestinian citizens.

Hussein allowed a democratic opening in 1989—instituting reforms that permitted new elections. To his displeasure, however, this gave a strong representation to radical Islamic groups. As a result he took a series of measures to reduce the number of Islamicists in parliament and introduced strict press laws.

Hussein died in 1999 and was succeeded by his son Abdullah ibn Hussein al-Hashemi. The British-educated Abdullah knew relatively little of Jordanian politics. But he soon had a considerable impact and embarked on a number of social and economic reforms. In 2002 his 'Jordan First' campaign envisaged reforms in the education system. Women's rights have also been a priority for Abdullah and his Palestinian wife, Queen Rania.

On the political front, Abdullah has to walk the same tightrope between keeping the Palestinians on side and maintaining good relations with the USA. He has been wary of the main Islamicist party, the Islamic Action Front, and twice postponed elections that were finally held in 2003. Tribal and independent candidates took two-thirds of the seats. The Islamicists won around 10% but complained the election was rigged.

Although generally popular, Abdullah lost some support in 2003 for assisting the USA in its war on Iraq. And in 2005 there were suicide bomb attacks on three Amman hotels.

162

Kazakhstan

Kazakhstan has yet to capitalize on its rich mineral resources, or the opportunity for democracy

Land area: 2,717,000 sq. km.
Population: 15 million—urban 56%
Capital city: Astana, 313,000
People: Kazakh 53%, Russian 30%, Ukrainian 4%, German 2%, Uzbek 2%, Tatar 2%, other 7%
Language: Kazakh, Russian
Religion: Muslim 47%, Russian Orthodox 44%, other 9%
Government: Republic
Life expectancy: 63 years
GDP per capita: $PPP 6,671
Currency: Tenge
Major exports: Oil, gas, metals

Kazakhstan is a landlocked country in central Asia that consists of a vast plain stretching from the Caspian Sea in the west to the mountainous border with Kyrgyzstan and China to the east. Most of the territory is grassy steppes or desert.

Kazakhs form around half the population in a country that has more than 100 ethnic groups. Russian settlers moved in from the 18th century and continued to do so during the Soviet era, along with ethnic Germans who were exiled here by Stalin in the 1930s. Most Kazakhs live in the east, while Russians are in the majority in the north.

Since independence, there have been efforts at 'Kazakhization', particularly the promotion of the Kazakh language. When Kazakhstan was effectively a Russian colony, Kazakhs had to learn Russian which then displaced Kazakh.

Demography, too, is starting to swing back in the Kazakhs' favour. This is partly due to emigration. Many emigrants were Russians: between 1989 and 2000, 1.6 million left. Around 500,000 ethnic Germans headed for Germany. The average age of Kazakhs is 20, while that of Russians is 46 and because of emigration and their higher birth rate Kazakhs are now in a majority again. After 2002, the population started to grow again, partly as a result of immigration from Uzbekistan.

In the past Kazakhstan's people had reasonable standards of human development, but health services deteriorated in the 1990s, and there were sharp rises in infectious diseases such as tuberculosis, as well as lifestyle diseases, particularly alcoholism. Services have now started to improve though health expenditure is still low.

Poverty has also started to fall: only around 10% of the population now live below the national poverty line. Nevertheless there are still wide regional gaps: the per capita income in the poorest province is only one-fifth that of the richest.

One-third of the workforce are employed in agriculture, though this now generates a shrinking proportion of GDP, under 7% in 2005. The soil is good but the climate is unpredictable and it swings between hostile extremes—fiercely hot summers and bitterly cold winters.

Agriculture has not developed much since the Soviet era. Privatization has been slow. Most farmers lease their land from the state and are often so short of credit, inputs, and equipment that they cannot complete their harvests.

Kazakhstan's other major source of employment—though more for Russians and Ukrainians—is heavy industry which centres on the production of oil and gas. Oil was first found here in 1899. Kazakhstan still has substantial oil deposits along the Caspian Sea coast. There is some dispute over just how big the proven reserves are but they appear to be growing. And despite the difficulties of exporting from this landlocked country, these deposits have proved

Oil companies have flocked to Kazakhstan
very attractive to companies from all over the world who are keen to engage in joint ventures. More than 80% of output now comes from foreign companies, the rest from the state-owned company Kazmunaigaz.

The country also has substantial deposits of metal ores, including chromium, lead, copper, zinc, and tungsten.

Kazakhstan's industrial past has bequeathed severe environmental problems. It was also a Soviet nuclear test site—between 1949 and 1989, 470 tests were carried out in the north, which now has to deal with a legacy of cancer and birth defects.

In the south, the main environmental problem is the drying up of the Aral Sea, whose water level has been falling as a result of withdrawals from feeder rivers elsewhere in Central Asia. The sea is also heavily polluted by fertilizer run-off—which is killing the fishing industry and is also contaminating water supplies.

When Kazakhstan achieved independence from the Soviet Union in 1991 the sole presidential candidate was Nursultan Nazarbaev—a former head of the Kazakh Communist Party, who initially had opposed independence. Like other leaders in Central Asia, he continued to rule in much the same autocratic fashion. And when it was time for re-election in 1999 he took no chances—suddenly bringing the election forward while the economy was in better shape, and banning his most serious rival. Standing against a communist candidate, he was duly re-elected with 82% of the vote.

Opposition started to surface in 2002 with the formation of the Democratic Choice of Kazakhstan (DVK). But the DVK and other parties have been effectively stifled by repression and electoral fraud and in the 2004 parliamentary elections were allowed to have only one seat between them. The process was similar for the presidential election which was arbitrarily brought forward to 2005. Unsurprisingly, Nazarbaev won with 91% of the vote.

Altynbek Sarsenbaev, an opposition figure, was murdered in February 2005 in what looks like the beginning of the struggle to succeed Nazarbaev.

Kenya

Kenya's 'Rainbow Coalition' is falling apart, undermined by continuing corruption

Land area: 580,000 sq. km.
Population: 33 million—urban 39%
Capital city: Nairobi, 1.3 million
People: Kikuyu 22%, Luhya 14%, Luo 13%, Kalenjin 12%, Kamba 11%, other 28%
Language: English, Swahili, and others
Religion: Protestant 45%, Roman Catholic 33%, Muslim 10%, other 12%
Government: Republic
Life expectancy: 47 years
GDP per capita: $PPP 1,037
Currency: Kenyan shilling
Major exports: Tea, coffee, horticultural products

Most of the eastern half of Kenya consists of a broad plateau with scattered hills that slope down to a narrow coastal strip. This is an area of uncertain rainfall, merging into the arid and semi-arid lands of the north and north-east. The western half has several mountain systems split in two by the snaking line of the Rift Valley, while to the south-west the mountains descend to a basin around the shores of Lake Victoria.

The population is concentrated in the western half of the country, particularly on the fertile central highlands and on the land to the west. The land in the north-east is occupied mostly by nomadic herding communities. Kenya has many different ethnic communities, but they can be divided into three main groups. The largest, with about two-thirds of the population, are the Bantu group which includes the Kikuyu and the

Luhya. A second group, the Nilotic, making up around a quarter of the population, includes the Luo and Kalenjin and the Masai nomads. A third, smaller group are the Cushitic. Others are of European, Asian, and Arab descent.

Kenya has had very rapid population growth though the rate has now fallen to 2.3% as a result of more intensive family planning as well as the death toll from AIDS—7% of adults are HIV-positive. Even so, the population density is already high in the areas that have better land, putting pressure on the environment and driving people into the cities, particularly to the slums of Nairobi.

Kenya also suffers from wide income disparities. At independence, much of the best land was taken and kept by the Kikuyu élite. And subsequent liberal free-market development led to rapid economic growth but also heightened inequality. The richest fifth of the population get half of national income. Some 40% of the population live below the national poverty line.

Agriculture accounts for around one-quarter of GNP and for around one-fifth of formal employment.

Kenya's small farmers grow food crops such as cassava and maize for subsistence, and also produce around 60% of two major export items, tea and coffee—the rest being grown on large estates. Kenya is the world's largest exporter of black teas.

Over recent decades Kenyan farmers have increasingly turned to horticulture for export, primarily fruit, vegetables, and cut flowers. These are air-freighted to Europe where they meet around 25% of the demand for off-season produce. Output continues to rise as Kenya finds new horticulture markets in Asia and North America.

Florist to the world

Manufacturing accounts for 10% of GDP, with the emphasis on food processing and small-scale consumer goods. Output has been constrained, however, by poor infrastructure, especially power supplies: the country faces frequent power cuts, following a badly managed liberalization of the electricity industry.

Kenya also has a vast informal sector working in small-scale manufacturing making everything from household goods to car parts—and contributing around one-fifth of GDP.

Tourism has been a further source of foreign exchange, and accounts for around one-fifth of GDP, as half a million visitors each year head for the game reserves and the beaches. However, the industry has suffered from political uncertainty and weak infrastructure as well as tougher competition from neighbouring countries.

For much of the period following independence in 1963, Kenya was effectively a one-party state—run by the Kenya African National Union (KANU). The first president was a Kikuyu, Jomo Kenyatta, who on his death in 1978 was succeeded by a Kalenjin, Daniel Arap Moi. In 1982, Moi made KANU the sole legal party, which it remained until 1992 when he was forced to concede multiparty elections. Moi and KANU won the 1992 elections and won again, though more narrowly, in 1997.

The final years of Moi's reign saw a steady descent into greater corruption, poverty and violence. Politics in Kenya has always had a strong ethnic element. KANU was originally a Kikuyu-dominated party, but Moi transformed it into one that primarily served his own group, the Kalenjin. Meanwhile, the economy was unravelling, the infrastructure was collapsing, and aid donors withdrew.

This dismal scene brightened at the end of 2002 when opposition groups united as the National Rainbow Coalition. They achieved a landslide victory in the presidential election—for Emilio Mwai Kibaki—and a corresponding win in the parliamentary election.

Kibaki, with a reputation for honesty, said he was determined to fight corruption and revive the economy. He introduced free primary education—which sent children flocking to school.

In November 2005, however, he suffered a setback when a referendum rejected a new constitution—after a campaign that stirred up tribal loyalties and split the coalition. In 2006, his credibility was further undermined by a report that revealed the extent of continuing corruption.

Korea, North

Increasingly unpredictable, North Korea now presents an alarming nuclear threat

0	Miles	150
0	Km	240

Land area: 121,000 sq. km.
Population: 22 million—urban 61%
Capital city: Pyongyang, 2.7 million
People: Korean
Language: Korean
Religion: Buddhism and Confucianism
Government: Communist
Life expectancy: 63 years
GDP per capita: $PPP 1,700
Currency: North Korean won
Major exports: Minerals, agricultural products

Officially, this is the Democratic People's Republic of Korea. Around four-fifths of the territory comprises mountains or high plateaux—the highest of which are to be found in the north-east. The main lowland areas, home to most of the population, and where most agriculture takes place, are to be found in the south-west.

The population is ethnically among the most homogenous in the world, a factor that has helped sustain national solidarity in the face of extreme hardship. There are also several million ethnic Koreans living across the border in China—including around 100,000 refugees.

One of the achievements of the communist regime has been education. Children have good access to pre-school activities, partly to allow their mothers to work, but also to expose them early to the ideals of the state. Health standards were also good until undermined by the famines of the 1990s. In 2004, more than one-third of children were thought to be malnourished. Estimates of the death toll from famine since 1995 range from one to three million. Since so many people have fled to China and the birth rate is low, the population is either static or declining.

Agriculture still employs one-third of the workforce, primarily growing rice, corn, and potatoes. Officially, this is produced by co-operatives and collectives—though in recent years much has also been traded through markets. Until the mid-1990s, agriculture was relatively successful, but from 1995 a series of climatic setbacks, including floods, droughts, and tidal waves, combined with mismanagement, crippled production and triggered a sequence of famines. Most people now rely on food aid.

North Korea once had a flourishing industrial sector because the partition of Korea in 1948 left it with most of the minerals and heavy industry. Though it has no oil, the country has plenty of brown coal and considerable hydroelectric potential. Industry, which is largely on the east coast, is dominated by iron and steel, machinery, textiles, and chemicals. But much of the equipment is now antiquated and production has shrunk. There is some light industry that includes textiles exports and TV manufacturing by South Korean firms such as Samsung and LG.

The Democratic People's Republic of Korea is a communist state which has been underpinned by a bizarre personality cult. The state was formed in 1948 from the territory of Korea above the 38th parallel which had been occupied by the Soviet Union. With Soviet help, the communist leader Kim Il-Sung rapidly took control. In 1950, he also attempted to seize South Korea, provoking the 1950–53 Korean War. Although he failed, Kim nevertheless entrenched his position and with the support of China, the Soviet Union, and a highly disciplined party, he managed to rebuild the country.

By the mid-1950s, however, he had drifted away from the Soviet sphere and developed his own philosophy 'Juche', an obscure mixture of communism and self-reliance, woven into a personality cult around Kim himself. The country

A bizarre personality cult

was never as self-reliant as it proclaimed. When the Soviet Union collapsed and China started to liberalize its economy and build stronger ties with the West, North Korea lost its two main props.

When Kim Il-Sung died suddenly in 1994 he left a confusing power vacuum. The situation became clearer in 1998 when the late Kim Il-Sung was posthumously elevated to 'president for eternity' while his son Kim Jong-Il became Chairman of the National Defence Committee and head of state. Even so, the real location of

power remains unclear—distributed between the state, the party, and the army. Kim Jong-Il is reported to have health problems and his sons Kim Jong-nam, Kim Jong-chul, and Kim Jong-woon are thought to be manoeuvring for the succession.

With persistent famines and a bankrupt government, North Korea seems permanently on the brink of implosion. South Korea does not want the economic burden of having North Korea fall into its lap and has been sending fertilizer and food aid. It has also established a fund of $450 million to help the north. China, nervous of a capitalist Korea on its doorstep, has been allowing hundreds of thousands of starving North Koreans to enter in search of food.

North Korea certainly frightens its neighbours. It has the world's fifth largest armed forces—1.2 million strong—which soak up around one-third of GDP.

Even more alarming are its nuclear threats. In 1994 North Korea and the US signed the General Framework Agreement, through which North Korea would cap its nuclear weapons programme in exchange for US aid for the construction of new light-water reactors. By 2002, however, having been described by the US president as belonging to an 'axis of evil', North Korea claimed it was again developing nuclear weapons. The USA has offered economic aid provided North Korea gives up its weapons, while also insisting on six-way talks that include South Korea, Russia, and Japan. But Kim Jong-Il seems unwilling to co-operate and in July 2006 carried out a series of missile tests.

Korea, South

South Korea's economic crisis exposed its over-reliance on huge and cosseted conglomerates

NORTH KOREA
Seoul
Inchon
Sea of Japan
Yellow Sea
Taejon
Taegu
Kwanju
Pusan

| 0 | Miles | 150 |
| 0 | Km | 240 |

Land area: *99,000 sq. km.*
Population: *48 million—urban 80%*
Capital city: *Seoul, 10.3 million*
People: *Korean*
Language: *Korean*
Religion: *Christianity, Buddhism*
Government: *Republic*
Life expectancy: *77 years*
GDP per capita: *$PPP 17,971*
Currency: *Won*
Major exports: *Electronic goods, textiles, cars, ships*

The country's official name is the Republic of Korea. Almost three-quarters of its territory is mountainous—the largest range being the T'aebaek, which occupies most of the eastern half of the country. Most of the population are to be found in the lowlands of the north-west and south-west and in the Naktong River basin in the south-east.

Almost everyone is ethnically Korean—and speaks the Korean language. Traditionally most Koreans have followed Confucian and Buddhist principles, but Christianity has had an increasing impact, particularly the evangelical protestant sects. One of these, though less significant now, is the controversial Unification Church of Reverend Sun Myung Moon, which has been exported around the world.

South Korea is one of the world's most densely populated countries. Until the early 1980s, South Korea sent emigrant workers to other countries, but the flow has now been reversed and South Korea has attracted more than 200,000 foreign workers, legal and illegal, to do the more menial work. Most come from South-East Asia and China and many are of Korean ethnic origin.

Koreans have good standards of health and are well educated. But the curriculum puts much emphasis on rote learning that does not encourage creativity.

Today few Koreans work as farmers—around 10%. Thanks to a ban on rice imports, the remaining farmers, who are mostly smallholders, receive high prices for their crops and are so productive that the country is usually self-sufficient in rice. Meanwhile consumer food prices are heavily subsidized.

Nowadays around one-third of the workforce are employed in manufacturing industry. The partition of Korea left the South worse off—since the North had most of the mineral resources and the heavy industry. Manufacturing in the South initially concentrated therefore on light industry for local consumption and then for export. But from the

1970s South Korea turned its attention to heavy industry, particularly steel and cars—largely in the Seoul-Inchon region—and later to advanced electronic products. The result was a phenomenal rate of growth: from the 1990s until 1997 it averaged around 9% per year. This led to steady rises in income, and in labour costs, so much of the simpler labour-intensive assembly work in footwear and toys has now moved overseas.

Service industries, on the other hand, have been less sophisticated than those of other countries of similar wealth—whether in retail, finance, or entertainment. This has started to change, however, particularly in retailing, with a relaxation in restrictions on foreign participation.

At the heart of South Korea's rapid economic expansion were the 'chaebol'—vast diversified family-owned conglomerates. In 1999, five of these—Hyundai, Samsung, Daewoo, LG, and the SK group—still accounted for one-third of sales and almost half of exports. They benefited from decades of state favours and easy credit but recklessly overstretched and were deep in debt by 1997 when the Asian financial crisis struck. The currency lost half its value, and in the year from June 1997 the stock market fell by 65%. In July 1999 the Daewoo group collapsed in the

The world's biggest bankruptcy

world's largest ever corporate bankruptcy. Meanwhile, foreign companies such as Hyundai Motors, Renault, and DaimlerChrysler took the opportunity to buy stakes in Korea's car industry.

Since then, the economy has bounced back and growth for 2006

could be 5%. Unemployment has come down to around 3%. North Korea, however, presents an economic threat since a collapse followed by a takeover could cost up to $1.2 trillion.

During most of its period of rapid growth, South Korea was in the grip of authoritarian military-backed regimes, which repressed opposition groups and especially the militant labour unions. The democratic opening appeared in 1987, when the government acceded to demands for democratic elections. South Korea now has a strong presidential system in which the president appoints the prime minister.

In 1997 the presidential election was won by a coalition led by the left-of-centre Kim Dae-Jung. He struggled to reform the economy and to fight corruption but his efforts were frustrated by the parliamentary elections in 2000 in which his Millennium Democratic Party (MDP) came second. He thus had to rule in coalition with several smaller parties. The largest party was the Grand National Party (GNP).

The 2002 presidential election was also won by the MDP's candidate, Roh Moo-Hyun. In 2003, however, part of the MDP split off to form a new party, Uri. Prior to the 2004 legislative elections Roh praised Uri, which as a public official he was not supposed to do. The rest of the MDP, along with the GNP, then launched impeachment proceedings against him. This evidently backfired, however, since Uri won an absolute majority.

Since then Uri's fortunes have declined, with a poor performance in the 2006 provincial elections.

Kuwait

Kuwait's hereditary monarchy retains power but faces greater demands for democracy

> **Land area:** 18,000 sq. km.
> **Population:** 3 million—urban 96%
> **Capital city:** Kuwait City, 388,000
> **People:** Kuwaiti 45%, other Arab 35%, South Asian and other 20%
> **Language:** Arabic, English
> **Religion:** Muslim 85%, Christian, Hindu, Parsi, and other 15%
> **Government:** Hereditary emirate
> **Life expectancy:** 77 years
> **GDP per capita:** $PPP 18,047
> **Currency:** Kuwaiti dinar
> **Major exports:** Oil, oil products

Kuwait is a small state at the north of the Persian Gulf between Iraq and Saudi Arabia. It has some oases and a few small fertile areas, but the country consists largely of a sloping plain that is almost entirely desert.

A couple of generations ago, Kuwaitis were nomadic tribesmen. But since 1946 oil has made Kuwait very wealthy. Kuwaitis are among the richest people in the world—and are also entitled to free education and subsidized health care, and are spared the inconvenience of income tax.

The government spends around 6% of its total budget on health, though it now requires expatriates to pay for services at public hospitals and clinics. Although education is compulsory, 3% of men and 15% of women are illiterate.

Most of the work is done by expatriates. In 2004 Kuwaitis made up only 35% of the population and

only 19% of the labour force—the rest being immigrant workers. The largest numbers nowadays are from Egypt, Palestine, and Lebanon, but there are also many from South Asia. Almost all employed Kuwaitis choose to work for the government or for public corporations.

Kuwait is still floating on a sea of oil which accounts for 45% of GDP and around 95% of export earnings. Kuwait has about 10% of world reserves, mostly in the Burgan field south of Kuwait City. At present rates of extraction, this should last for 100 years or more.

Since the 1970s, the oil industry has been nationalized. The government owns local refineries and also markets around 40% of its own oil. Kuwait Petroleum International operates more than 5,000 petrol stations using the 'Q8' brand name in 10 European countries. It also now has more than 100 stations in Thailand. In addition, Kuwait has large reserves of gas.

Kuwait has some manufacturing industry—also linked to oil. In the past this has typically involved such low-value products as ammonia, urea, and fertilizers. But there have been efforts to move to more profitable products such as polypropylene. One of the largest petrochemicals complexes, Equate I, is a joint venture between Kuwait's Petrochemical Industries Company and Union Carbide, and there are now plans for an Equate II.

With scarcely any arable land or water for irrigation, Kuwait's agriculture is very limited. Some farms grow vegetables and fruits but almost all the country's food has to be imported. There is also a fishing industry whose catches are dedicated to the local market.

Much of the oil infrastructure has been restored since the 1990 Gulf War, but Kuwait's economy will never be the same. To pay for the war the government had to sell off around half its $100-billion overseas investments.

The government also responded by introducing some charges for health care, for example, and there have been suggestions for income or consumption taxes. But there is stiff opposition from conservatives who argue that the government should instead attack corruption and reduce defence expenditure.

Disturbed by the thought of taxation

One of the most important external influences has been neighbouring Iraq. In 1961, when Kuwait achieved full independence, Iraq claimed the territory as part of Iraq. Finally in August 1990 Iraq led by Saddam Hussein occupied Kuwait until it was expelled by a US-led coalition in Operation Desert Storm. Kuwait was naturally also a strong supporter of the US invasion of Iraq in 2003 and of the fall of Saddam Hussein.

Kuwait is a hereditary emirate and the ruling al-Sabah family holds all the key government posts. When the ailing Sheikh Saad Abdullah al-Salem al-Sabah was forced by the family to step down he was replaced in 2006 by Sheikh Sabah al-Ahmed al-Jabr al-Sabah. He then appointed his half-brother, Sheikh Nawaf al-Ahmed al-Jabr al-Sabah, as crown prince, and his nephew, Sheikh Nasser Mohammed al-Ahmed al-Sabah, as prime minister.

Kuwait does have a parliament which consists of 50 members elected for four-year terms plus 15 cabinet members appointed by the emir. Its powers are limited but it can at least serve as a place to air grievances, since members have the right to question ministers. Kuwait also has a relatively free press.

Formal political parties are banned, but there are more informal groupings both inside and outside the parliament. On the one hand, there are the liberals who broadly support the government and are concentrated in the centre of Kuwait City; on the other, there are the more oppositional conservative tribal-based Islamic groupings.

The July 2006 elections represented something of a landmark. First, because for the first time women were allowed to vote and stand for election, though none succeeded. Second, because these elections resulted in a large majority for the opposition forces. These form a fairly disparate group, united only by their desire for electoral reform, though they may try to block the appointment of ministers and they aim to combat corruption. Nevertheless, with high oil prices Kuwait's rulers can muffle dissent with higher public spending.

Kyrgyzstan

The latest of the former Soviet republics to experience a popular revolution

KAZAKHSTAN

■ Bishkek

Lake Ysyk-Köl

Osh

TAJIKISTAN

CHINA

| 0 | Miles | 300 |
| 0 | Km | 480 |

Land area: 199,000 sq. km.
Population: 5 million—urban 34%
Capital city: Bishkek, 825,000
People: Kyrgyz 65%, Uzbek 14%, Russian 13%, Dungan 1%, other 7%
Language: Kyrgyz, Russian
Religion: Muslim 75%,
Russian Orthodox 20%, other 5%
Government: Republic
Life expectancy: 67 years
GDP per capita: $PPP 1,751
Currency: Som
Major exports: Food, metals, electricity

Kyrgyzstan is a landlocked country with spectacular scenery. It is almost entirely mountainous, lying at the junction of several ranges that run mostly east to west and whose peaks are permanently covered in snow and ice. More than half the territory is above 2,500 metres. Some of the main lowland basins are in the south-west and there is one in the north where the capital Bishkek is to be found. The main river, flowing east to west, and a major source of irrigation, is the Naryn. The country also has many lakes, the largest of which, Ysyk-Köl in the north-east, is 1,500 metres above sea level.

The Kyrgyz people make up around two-thirds of the population—a proportion that has risen as a result of the exodus of Russians and ethnic Germans: between 1990 and 1995 half a million departed and each year around 15,000 more leave.

The majority of Kyrgyz live in the rural areas. The second largest minority are the Uzbeks, who are concentrated in the south, around the city of Osh. Ethnic relations have at times been tense.

While standards of health have generally been good, health services are deteriorating and there have been outbreaks of contagious diseases. Education services too have suffered from budget cuts and school enrolment has fallen. Around 40% of the population are poor with those in the south being worst off.

Agriculture is the largest source of income, employing one-third of the workforce. Livestock has long been a mainstay of the economy—producing wool, hides, and meat. And many people still work as nomadic herders, living in their traditional round felt tents called 'gers'.

Only a small proportion of the land is arable, but more of the grazing land is now being used for crops—benefiting from extensive irrigation from rivers rushing down from mountain peaks. Most of the previous collective farms are now joint-stock companies and relatively efficient. Crops include cotton, wheat, and vegetables and the country is now self-sufficient in grains.

During the Soviet period, Kyrgyzstan underwent a rapid process of industrialization, chiefly based on the exploitation of mineral deposits. The country has extensive coal reserves, as well as non-ferrous metal deposits, including gold, uranium, mercury, and tin. Gold production, for example, has been stepped up following the start of mining at Kumtor, near Bishkek. Although the government owns two-thirds of Kumtor, the mine is operated by a Canadian company.

Much of the heavy industrial manufacturing has been fairly inefficient and output has declined sharply, largely as a result of low investment.

Kyrgyzstan was one of the faster reforming countries of the ex-Soviet republics in Central Asia. More than 60% of eligible enterprises were rapidly privatized, though the state retained 'strategic enterprises'. This, combined with tight macroeconomic policies, ensured considerable support from international donors and financial institutions. But the economic and social cost was high and the economy has been slow to respond—hampered by the exodus of skilled Russian workers.

Kyrgyzstan has lost many skilled Russians

Growth only really resumed after 1995 and has since been erratic, depending largely on output from Kumtor and to a lesser extent on agriculture. The economic outlook remains problematic—the country's relative isolation has been deterring foreign investment in anything other than mining.

Kyrgyzstan's economic reforms were driven largely by former President Askar Akaev, who was in office between 1990 and 2005. The riots in 1990 had fatally weakened the Communist Party and opened the way for reformists to elect Akaev, a former physicist, as president. He subsequently banned the Communist Party and, following independence in 1991, he was elected unopposed.

At that point, the parliament had three main groups: the communists, who were now legal again but were hostile to reform, Akaev's Union of Democratic Forces (SDS), and a social democratic grouping that tried to resist the increasing presidential powers.

In 1995, Akaev was convincingly re-elected and took the opportunity to push through yet another constitutional amendment to tilt power more in his favour. Akaev and the SDS duly won the elections in 2002. But since Akaev had banned several opposition groups and jailed one of his main challengers, the opposition became more vocal.

Akaev's administration came to a sudden, and rather surprising, end in March 2005 after a blatantly fraudulent election provoked a full-scale insurrection. Thousands of people took to the streets, eventually causing Akaev to flee the country.

One of the leaders of the 'tulip revolution', Kurmanbek Bakiev, then won a landslide victory in the presidential election in July 2005 and another of the leaders, Feliks Kulov, became prime minister. Since then the political situation has remained unstable—two members of parliament have been assassinated.

Laos

An isolated communist state that is slowly opening up to the rest of the world

| 0 | Miles | 200 |
| 0 | Km | 320 |

Land area: 237,000 sq. km.
Population: 8 million—urban 21%
Capital city: Vientiane, 319,000
People: Lao-Loum 68%, Lao-Theung 22%, Lao-Soung including Hmong and Yao 9%, Vietnamese/Chinese 1%
Language: Lao, French, English, and various ethnic languages
Religion: Buddhist 60%, animist and other 40%
Government: Communist
Life expectancy: 55 years
GDP per capita: $PPP 1,759
Currency: Kip
Major exports: Wood, coffee, electricity

Laos is a landlocked and almost entirely mountainous country. The highest mountains are to the north, rising to 2,800 metres above sea level, while another range lies to the east along the border with Vietnam. The only extensive lowland areas are to the south in the floodplains of the Mekong River—where around half the population live. Only around 5% of the territory is suitable for agriculture and most of the rest is covered with forests—broadleaf in the north, but tropical hardwoods in the south.

Laos is sparsely populated but it has, according to the government, around 68 'nationalities' divided into three main groups. The largest, and politically the most dominant, are the Lao-Loum, who live in the cities and in the lowlands of the Mekong delta. The Lao-Theung, also called the Mon-Khmer, live in the uplands throughout Laos and can also be found in neighbouring countries. Finally, there are the highland Lao-Soung, who include the Hmong (Meo) and the Yao (Mien).

Levels of human development are low. Around 40% of the population live below the poverty line, consuming less than 2,100 calories per day, though with much higher poverty rates in the rural areas—40% of the population live more than six kilometres from a main road. Laos has more than 80% of children enrolled in primary school, but the quality of education is low. Health standards are also poor, partly because of inadequate public services. Many people are also addicted to opium, and the rate of HIV infection has been increasing.

More than 80% of Laotians depend for survival on agriculture, which accounts for more than half of GDP. But farming is not very efficient. Upland farmers rely on slash-and-burn techniques and even in the more fertile lowlands rice yields are low—vulnerable to droughts and floods and hampered by lack of credit and inputs, as well as by crumbling infrastructure that has suffered from decades of war and low investment.

More recently, Laos has been trying to produce more cash crops for export—including cotton, groundnuts, sugar cane, and tobacco. But the most notable export success has been coffee. This was originally introduced by the French colonialists and is now being revived by small farmers, who grow good-quality robusta coffee for export to Europe.

Nevertheless, the country's leading export earners remain timber and wood products. Laos has valuable forest resources, including teak, rosewood, and ebony. IKEA, for example, has a contract to buy wood products. A less official export is drugs: Laos is thought to be the world's third largest producer of opium and heroin. The country also serves as a transit route for heroin from Burma.

Laos has no heavy industry but does have some small-scale manufacturing around Vientiane.

Laos exports hydro-electricity
Some of the more successful products include handicrafts, garments, foodstuffs, and wood products. Another export prospect is hydroelectric power. Laos's rugged terrain is cut through with fast-flowing rivers, some of which are dammed to produce power for export to Thailand and in future to Vietnam. But projects for dams funded by World Bank loans have fallen foul of international environmental protests.

Political life in Laos is controlled by the Lao People's Revolutionary Party (LPRP). Laos has been a communist country since 1975 and the most recent constitution in 1991 confirmed that power would remain exclusively in the hands of the governing party. But whether Laotian communism amounts to socialism is more questionable. Politics is based less on class and more on regional, family, and ethnic loyalties. From the outset, it looked as though socialism would remain a long-term goal, and it seems more distant than ever.

Since the mid-1980s the government has pursued a more liberal economic path. In 1986 it abandoned collectivization of farms and introduced the 'New Economic Mechanism', which also removed many restrictions on private enterprise. Since 1988, the government has also been privatizing state-owned enterprises: between the early 1990s and 2000 the number of state-owned enterprises fell from 800 to 29—though those surviving tend to be the largest undertakings.

Politically, the situation is less fluid and the party hardliners in Laos retain their grip. Non-communist political parties are proscribed and the government clamps down on all forms of dissent. In the National Assembly elections in 2006 only two non-party members won seats. There seems little immediate prospect of political liberalization.

The most visible opposition comes from insurgent Hmong tribesmen, known as the 'Chao Fa'. They are the remnants of a guerrilla Hmong army that the CIA financed during the Vietnam War. As a result of increasing discontent with the government, they seem at times to be getting more popular support. In 2004 there was renewed fighting though since then things have been quieter.

Latvia

Latvia has now joined the EU as its poorest member

Latvia consists of gently rolling plains interspersed with low hills. Around one-quarter of the land is forested and there are numerous lakes, marshes, and peat bogs. The long coastline includes major ports as well as holiday resorts and extensive beaches.

Latvia's people, the Letts, have come close to being a minority in their own country. Before Latvia was annexed by the Soviet Union in 1940 the Latvians were 77% of the population, but a combination of deportation of Latvians and immigration from other Soviet republics, particularly Russia, dramatically altered the ethnic balance. By 1989, the Letts made up only 52% of the population.

After Latvia regained independence in 1991 the government was determined to reassert Latvian control and in 1994 introduced a rigorous Citizenship Law. Although, under pressure from Russia, this

> **Land area:** 65,000 sq. km.
> **Population:** 2 million—urban 66%
> **Capital city:** Riga, 804,000
> **People:** Latvian 58%, Russian 30%, Belarusian 4%, Ukrainian 3%, other 5%
> **Language:** Latvian, Lithuanian, Russian
> **Religion:** Lutheran, Roman Catholic, Russian Orthodox
> **Government:** Republic
> **Life expectancy:** 72 years
> **GDP per capita:** $PPP 10,270
> **Currency:** Lat
> **Major exports:** Timber, wood products, textiles, metals

was relaxed somewhat in 1998 it still requires people taking Latvian citizenship to pass stiff tests in language and history. And unless you are a citizen you cannot vote, buy land, or enter a number of professions. Russians who arrived after 1940 are regarded as 'occupiers' and do not get automatic citizenship. Although many have now naturalized, around 20% of the population were stateless in 2005.

Latvia has been one of the more successful ex-Soviet republics, but it still has high levels of poverty. Living conditions and standards of health deteriorated sharply in the first years after independence, with a rise in diseases of poverty such as TB, though the situation has gradually started to improve.

Poverty tends to be greater in the eastern regions where unemployment can be as high as 20%, compared with 5% in the capital, Riga.

Industry in Latvia had been geared to producing industrial and consumer goods, including processed food, for the Soviet Union. After independence, when the country was opened up to foreign competition, many industries such as machine building went into steep decline, but have since revived.

As a result of a steady process of privatization, by 2005 70% of the economy was in the hands of the private sector which provided three-quarters of employment.

Although agriculture represents a small and declining proportion of the economy it is still a major source of employment. Most of the land was collectivized during the Soviet era. Much has now been restored to private hands but in many cases the ownership of land remains unclear. The main activities are dairy farming and livestock raising, along with cereal production. It seems likely that accession to the EU, along with increased competition from other countries, will further reduce agricultural output.

Probably the most important rural activity is timber production. Forests cover around 45% of the land area. Since 1996, the output of timber-related industries has more than doubled and timber and wood products make up around one-third of exports.

Forestry is one of Latvia's boom industries

Another striking economic development in recent years has been the rise of the service sector, which now accounts for around two-thirds of GDP. Much of this is connected with trade since Latvia, with three important ports, is a major transit route for goods in and out of Russia.

Following a positive result in a referendum in 2003, Latvia entered the EU in May 2004, where it is now the poorest country. It is also probably the most corrupt, in a country where political parties are closely aligned with business.

Since independence in 1991, politics has been a shifting and unstable affair, producing more than a dozen weak and transient governments. This is partly because of the diverse and shifting nature of political parties of whom 20 competed in 2002 for the elections to the Saeima, the unicameral parliament. This resulted in a governing coalition three of whose members were new.

The most prominent is the New Era Party, which was formed in 2002, led by a former physicist and central bank chairman, Einars Repse. Its partners, also formed in 2002, included the eurosceptic Union of Greens and Farmers and the First Party, which aims to promote Christian values.

Among the other largest parties in the parliament are the People's Party which was formed in 1998, largely as a vehicle for Andris Skele who twice became prime minister, but has since resigned from the party. When in government the People's Party has been associated with vested interests and corruption.

To no-one's great surprise the coalition collapsed in March 2004. Its replacement duly fell apart in December 2004 to be replaced by one comprising four centre-right parties headed by Aigars Kalvitis of the People's Party. In 2005 this coalition seemed under threat when the anti-corruption bureau investigated Mr Repse, but it has survived.

The president is elected by the Saeima. The president has some political authority and appoints the prime minister. Since 1999 the president has been the well-respected former academic Vaira Vike-Freiberga, who had previously been exiled in Canada.

Lebanon

Still recovering from 15 years of civil war, Lebanon found itself under attack from Israel

Land area: *10,000 sq. km.*
Population: *4 million—urban 88%*
Capital city: *Beirut, 403,000*
People: *Arab 95%, Armenian 4%, other 1%*
Language: *Arabic, French*
Religion: *Muslim 60%, Christian 40%*
Government: *Republic*
Life expectancy: *72 years*
GDP per capita: *$PPP 5,074*
Currency: *Lebanese pound*
Major exports: *Paper products, foodstuffs, textiles*

Along its Mediterranean coast Lebanon has a narrow, flat coastal plain. Inland there are two parallel mountain ranges running north-east to south-west: the Lebanon Mountains, which ascend abruptly from the plain; and the Anti-Lebanon Mountains along the border with Syria. Between these two ranges is the fertile Bekaa Valley.

Almost all Lebanese are ethnically Arab, but they are sharply divided into 'confessional' groups—Muslim and Christian sects. At independence in 1943, political leaders established an unwritten National Pact that shared political power according to the proportions of each sect in the 1932 census. At that point, the most numerous were the Maronite Christians, followed by Sunni Muslims, Shia Muslims, Greek Orthodox Christians, Druze (Muslims), and Greek Catholics.

Overall, Christians were in a narrow majority and for many years were politically dominant. Since then, waves of immigration, notably of Palestinians, have altered the balance. Of today's four million population, around 60% are Muslim, including around 200,000 Palestinian refugees living in camps as well as 300,000 migrant workers, mostly from Syria.

Prior to the civil war of 1975–90, Lebanon was one of the most developed Arab states—a major centre for trade, finance and tourism. Much of that was wrecked in the civil war which virtually levelled Beirut. Today, the Lebanese population remains highly educated with good standards of health, but many people are unemployed, income disparities are widening, and one-quarter of the population live in poverty.

Until the 2006 war, Lebanon had been recovering slowly. Unsurprisingly, one of the most vibrant sectors has been construction. From 1992, money flooded in for the rebuilding of Beirut, often from Lebanese expatriates.

But as before, the main sources of employment are services of various kinds, many connected with trade, which account for around 70% of GDP. The financial sector is also being redeveloped and tourists had been coming back—arrivals were rising at 5% per year. Many new manufacturing enterprises have also sprung up, most of which are small, engaged, for example, in food production, furniture-making, and in textiles and clothing.

Agriculture, despite favourable soil and climate, has been less significant: it now accounts for only 13% of GDP and also employs 13% of the labour force. Farmers grow fruit and vegetables on the coastal plain, and wheat and barley in the Bekaa Valley, but there has **Lebanese** been relatively little **farmers are** investment and many **turning to** farmers in the Bekaa **cannabis** are turning again to cannabis, which was a major source of militia income during the civil war.

Lebanon's political troubles date back to the late 1960s when the Palestine Liberation Organization (PLO) started using Lebanon as a base. The Muslims, many of whom supported the PLO, had long resented Christian political domination and from 1975 their disagreements erupted into a full scale civil war. In 1978, as part of its campaign against the PLO, Israel invaded and seized a 'security zone' on Lebanon's southern border.

The war stopped in 1990—after 150,000 deaths and $25 billion-worth of damage. The end came when Syrian forces took control, helping the government to disarm the militias. Israel finally withdrew from southern Lebanon during 2000, though the Syrians remained and Hizbullah continued to strike Israeli targets.

Political loyalties in Lebanon align not behind parties but more along personal or religious lines. The largest group now are the Shia, who include the militant Hizbullah. The Sunni are a less coherent group and during the war had weaker militias.

In the deal that ended the civil war, Lebanon's National Pact was amended to increase the powers of the prime minister (always a Sunni Muslim) and the speaker of parliament (a Shia Muslim), and to reduce those of the president (a Maronite Christian).

In 1992 Rafiq Hariri, a Saudi-Lebanese billionaire, emerged as prime minister. He set about the reconstruction of Beirut, but the economy failed to revive and Lebanon found itself deep in debt.

In 1998 Syria supported the election as president of a Christian former army commander, Emile Lahou. Hariri resigned. Following parliamentary elections in 2000, however, Hariri received so much support that Lahou was obliged to reinstate him. In February 2005, Hariri was killed in a bomb blast. One million people took to the streets blaming Syria and demanding that Syrian troops withdraw—which they did in May 2005.

In June 2005, parliamentary elections gave the most seats to a coalition led by Hariri's son Saad. Fouad Siniora became prime minister.

Meanwhile Hizbullah had sustained its attacks on Israel. In July 2006, after the capture of two of its soldiers, Israel responded with a massive air and ground attack on southern Lebanon, while Hizbullah rained rockets on northern Israel. Before a ceasefire the month-long war had killed 1,100 Lebanese and caused $7 billion-worth of damage.

Lesotho

Lesotho's democracy is fragile—and dependent on South Africa. Now it faces devastation from HIV/AIDS

0	Miles	100
0	Km	160

Land area: 30,000 sq. km.
Population: 2 million—urban 18%
Capital city: Maseru, 120,000
People: Basotho
Language: Sesotho, English, Zulu, Xhosa
Religion: Christian 80%, indigenous 20%
Government: Constitutional monarchy
Life expectancy: 36 years
GDP per capita: $PPP 2,561
Currency: Loti
Major exports: Garments, footwear

The landlocked kingdom of Lesotho is embedded in South Africa. Most of the territory is mountainous with its highest points in the east and north-east. The lowest and most densely populated areas, which have the best land, are in the north-west, particularly in the valley of the Caledon River.

The kingdom was formed in the 19th century when the Basotho ethnic group took refuge in the mountains to avoid being killed by the more aggressive Zulus and Boers. Today, almost all the people are still Basotho, though this includes a number of subgroups.

Lesotho is a poor country but by African standards has relatively high levels of human development. More than 80% of the population are literate, and 86% of children are enrolled in secondary schools.

Until the arrival of HIV/AIDS health standards too had been quite good, partly as a result of the fresh non-tropical climate, but also because of well-distributed health services.

Around 250,000 of Lesotho's people live temporarily or permanently in South Africa, many working on farms, but fewer are employed as miners because South African mines have become less profitable. By 2005 the number of miners had dropped to 47,000.

However, this still means that around 10% of the labour force is in South Africa. This has had a major social impact. Family structures are very disrupted, and most of the agricultural work in Lesotho is done by women.

A large migrant population also exposes Lesotho to HIV/AIDS: by 2003 almost one-third of the adult population were HIV-positive, with an infection rate above 40% in Maseru.

For more than half the population, however, the main source of income is still farming. Only around one-tenth of the land is suitable for agriculture, and even this has become less productive as a result of over-exploitation and soil erosion, and agriculture contributes only 16% of GDP. The main crops are maize, wheat, and sorghum, but even in good years around one-quarter of food needs to be imported. Many people raise cattle which they buy with migrant remittances. They also raise sheep and goats which generate some export income from wool and mohair.

In recent years, Lesotho has seen a surge in manufacturing industry. Investors from Taiwan, Hong Kong, and Singapore have

Making garments and shoes for the USA established a series of joint ventures with the Lesotho National Development Corporation to manufacture clothing and footwear for export. Thanks to preferential access, more than half of these goods go to the United States. Other companies are now making handicrafts and furniture for export. More companies are also engaged in food processing and making other goods for local consumption.

Lesotho's political relationship with South Africa has always been close. During the colonial era, South Africa had wanted to incorporate what was then Basutoland. The Basotho successfully resisted and in 1966 Lesotho was born as a constitutional monarchy. Elections the previous year had narrowly been won by the Basotho National Party (BNP), led by Chief Leabua Jonathan who became prime minister. Chief Jonathan was soon in conflict with King Moshoeshoe II, who had tried to extend his authority.

But Jonathan was no democrat. In 1970, when it looked as though the opposition Basotho Congress Party (BCP) had won the election, he suspended the constitution, dispatched the king temporarily into exile, arrested opposition politicians, and continued to rule by force.

In 1974 the BCP, following a failed uprising, formed the Lesotho National Liberation Army. But Jonathan was not overthrown until 1986, when the military took over and re-established King Moshoeshoe as head of state. In 1990, the army quarrelled with the king, exiled him again, and installed his son as King Letsie III.

This regime ended after a bloodless coup in 1991, which led in 1993 to new elections. These were convincingly won by the BCP, led by Ntsu Mokhehle. He reinstalled Moshoeshoe—who died in 1996 and was again succeeded by Letsie.

The BCP tried to promote national dialogue but eventually quarrelled within its own ranks and in 1997 fell apart. Mokhehle himself quit to form the Lesotho Congress for Democracy (LCD), though Mokhehle did not stay as leader, being replaced by Pakalitha Mosisili. In the 1998 election, the LCD won only 61% of the vote but gained 78 of the 79 seats.

In September 1998, an electoral commission refused to overturn the result, provoking riots and a revolt by junior officers. The government asked for intervention from South Africa and Botswana—under the rules of the Southern African Development Community—though most of the troops came from South Africa. The next election, it was decided, would have an element of proportional representation.

The election was held in May 2002 with virtually the same result, and Mosisili returned as prime minister. The BNP, the main opposition, contested the result. The LCD has also won subsequent by-elections.

Liberia

Liberia has yet another peace agreement that will hopefully end decades of warfare

Land area: 98,000 sq. km.
Population: 3 million—urban 47%
Capital city: Monrovia, 1.3 million
People: African 95%, Americo-Liberian 3%, other 2%
Language: English and about 20 others
Religion: Traditional 40%, Christian 40%, Muslim 20%
Government: Republic
Life expectancy: 42 years
GDP per capita: $PPP 1,000
Currency: Liberian dollar
Major exports: Iron ore, timber, diamonds

The most developed part of Liberia is the narrow coastal strip that extends around 50 kilometres inland. Behind this rises a belt of low hills and plateaux, and beyond this is the densely forested mountainous interior.

Liberia's 15 or more ethnic groups were left relatively undisturbed by the colonial powers until freed American slaves arrived in the mid-19th century and established the Republic of Liberia. Nowadays, the Americo-Liberians make up only around 3% of the population but until the early 1980s they dominated political life. The wars have partly been a struggle between this élite group and the rest of the population, as well as between other ethnic groups.

The wars are thought to have killed around 200,000 people, displaced around 700,000 within Liberia, and sent 350,000 refugees into neighbouring countries. Even before

the wars, this was one of Africa's most urbanized countries, but the fighting and the flows of refugees will probably have raised the proportion to 60% or more.

Levels of human development have fallen steeply over the past few years. Infant mortality is high and the health system has lost many of its staff. The education system too collapsed in many parts of the country as more than 40,000 children were recruited into the different militias.

In the past agriculture has provided work to around three-quarters of the workforce—the staple foods being rice and cassava. But the fighting drove many subsistence farmers from the land, and the population remains heavily dependent on food imports.

Many small farmers were also responsible for the country's major cash crop, rubber. One-third of this was produced by smallholders and the rest on large plantations now owned by the Japanese Bridgestone Company. The rubber industry now needs to be rebuilt—with investment not just in plantations but also in roads and in employee housing, much of which was destroyed in the war.

Logging has also been an important source of income. About

50% of Liberia is covered with trees and there are an estimated 3.8 million hectares of productive forests. During the war there was a steep rise in timber exports as the government liberally granted concessions to Asian companies.

Liberia is well endowed with minerals, including iron ore, gold, and diamonds. Indeed much of the fighting involved struggles for control over these resources. Iron ore was the most significant in terms of export income. Production had already fallen in the

One billion tons of high-grade iron ore mid-1980s as a result of a global slump in the demand for steel. But the war brought the industry to a rapid halt. Again, heavy investment will be needed to replace the damaged equipment. But with proven reserves of over one billion tons of high-grade ore this should be profitable.

Gold and diamond production will revive more quickly, since these are small-scale operations with about five thousand mining and dealing operations. But since much of the output is smuggled out of the country these make a limited contribution to the national economy.

The seeds for Liberia's civil war were sown in 1980 when a military coup led by Master-Sergeant Samuel Doe ended 150 years of domination by the Americo-Liberians. His regime promoted the interests of his own ethnic group, the Kahn, and rapidly degenerated into ruthless repression. After a rigged election in 1985 and attempted coup by the Gio group, full-scale civil war erupted in 1989 when Charles Taylor (an Americo-Liberian) invaded with rebel forces.

In 1990 a group of neighbouring countries through the Economic Community of West African States sent in a peacekeeping force, but this in turn became involved in offensive operations. Doe was assassinated in November that year. The ensuing years of anarchy saw many failed peace deals and a bewildering succession of alliances and splits between warring factions.

In 1995, another peace deal led to elections in 1997 which were won by Taylor. He ruled in lavish style, with a heavy hand, and with armed guards, as well as with an 'Anti-Terrorist Unit' commanded by his son.

Most opposition politicians fled, some of them to form 'Liberians United for Reconciliation and Democracy' (LURD), a rebel group that entered the country in 1999 launching a new round of civil wars. Faced with US pressure and advances by LURD and another group, Taylor, by then an indicted war criminal, finally resigned in August 2003 and left for Nigeria.

The ensuing Comprehensive Peace Agreement, which was backed by 15,000 peacekeepers, led to a two-year provisional administration.

In October 2005 presidential elections were held, and won by a former UN and World Bank civil servant, Ellen Johnson-Sirleaf, who defeated former footballer George Weah. She is well respected across the continent and has a good relationship with donors but has a huge task ahead, particularly in reviving public services and tackling corruption. One potential source of trouble was removed in 2006 when Taylor's trial for war crimes was transferred to the Hague.

Libya

Libya is ending its diplomatic isolation, but democracy is still not on the agenda

Land area: 1,760,000 sq. km.
Population: 6 million—urban 86%
Capital city: Tripoli, 1.1 million
People: Berber and Arab
Language: Arabic, Italian, English
Religion: Muslim
Government: Military dictatorship
Life expectancy: 74 years
GDP per capita: $PPP 11,400
Currency: Libyan dinar
Major exports: Oil, petroleum products, natural gas

Most of Libya is a vast barren plain of rocks and sand. The majority of the population live in Tripolitania, in the north-west of the country, a region that includes the capital and a number of coastal oases. The other main inhabited area is Cyrenaica in the north-east.

Many of Libya's people descend from Bedouin Arab tribes—and the tribe is still often the basic social unit. The original Berber population has also largely been absorbed into Arab culture. Nowadays, however, most people live in the cities and the population is very young—60% are under 20 years old.

Libya's oil wealth has not been widely distributed. Arround 70% of Libyan workers are employed in one way or another for the government on salaries that average only $190 per month. Unemployment is high, at around 30%.

But Libyans do benefit from subsidized housing and largely free education and health care. Notably for an Arab country, women are at least as well educated as men.

In addition to the native population, there are also more than one million immigrant workers, chiefly from Egypt, often doing jobs that local workers refuse. Sometimes there have been tensions with the local population, and even riots.

Since 1959, when oil was first discovered, Libya has become one of the world's largest producers. Oil makes up more than 90% of exports and provides over 80% of government income. At current rates of extraction the oil should last 50 years or more. Most oil production is controlled by the National Oil Company but there are many exploration and co-production contracts with European oil companies. Libya also has huge reserves of natural gas that have yet to be exploited.

Most other industrial operations are run by the state. They do now produce many consumer goods, though companies from Italy and Korea are manufacturing some items like refrigerators and video recorders.

Less than 2% of the land is suitable for farming so agriculture is limited and employs only 6% of the workforce. Around 80% of cereals must be imported. The main problem is water. Along the coast, groundwater has been over-extracted so sea water may seep in to replace it. The government's solution is the $20-billion 'great man-made river' scheme which will pump groundwater from aquifers in the south-east to the north. This project already delivers water, at great expense, to Tripoli and Benghazi, but has hit technical problems and is only running at 15% of capacity.

For the past 35 years, Libyan politics has revolved around Muammar Qaddafi. Following a military coup in 1969, Libya's monarchy was overthrown and a group of military officers led by the then 29-year-old Captain Qaddafi seized control and redirected Libya towards Arab socialism.

In 1977 he changed the country's name to the Socialist People's Libyan Arab Jamahiraya ('state of the masses'). In theory this is popular government, with everyone participating in Basic People's Congresses and choosing representatives to a General People's Congress which selects members of the

Libya is a 'socialist state of the masses'

government. The philosophy is spelled out in Qaddafi's 'Green Book' —a mixture of Islam and socialism.

In reality, all power resides with the idiosyncratic Colonel Qaddafi, whose picture is ubiquitous. He is assisted by 'revolutionary committees' who control most institutions, including the media. Citizens also find their activities scrutinized by 'purification committees'. Human rights abuses range from arbitrary arrest and torture to extra-judicial executions and disappearances.

Qaddafi's main fear is an Islamic uprising. In response, he has appropriated Islamic symbols and set out an Islamic 'third path' between communism and capitalism, but this does not convince his more radical Islamic enemies.

Qaddafi has also been a thorn in the side of Western governments, particularly the USA. In the past he used Libya's oil wealth to bankroll a range of guerrilla groups from the PLO to the IRA. A series of terrorist incidents culminated in the bombing in 1988 of a Pan Am jet over Lockerbie in Scotland, killing 270 people. Libya originally denied involvement but in 1999 handed over the suspects for trial—one of whom was found guilty—and in 2003 admitted 'civil responsibility' and agreed to pay $2.7 billion to the victims' families.

Relations with the UK and the USA also improved when Qaddafi condemned the 9/11 attack. And in an even more remarkable move, perhaps reflecting on the US invasion of Iraq, in December 2003 Qaddafi promised to give up Libya's weapons of mass destruction and open secret facilities for inspection. As a result many of the economic and diplomatic sanctions have now been lifted.

Following Libya's international rehabilitation, Qaddafi's position seems even more secure. In 2006 he was still only 64, so will probably be around for some time; in any case his sons already hold public office, so there is a potential dynasty.

Lithuania

The most ethnically homogenous Baltic state, but with a shrinking population

Land area: 65,000 sq. km.
Population: 4 million—urban 67%
Capital city: Vilnius, 553,000
People: Lithuanian 83%, Polish 7%, Russian 6%, other 4%
Language: Lithuanian, Polish, Russian
Religion: Roman Catholic
Government: Republic
Life expectancy: 72 years
GDP per capita: $PPP 11,702
Currency: Litas
Major exports: Mineral products, textiles, machinery

Lithuania's terrain is mostly flat—a plain dotted with low hills and around 3,000 small lakes. The highest points are in the Baltic Highlands in the east and south-east. A feature of the sandy coastline is the narrow 100-kilometre sand spit that creates a distinctive lagoon.

Compared with the other Baltic states—Estonia and Latvia—Lithuania's population has a higher proportion of its own national group: more than 80% of people are ethnic Lithuanian. As a result, when it regained independence in 1991 Lithuania was less nervous than the other Baltic states about offering automatic citizenship to all—an offer that most people took up. Lithuanians also differ in that, like the Poles to the south, they maintain a strong Roman Catholic tradition.

The population has shrunk since 1991, initially as a result of emigration to Russia, Poland, and elsewhere, but more recently because of falling birth rates and rising death rates. Mortality rates, particularly from heart disease and cancers linked to poor diets, rose steeply until 1995 though since then have they levelled off. Many men's health problems are linked to high consumption of alcohol.

The transition also caused an increase in poverty: beggars and street children are becoming part of a new underclass. In 2005 official unemployment was only 8% but the real figure is certainly much higher. And in any case most workers have to do a second job in order to survive.

Lithuania had industrialized rapidly in the Soviet era, but after 1991 the country lost its major markets and its sources of cheap energy. By 1995, industrial output had halved. Having reoriented more towards the West, some industries, particularly textiles, chemicals, and wood products, have since recovered. By 2004 manufacturing accounted for 18% of GDP and employed 17% of the workforce.

But with only small amounts of domestic oil and gas, the country remains very dependent on energy imports. Two Chernobyl-style reactors at Ignalia provide over 80% of electricity—making this probably the most nuclear-dependent country in the world. Under EU pressure Lithuania has promised to close them by 2009—which means it will have to buy more gas from Russia or build a new reactor.

Independence boosted the service sector, which by 2004 accounted for 70% of GDP. Much

Lithuania embraces an orphaned Russian province

of this is concerned with transport and communications with Russia. A significant proportion of exports are actually re-exports to Russia of imported cars and consumer goods. Moreover, Russians have to cross Lithuania to reach their 'orphaned' province of Kaliningrad.

Agriculture continues to make an important if declining contribution—chiefly livestock and cereals. By 2004 it accounted for 6% of GDP and 16% of employment. One of the first steps after independence was to break up the collective farms and co-operatives and return land to the descendants of former owners. Since most of this was in fairly small plots, efficiency and output fell.

In May 2001 Lithuania joined the EU which has further unsettled farmers who, although they will benefit from subsidies, also have to meet higher standards of hygiene and inspection.

One-quarter of the country is covered by forests which supply raw materials for wood and paper industries. With a strong world market the industry has been growing rapidly.

Lithuania's independence struggle had been led in the late 1980s by Sajudis—the Lithuanian Movement for Restructuring. But in 1992 in the first free election to the unicameral parliament, the Seimas, the voters reverted to the former communist party which had reinvented itself as the Democratic Labour Party (DLP). In 1993, the DLP candidate Algirdas Brazauskas was elected president. The president has considerable powers—including the appointment of ministers.

The DLP government made some progress but was rocked by a banking crisis and following the 1996 election Sajudis, now renamed the Homeland Union Party (HU), formed a coalition government with the Christian Democratic party.

The 1997 presidential election was surprisingly won by an independent candidate, a returning US emigré, Valdas Adamkus, who at that point was still a US national.

The general election in October 2000 produced another coalition government, New Policy. The main partner was the Liberal Union, led by Rolandas Paksas. The other partner was New Union–Social Liberals (NU). The coalition collapsed in June 2001 and the NU formed another with the Social Democratic Party whose leader, former president Algirdas Brazauskas, took over as prime minister.

Paksas was to return, as president, in January 2003 when he defeated the incumbent Adamkus. But he was impeached and dismissed after a financial scandal and Adamkus returned in the ensuing election in June 2004. In the general election in October that year Brazauskas was returned as prime minister at the head of a four-party coalition.

Luxembourg

Luxembourg is the world's richest country—but with a lot of help from foreigners

The northern one-third of Luxembourg is mountainous, with dense forests. The more fertile part of the country is in the southern two-thirds—the rolling plains of the 'Bon Pays'.

Luxembourg's people have their own dialect and a strong national identity, but most also speak French and German. They have the world's highest per capita income—and usually Europe's lowest rate of unemployment.

But the birth rate is also low, so immigrants are needed to fill the labour gaps. Foreign-born residents make up 40% of the population, and more than half the labour force. Portugal is the main provider. In addition there are some 60,000 'frontaliers'—cross-border workers who commute daily from France, Belgium, and Germany.

Most immigrants used to come to work in Luxembourg's heavy industry, particularly steel which took advantage of iron-ore deposits along the southern border. Steel is still an important industry, and the main export. And following an international merger in 2002 Luxembourg is the base for the world's second largest steel company: Acelor. There are also other major manufacturing industries, including a Goodyear tyre plant and a DuPont chemical plant.

Land area: 3,000 sq. km.
Population: 0.5 million—urban 92%
Capital city: Luxembourg, 76,000
People: Luxembourgish and immigrants
Language: Luxembourgian, French, German, English
Religion: Roman Catholic 87%, other 13%
Government: Constitutional monarchy
Life expectancy: 72 years
GDP per capita: $PPP 62,298
Currency: Euro
Major exports: Steel, chemicals, rubber products

Today, however, the core of Luxembourg's economy is formed by financial and business services, which contribute around one-third of government revenue. Luxembourg is one of Europe's leading financial centres. Its low tax rates attract funds from Germany, and it controls more than 90% of Europe's offshore investment funds.

Luxembourg is a constitutional monarchy—a Grand Duchy. In September 2000, Grand Duke Jean abdicated in favour of his son Henri.

All governments since 1919 have been coalitions, usually led by the centre-right Christian Socialist Party (CSV). The CSV were the major party after the 1984 election. They governed in coalition with the Socialist Party (LSAP), with Jacques Santer as prime minister, and were re-elected in 1989 and 1994. In 1995, Santer left for an ill-fated presidency of the European Commission. He was replaced in Luxembourg by Jean-Claude Juncker. In the 1999 elections, however, the LSAP were pushed to third place by the right-wing Democratic Party which became the coalition's junior partner. The elections in June 2004 also produced a CSV-led coalition with Juncker as prime minister.

Macedonia

Macedonia avoided the worst of the Balkan wars, but has persistent ethnic tensions

Land area: 26,000 sq. km.
Population: 2 million—urban 60%
Capital city: Skopje, 467,000
People: Macedonian 64%, Albanian 25%, Turkish 4%, Roma 3%, Serb 2%, other 2%
Language: Macedonian 70%, Albanian 21%, Turkish 3%, Serbo-Croatian 3%, other 3%
Religion: Macedonian Orthodox 64%, Muslim 33%, other 3%
Government: Republic
Life expectancy: 74 years
GDP per capita: $PPP 6,794
Currency: Macedonian denar
Major exports: Food, tobacco, machinery

This landlocked country is largely mountainous. In the past it was also heavily forested, and though forests remain, particularly in the west, the cleared land revealed thin soil, much of which has now been eroded. The best land is to be found in the valley of the Vardar River, which runs through the centre of the country from north to south. Tectonic fault lines run in the same direction and the country remains vulnerable to earthquakes.

Macedonia's ethnic composition is a source of political tension. The 1994 census showed two-thirds of the population as ethnic Macedonian who belong to the Eastern Orthodox Church. There are also minorities of Turks, Romanians, and Serbs.

But the most significant minority were ethnic Albanians, who made up one-quarter of the population. The Albanians, who are Sunni Muslims, live largely in the north-west. They claim they were under-counted and that in fact they represent around one-third of the population. Albanians are often treated as second-class citizens. In 2001 guerrillas from the Albanian National Liberation Army (NLA) attacked the government. After a peace agreement in Ohrid in August 2001, the government made a number of concessions, concerning power-sharing, the use of the Albanian language, Albanian education, and hiring more Albanian policemen.

Macedonia was the poorest of the constituent republics of former Yugoslavia. It had developed heavy industries such as steel, chemicals, and metal-based manufacturing, and was mining its deposits of lead, zinc, copper, chromium, and coal. It also had light industries such as textiles. But the break-up of Yugoslavia and subsequent trade embargoes sent industry into a steep decline. By 1995, output had dropped by more than one-third. There has been limited foreign investment in mining. And manufacturing industries have since recovered but between 1998 and 2005 employment in manufacturing continued to fall.

Officially unemployment in 2004 was 37% but many people actually work in the black economy which accounts for around one-third of GDP.

Agriculture also suffered in the early years but has recovered more strongly. Even as part of Yugoslavia, Macedonian agriculture depended largely on small private farms which produce most of the food as well as important cash crops such as tobacco. Tobacco remains an important crop since it supports around 40,000 households. The country could be self-sufficient in food, and even export vegetables and fruit, but at present has to import much of its requirements. One problem is that the state is frequently late in paying for its purchases of wheat and tobacco so farmers have little incentive to increase output.

Macedonia's emergence as an independent republic in 1991 was a tortuous process. Neighbouring Greece immediately objected to the name on the grounds that this and the constitution implied sovereignty over the adjacent Greek province of the

Objections to the country's name

same name. Greece's objections for a time blocked international recognition. The name problem was solved by calling the country the 'Former Yugoslav Republic of Macedonia'—though few people use this. Macedonia joined the UN in 1993 and almost immediately requested a UN peacekeeping force which successfully prevented annexation by Serbia.

The 1992 parliamentary election was won by the ex-communist Social Democratic Alliance of Macedonia (SDSM), which entered into a coalition with the moderate Albanian Party for Democratic Prosperity. The SDSM increased its representation in the 1994 parliamentary elections which were boycotted by the main opposition party, the Internal Macedonian Revolutionary Organization (VMRO). But the government was beset by pyramid scheme scandals and by riots in the Albanian-dominated city of Gostivar.

In the general election in 1998, the VMRO, led by Ljubcho Georgievski, did stand, gained the largest number of seats, and formed a centre-right coalition government. Its main partner was the new pro-business Democratic Alternative (DA) Party, along with the Democratic Party of Albanians (DPA), a radical separatist group. The government proved an unstable coalition and grew increasingly unpopular as a result of widespread corruption. It did deal successfully with the Albanian insurgent campaign of 2001 but lost the parliamentary election of 2002.

This saw a return for the SDSM who formed a coalition with the ethnic Albanian party Democratic Union for Integration (DUI), which also represented the NLA, with the SDSM's Branko Crvenkovski as prime minister. This raised the hope that ethnic tensions might be reduced though the respite may only be temporary. In February 2004 the president, the SDSM's Boris Trajkovski, who had been an important conciliator, was killed in a plane crash. The subsequent election in April 2004 saw Crvenkovski chosen as president. He was replaced as prime minister first by Hari Kostov, and later by Vlado Buckovski.

Madagascar

Madagascar's drive for foreign investment is doing little for the rural poor

Land area: 587,000 sq. km.
Population: 18 million—urban 27%
Capital city: Antananarivo, 1.2 million
People: Merina, Betsileo, Betsimisaraka, and others
Language: French, Malagasy
Religion: Indigenous beliefs 52%, Christian 41%, Muslim 7%
Government: Republic
Life expectancy: 55 years
GDP per capita: $PPP 809
Currency: Malagasy franc
Major exports: Coffee, vanilla, cloves

Madagascar is dominated by a central mountainous plateau that covers around two-thirds of the island—the highest peak is 2,800 metres. To the east, the mountains drop steeply to a narrow coastal strip. To the west the slope is more gentle to broad fertile valleys. Most of the original forest cover on the mountains has been lost and Madagascar now suffers from severe erosion as millions of tons of soil are washed each year into the sea. Deforestation has also stripped the natural habitats of many rare plant and animal species.

Madagascar is frequently battered by cyclones—and in early 2000 was hit by three in fairly rapid succession, with others in 2003 and 2004.

Most Malagasy are descendants of immigrants who arrived more than 2,000 years ago from South-East Asia. Today, these consist of around 20 ethnic groups, of whom the largest are the Merina (around one-quarter of the population) and the Betsileo. Some groups such as the Betsimisaraka are of mixed African, Malayo-Indonesian, and Arab ancestry. All speak dialects of Malagasy, though the education system also works in French.

Levels of human development in Madagascar remain low. Two-thirds of the population live in poverty—with an increasing disparity between urban and rural areas. Life expectancy, however, is a little higher than in other countries of Sub-Saharan Africa, and the island's relative isolation offers some protection from the rapid spread of HIV/AIDS. Nevertheless, one-third of children are malnourished and only one-third of the rural population have access to safe water.

Madagascar's poverty is concentrated in the countryside where most people struggle to survive through agriculture. Much of this is at a subsistence level, primarily growing rice, along with cassava, sweet potatoes, maize, and other crops. But farms are generally small, yields are low, and food has to be imported. The main cash crops are coffee, vanilla, sugar, and cloves, but here too output has been stagnant. Some of the problems result from underinvestment and there is relatively little irrigation. In fact much more land could be brought under cultivation to grow food that might displace imports. Livestock rearing which occupies more than half the land is also an important source of income—indeed it often takes up land that might better be used for growing crops.

Prices for cash crops have been liberalized in recent years, so farmers' rewards are linked more to world prices. But with poor infrastructure and inadequate systems for quality control they are at a disadvantage against international competitors.

Fishing, especially for prawns, should be an important source of income but the local industry is relatively underdeveloped, so much of the income actually comes from licences sold to European and Japanese fleets that fish for tuna in Madagascan waters.

Selling fishing licenses to the EU and Japan

Madagascar has some basic industries producing goods for local consumption, including plastics, pharmaceuticals, and footwear. It also had some success in developing export-processing zones which account for around 80% of foreign investment. These include a textiles industry which is based on locally-grown cotton.

Madagascar's dominant political figure until 2001 was Didier Ratsiraka. He first came to power at the head of a military government in 1975 and held a referendum that approved a new constitution allowing for only one political organization. This was to be a 'national front' that embraced all the political parties, though it was actually controlled by Ratsiraka's party which was subsequently renamed the Association pour la renaissance de Madagascar (Arema).

Ratsiraka embarked on a socialist transformation that involved nationalizing banks and other major companies, but by 1980 was seeking the support of the IMF. The ensuing austerity measures contributed to growing political unrest. By 1990, he had agreed to allow other political groups to operate.

This led to the formation of a new opposition front, the Forces vives (FV), led by Albert Zafy. After further widespread protests and a fresh constitution Zafy and the FV won the elections in 1993. But Zafy's chaotic presidency was fraught with problems and in 1996 he was impeached for corruption and abuse of power. In the ensuing presidential election Ratsiraka was voted back into office. By this time, however, he had abandoned socialism for a 'humanist and ecological republic'.

For the 2001 presidential election the opposition united behind businessman Marc Ravalomanana of the Tiako-I-Madagasikara (TIM) party and mayor of Antananarivo. After the first round Ravalomanana was just short of 50% of the vote but said the election had been rigged and took control of the administration while Ratsiraka and his supporters blockaded the capital. A recount confirmed Ravalomanana's victory and Ratsiraka was eventually exiled.

Ravalomanana has consolidated his position, though in an increasingly autocratic style. TIM also achieved a comfortable majority in the 2002 National Assembly elections.

Malawi

Malawians now have a democracy, but they remain desperately poor and millions are dying from AIDS

Land area: 118,000 sq. km.
Population: 12 million—urban 16%
Capital city: Lilongwe, 464,000
People: Chewa, Nyanja, Tumbuko, and others
Language: English, Chichewa, and others
Religion: Christian 80%, Muslim 13%, other 7%
Government: Republic
Life expectancy: 40 years
GDP per capita: $PPP 605
Currency: Kwacha
Major exports: Tobacco, tea, sugar, coffee

Malawi's dominant geographical feature is the East African Rift Valley, which cuts through the country from north to south. This includes Lake Malawi, through which runs most of the eastern border, and the Shire River valley, which drains the lake southwards into Mozambique. To the west of the Rift Valley in the centre of the country, there is an extensive plateau area. There are also high plateaux in the north and intensively cultivated highlands in the south.

The main ethnic groups are the Chewa, who are the majority of people in the centre of the country and whose language is the most widely spoken. The Nyanja are mostly in the south, and the Tumbuko in the north.

By African standards, Malawi is densely populated and, with few people using contraception, the population growth rate prior to the AIDS epidemic was 3.5% per year, but could now be around 2%.

Malawi is a poor country. Around two-fifths of people live below the poverty line. Its education levels are higher than in some African countries and it even has its own version of an élite English public school, the Kamuzu Academy. But most children are less fortunate and drop-out rates at ordinary government schools are high. Half the children are malnourished and almost one-fifth do not live beyond their fifth birthday.

Mortality rates will climb even more steeply as a result of AIDS. By 2003, 15% of the population were HIV-positive. By the end of 2005 it was estimated that 40% of teachers, for example, had died. AIDS deaths are pushing the survivors even deeper into poverty, since those who will die are the main breadwinners, and the country will also face the heavy cost of caring for AIDS patients.

With 85% of people living in the rural areas, this is one of the world's least urbanized countries. Farmers have cultivated virtually all the arable land, and in the mid-1980s more than half of rural households had less than one hectare each. Most concentrate on subsistence crops like maize, sorghum, or millet, and the country is largely self-sufficient, though at low levels of consumption. However, farmers are vulnerable to droughts: in 2002 Malawi suffered its worst famine in 50 years.

Cash crops have also been important, particularly tobacco, which has provided 60% of export income, along with tea, sugar, and cotton. Previously the policy was to concentrate cash crops on large commercial estates (which still occupy around 20% of arable land) but the government is now encouraging smallholders by offering better prices. Even so, prospects for both food and cash crops are limited since many farmers cannot afford the inputs, and much of their land has been over-exploited.

Malawi relies heavily on exports of tobacco

Another important source of food is Lake Malawi, whose fish provide more than two-thirds of animal protein consumption—though catches have fallen in recent years as a result of over-fishing and pollution.

Malawi has some mineral deposits but no substantial mining industry. Now the government is trying to encourage foreign companies to develop bauxite and titanium mines.

For the first three decades following independence in 1962, political life in Malawi was overshadowed by the diminutive but dictatorial figure of Dr Hastings Banda at the head of his Malawi Congress Party (MCP). He was first prime minister and then president, and finally in 1970 declared himself 'president for life'. Banda's rule became increasingly despotic. He had an extensive network of spies and secret police to enable him to jail and torture opponents.

Banda's regime started to fall apart in the early 1990s following strikes and riots. Eventually he was forced to concede a 1993 referendum which approved a return to multiparty democracy.

In the ensuing presidential election in 1994, Banda was defeated by a former cabinet minister, Bakili Muluzi, at the head of the United Democratic Front (UDF). The legislative election also gave the UDF the most seats, ahead of the MCP, but without a majority it initially entered into a coalition with the third party, the Alliance for Democracy (Aford).

After his election Muluzi released a number of political prisoners. He also had Banda arrested on a murder charge, though he was acquitted, and died (101 years old) in 1997. In the 1999 elections, Muluzi scraped through to a second five-year term.

The second UDF government was marred by increasing problems of corruption, and aid donors withheld funds in protest.

In July 2002, the parliament rejected a motion to remove the limit on the number of presidential terms. Muluzi accepted this and backed Bingu wa Mutharika as the UDF candidate for the May 2004 election, which he duly won.

However Mutharika soon fell out with Muluzi who had remained as UDF chairman. Mutharika formed his own Democratic Progressive Party (DPP) and started an anti-corruption drive focusing among others on Muluzi, who has responded by trying to get Mutharika impeached.

Malaysia

Malaysia has a new leader but the same autocratic style of government

Land area: 330,000 sq. km.
Population: 24 million—urban 64%
Capital city: Kuala Lumpur, 1.4 million
People: Malay 51%, Chinese 24%, indigenous 11%, Indian 8%, other 6%
Language: Bahasa Melayu, English, Chinese, Tamil, and tribal dialects
Religion: Muslim, Buddhist, Hindu, Confucian, Christian, tribal religions
Government: Constitutional monarchy
Life expectancy: 73 years
GDP per capita: $PPP 9,512
Currency: Ringgit
Major exports: Manufactured goods, rubber, palm oil

Malaysia has an unusual geographical composition. Its territory is evenly divided between a portion on the Asian mainland and a similar area on the north-west of the island of Borneo. On the mainland, peninsular Malaysia, which has around 80% of the population, is largely mountainous in the north, with coastal lowlands to the west and south. Eastern Malaysia, 600 kilometres away across the South China Sea, is more sparsely populated, with a swampy coastal plain rising to high mountains that form the border with Indonesia.

Malaysia also has a distinctive racial composition. The majority are classified as ethnic Malays, most of whom are Muslims. But around one-third of the population are of Chinese origin, who live chiefly in the urban areas. There are also a number of South Asians, as well as small tribal groups who are found particularly in Eastern Malaysia. Standards of education and health have improved rapidly though there remain wide disparities between urban and rural areas and between richer and poorer states. There are also notable differences in population growth between ethnic groups. Over the period 1996–2000 the growth rate for Malays was 3.2% while for Indians it was 1.8% and for the Chinese 1.4%.

Since the early 1970s, Malaysia has transformed itself into an export-oriented industrial country. More than one-quarter of the population now work in manufacturing industry, which until recently was largely concentrated in states on the west of the peninsula. However the government has now created 14 'free industrial zones' along with 200 other industrial estates across the country. Much of this has been driven by electronics multinationals that have used Malaysia as an assembly base.

The government has recently been making efforts to have more manufacturing take place in Malaysia and use higher levels of technology. The most ambitious project is a 'multimedia super-corridor', carved out of land that previously was either jungle or palm-oil plantations. This was slow to take off but by mid-2005 had attracted 1,112 companies employing 23,000 workers.

Attracted by better-paid jobs in the cities, many people have been leaving the land. As a result, agricultural production has fallen and Malaysia is now a net importer of rice. Production of rubber, long a major export, has also been declining—though partly because plantations have switched to the more lucrative palm oil of which output continues to expand.

Malaysia is also reliant on immigrant labour. In 2004, there were an estimated 1.3 million migrant workers, 6% of the population. There have been pressures for them to return—though there is little sign that Malaysians want those jobs again.

Malaysia's distinctive ethnic mix has also had a profound impact on its political processes. The country is a federal constitutional monarchy, and formally it is an Islamic state. Each of the 13 states has a state assembly and an executive council that deals with state issues. Nine of the states have hereditary rulers who take turns to serve a five-year term as king.

Rulers take turns as king of Malaysia

Since independence in 1957, political power at the federal level has been in the hands of the main Malay political party, the United Malays' National Organization (UMNO), which has ruled in a National Front coalition with parties representing other racial groups.

The political system in Malaysia turned decisively in a new direction following riots in 1969 when Malays protested against the pervasive economic power of the Chinese minority. From 1971, the government embarked on a 'new economic policy' that would deliberately favour Malays and indigenous ethnic minorities, referred to as 'bumiputra' (sons of the soil). This included preferential access to government jobs, to higher education, and to a share of equity in national and foreign investment.

After becoming prime minister in 1981, UMNO leader Dr Mahathir Mohammed built a dominant position within Malaysia and became an outspoken international figure, notably accusing foreign currency speculators of provoking Asia's financial crisis, and suggesting that terrorist attacks against the USA were justified.

Mahathir's position was weakened somewhat in 1998 when he had his obvious successor, Anwar Ibrahim, jailed on charges of corruption and sexual misconduct. And the main opposition group, the Parti Islam sa-Malaysia (PAS), seemed to be gaining ground. Nevertheless, Mahathir remained in control until he stepped down after 22 years in 2003.

His chosen replacement as prime minister was Abdullah Badawi whose position was confirmed with a striking victory in the parliamentary elections in 2004, when the National Front gained 90% of the seats. This was a crushing defeat for PAS, and a rejection of their fundamentalist approach. Badawi offers a more progressive vision of Islam, while also promising a less corrupt and more responsive style of administration. He has also fallen out with Mahathir, having cancelled one of his more grandiose construction projects.

Maldives

One of the most prosperous, if least democratic, countries in South Asia

Land area: 300 sq. km.
Population: 0.3 million—urban 29%
Capital city: Male', 62,000
People: Sinhalese, Dravidian, Arab, African
Language: Maldivian Divehi, English
Religion: Muslim
Government: Republic
Life expectancy: 67 years
GDP per capita: $PPP 3,900
Currency: Rufiyaa
Major exports: Marine products

The Maldives consists of 1,190 coral islands grouped in atolls, spread out over a wide area. Of these 200 are inhabited, while 80 others are tourist resorts. Since few rise more than one metre above sea level, global warming could swamp them. In 2004 the Asian tsunami washed across the islands, though only killed 82 people.

Culturally, the Maldives is a blend of influences from India, Sri Lanka, Arabic countries, and elsewhere. All Maldivians speak the national language, Divehi, and are Muslims. In addition migrant workers from Bangladesh, India, and Sri Lanka make up more than one-quarter of the labour force, providing professionals, such as teachers, as well as labourers.

By South Asian standards, Maldivians are healthy and well off, but rapid development has also been socially disruptive. The Maldives has the world's highest divorce rate and many young people are abandoning the atolls for Male' and some have fallen victim to heroin peddlers.

The country's major source of income is tourism, which brings in around half a million visitors per year at the top end of the market—up to $10,000 per night—providing one-third of government revenue and 70% of foreign exchange. The hedonistic tourists are however mostly kept on their own islands, well away from the conservative Muslim population.

Fishing is also an important source of income and employment. The main catch is tuna, which is processed locally and exported frozen or canned.

The Maldives has a distinctive constitution. Citizens elect members to the 50-seat parliament, the Majlis, which chooses a president whose appointment then has to be confirmed in a national referendum.

Since 1978, the winner, through a mixture of guile and repression, has been Moumoon Abdul Gayoom, who in 2003 gained a sixth successive term with more than 90% of the vote.

Under international pressure, Gayoom has been edging towards democracy. In June 2004, he allowed the registration of political parties, and established a special people's Majlis to draw up a new constitution. Not long afterwards, however, he had opposition demonstrations broken up with baton charges, and the leaders arrested and tortured.

The main opposition comes from the Maldivian Democratic Party, which has been run from Sri Lanka by Mohamed Nasheed. He returned in 2004 only to be arrested, though subsequently released. Similarly, in 2005, journalist Jennifer Latheef was sentenced to ten years for 'terrorism'.

Mali

Mali is one of the poorest developing countries, but also one of the more democratic

0	Miles	400
0	Km	640

ALGERIA

MAURITANIA

Timbuktu

Mopti

Kayes

Segou

Bamako

BURKINA FASO

Land area: 1,240,000 sq. km.
Population: 13 million—urban 32%
Capital city: Bamako, 1.3 million
People: Mandé 50%, Peuhl 17%, Voltaic 12%, Songhai 6%, Tuareg and Moor 10%, other 5%
Language: French, Bambara, other African languages
Religion: Muslim 90%, indigenous beliefs 9%, Christian 1%
Government: Republic
Life expectancy: 48 years
GDP per capita: $PPP 994
Currency: CFA franc
Major exports: Cotton, gold, livestock

Mali is almost entirely flat with only occasional broken rocky hills. The northern one-third of the country falls within the Sahara Desert. The centre is the semi-arid Sahel belt, which gives way further south and south-west to grasslands and to the valley of the Niger—a broad river that periodically floods the land, depositing rich alluvial soil.

Mali has a diverse collection of ethnic groups and clans that tend to be concentrated in specific areas. The main distinction is between the Berbers and the black groups. Berbers include the Tuareg and the Moors, who are nomadic herders and occupy the Sahelian zone. Of the black Mandé group, the largest are the Bambara who are farmers along the Niger River. The Peuhl are another nomadic herding group in the Sahel. Most Malians are Muslims, though religious affiliations are also divided along ethnic lines. Islam is not as strong a political force as in other Islamic countries but there are a number of fundamentalist groups. Mali's ethnic diversity has frequently led to conflict between the nomadic and sedentary groups.

The Tuareg in particular have felt marginalized and in 1991 they rebelled, demanding autonomy from the central government. The army responded forcefully, driving more than 80,000 Tuareg into neighbouring countries. Hostilities have ended and most refugees have returned, but there are still occasional conflicts.

There are also around three million Malians working as emigrants, primarily in neighbouring countries such as Côte d'Ivoire, but there are also many in France.

Mali is one of the world's poorest countries: two-thirds of people are below the poverty line. More than 40% of children are malnourished. Life expectancy is short even though HIV infection rates are low by African standards at around 2%. Half of children do not attend school and as a result four-fifths of the population are illiterate.

Most people depend for survival on subsistence agriculture and livestock. They grow food crops in the south on irrigated land around the Niger and its tributaries, their main crops being millet, rice, wheat, and corn. This is also where they grow the principal cash crops: peanuts and particularly cotton which is a major source of export income. Most cotton is grown by small farmers and co-operatives; quality is high and output has been increasing. Although there are some plans to process cotton locally, most is exported raw.

Livestock production accounts for around one-fifth of GDP. This used to be the preserve of nomadic herding communities, but nowadays most of the cattle are on small farms. Fishing in the Niger river is also an important source of income and around one-fifth of the catch is exported to Côte d'Ivoire.

Mali has little industry beyond processing agricultural output. Most consumer goods are imported, or smuggled in, from neighbouring countries. But one increasingly important activity is gold mining. Output in 2002 was around 66 tons, mostly through South African companies working with the government. In 2000 gold overtook cotton as the leading export earner. The country's reserves are equivalent to at least 500 tons. Production costs are low, and world prices have been rising so Mali is likely to rely more on gold exports.

Mali has around 500 tons of gold

For 23 years, Mali was under the autocratic rule of Moussa Traoré, who became president in 1969 following a military coup. In 1979 the country returned to single-party civilian rule. This made little difference since Traoré was elected president at the head of a military-backed party and re-elected in 1985—as the only candidate. Throughout this period there had been a number of protests, with a former minister Alpha Oumar Konaré as one of the leading figures. These, along with several coup attempts, were violently suppressed.

Traoré survived until 1991 when, after mass protests, he was arrested and replaced by an interim administration headed by Lieutenant-Colonel Amadou Toumani Touré.

In 1992, a new constitution led to multiparty elections and a victory in the National Assembly for a coalition led by Konaré and the Alliance pour la démocratie au Mali (Adema). Konaré was elected president.

Konaré was a veteran pro-democracy campaigner and since he took human rights and clean government fairly seriously he was rewarded with support from donors. The opposition parties, however, accused Konaré of trying to recreate a one-party state and boycotted the 1997 election which Konaré duly won, albeit in a low turnout.

The constitution prevented Konaré standing again in the 2002 presidential election, prompting a lot of infighting in his party for the candidature. This opened the way for former transitional president Touré, who won as an independent.

The National Assembly elections in the same year spread the seats out over a large number of parties in four main coalitions, so Touré's main task has been to build a consensus in favour of his policies.

Malta

Malta is now a bridge between the EU and North Africa

Malta comprises a small group of islands in the Mediterranean, of which the largest are Malta, Gozo, and Comino. The land is largely low-lying and dry, without permanent rivers or lakes, so around half of water comes from desalination plants.

Malta is a bridge between Europe and North Africa, and its language has both Latin and Arabic components. In the past, young Maltese often chose to emigrate, creating a Maltese diaspora around the world, notably in Australia. Nowadays people tend to stay, and the population continues to grow, if only slowly. The Maltese enjoy an extensive welfare state with free education and health care.

Not many work in agriculture, which is restricted by thin soil and the lack of water, so although farmers do have crops of cereals, fruits, and vegetables, most food has to be imported.

One-fifth of the workforce are in manufacturing industry, either in small-scale factories making goods for local consumption or in the export factories established as a result of foreign investment—typically in clothing and light engineering, and a large SGS-Thomson electronics factory. However the more labour-intensive industries are declining in importance.

Until 1979, Malta had a major

Land area: 316 sq. km.
Population: 0.4 million—urban 92%
Capital city: Valletta, 9,100
People: Maltese
Language: Maltese, English
Religion: Roman Catholic
Government: Republic
Life expectancy: 78 years
GDP per capita: $PPP 17,633
Currency: Maltese lira
Major exports: Clothing, electronics

British naval base and still has a ship-repair industry. More important to the economy now is tourism, which accounts for 17% of GDP. Around 40% of the 1.2 million visitors are from the UK. Many others arrive from elsewhere on cruise liners.

Maltese politics is frequently confrontational and animated, and as a result electoral turnouts above 90% are common. For much of the 1970s and 1980s, Malta was governed by the then socialist Maltese Labour Party (MLP) led by Dom Mintoff, who built up a welfare state. Even today, a high proportion of the workforce are employed in the public sector.

Since 1987, however, with a brief 1996–98 Labour interruption, the country has been in the hands of the centre-right Nationalist Party (PN) which has been trying to privatize and to cut down the public sector. The PN was returned to power in the 2003 general election, led by Eddie Fenech Adami, though he was in 2004 appointed president and replaced as prime minister by Lawrence Gonzi.

Ideologically, the two parties have moved closer together. The most divisive issue has been EU membership; the PN was pro and the MLP against. The issue was finally settled in favour of the EU in 2003 and Malta joined in May 2004.

Martinique

A French dependency with hopes for independence

Martinique is a mountainous island in the eastern Caribbean with an active volcano, Mont Pelée, though it does have some flatter parts in the south-west. Abundant rainfall for most of the year sustains lush vegetation: more than one-third of the island is covered with tropical rainforests. There are also many sandy beaches.

Land area: 1,000 sq. km.
Population: 0.4 million—urban 79%
Capital city: Fort-de-France, 100,000
People: African and African-white-Indian mixture 90%, white 5%, other 5%
Language: French, creole
Religion: Roman Catholic 85%, other 15%
Government: Overseas department of France
Life expectancy: 79 years
GDP per capita: $PPP 14,400
Currency: Euro
Major exports: Petroleum products, bananas, rum

The people of Martinique are a rich racial mixture of black, white, and Indian. Thanks to steady flows of French aid, their standard of living is fairly high. The capital, Fort-de-France, is a smart cosmopolitan city, combining French and West Indian cultures, though its attractions have also encouraged many younger people to emigrate onwards to metropolitan France. The descendants of the white settlers, the Béké, still own plantations but most political and economic power nowadays is in the hands of the creole élite.

Historically, Martinique was a major sugar producer and there are still sugar plantations, which supply the raw material for the rum distilleries, but farmers have been diversifying more into tropical fruits including pineapples, avocados, and bananas. Agriculture only contributes around 5% of GDP and most food has to be imported.

Today, the main industry is tourism. More than three-quarters of the workforce are employed in services, working in hotels and restaurants, as well as in government.

There is also a refinery which produces oil products for local consumption and for other French dependencies in the Caribbean.

As an overseas department of France, Martinique sends two representatives to the French Senate and four to the French National Assembly. For local consultation, it also has a general council and a regional council.

Martinique has typically voted for left-wing parties and since 1992 the president of the general council has been Claude Lise of the Progressive Martinique Party. Of the French dependencies, Martinique has had the most persistent agitation for independence. While some political groups just want greater autonomy, and to attract more investment, others have been elected on pro-independence platforms.

Given the high levels of unemployment, however, and the dependence of many people on welfare payments and other transfers from France, in the short term a vote for independence seems unlikely.

Mauritania

Nomadic communities have settled in the towns, but many traditional practices remain—including slavery

Land area: 1,026,000 sq. km.	
Population: 3 million—urban 62%	
Capital city: Nouakchott, 611,000	
People: Mixed Maur/black 40%, Maur 30%, black 30%	
Language: Hasaniya Arabic, Pular, Soninke, Wolof	
Religion: Muslim	
Government: Military transitional	
Life expectancy: 53 years	
GDP per capita: $PPP 1,766	
Currency: Ouguiya	
Major exports: Fish, iron ore	

Mauritania is essentially the western section of the Sahara Desert. It does have a strip of arable land alongside the southern border that is formed by the Sénégal river, and the territory to the north of this consists of dry grasslands with occasional bushes, but more than half the country is covered with sand dunes.

The population represents an overlap between ethnic groups from the north and south. Those from the north are lighter coloured and of mixed Arab and Berber descent. Those from the south are black, from a number of groups, including the Fulani, Soninke, and Wolof. But there has also been extensive intermarriage, so most of the population is a mixture of the two groups.

Mauritanian society has traditionally been very hierarchical—with noble families at the top and then a series of different castes extending down to servants and slaves. Slavery was abolished during the French colonial period, and the prohibition was reiterated with new laws in 1980. However, Amnesty International reports that slavery and serfdom continue—even if owners can no longer call on the law to pursue runaway slaves. The slaves may belong to any ethnic group, but particularly the blacks.

Living standards are very low. Around one-third of children are malnourished and infant mortality is high—especially for those outside the towns and cities who have few facilities for health care. Half the population live in poverty.

In the past, most Mauritanians lived in nomadic herding communities, and livestock still contributes around one-sixth of GNP—one million each of cattle and camels in the south and 14 million goats and sheep further north. But the numbers of nomadic herders have been falling. To some extent this is a response to desertification—a result in part of over-grazing. A series of droughts also wiped out herds, driving destitute families to the cities. Moreover, the urban life in general is attractive to younger generations.

Mauritania's little available agricultural land, around 1% of the territory, is used mostly to grow millet and sorghum. Some rice is produced on irrigated land—usually by agricultural labourers working for richer landlords. Nevertheless, most cereals have to be imported. Many rural communities also maintain herds of cattle, camels, and goats.

The Atlantic coastline opens up opportunities for fishing—which provides around half of export revenue. This is one of the richest fishing areas in the world but over-fishing by local and foreign boats, the 'supertrawlers', has depleted the stock. The government has taken measures to combat over-fishing—including tighter surveillance and new agreements for sharing production with fleets from the EU.

Fishing grounds depleted by foreign trawlers

With limited agricultural potential, Mauritania's economic survival has rested on the exploitation of mineral resources which have generated 12% of GDP and half of export revenue. The most important is iron ore. The Zouérat mine near the border with Western Sahara has rich deposits of iron ore that have been extracted since the early 1960s. Most of the output goes to the EU.

One promising development is the discovery of offshore oil and gas which Australian and British companies expect to start extracting sometime in 2006.

Mauritania's poverty and vulnerability to shocks, whether from the climate or international commodity markets, have frequently obliged it to turn to the IMF. Popular protests against the ensuing austerity measures have usually been dealt with fairly harshly. Mauritania is, however, a major aid recipient and has also qualified for debt relief—which has allowed it to increase social spending.

Since independence in 1960, political life in Mauritania has been controlled by the Arabic speakers from the north who have periodically made efforts to further 'Arabize' the rest of the country. In 1984 the one-party government that had been in power since independence fell to a military coup led by Colonel Maaouya Ould Sid'Ahmed Taya.

Under pressure from both inside and outside the country, Taya brought in a new constitution in 1991 that legalized political parties, including his own Parti républicain démocratique et sociale (PRDS).

Taya won the first presidential election in 1992. The main opposition party charged that this was fraudulent and boycotted the 1997 presidential election, which Taya duly won with 90% of the vote. Taya was no democrat and has frequently harassed, and several times banned, opposition parties. Taya survived a coup attempt by officers from an eastern clan in 2003 and went on to win a further six-year term.

Never very popular, Taya was creating ever greater discontent by suppressing Islamic militants and having friendly relations with Israel. In August 2005 he was overthrown in a bloodless coup by army officers led by Colonel Ely Ould Mohamed Vall. They say they will return the country to democracy but do not appear in any great hurry to do so.

Mauritius

A human development success story with a booming economy

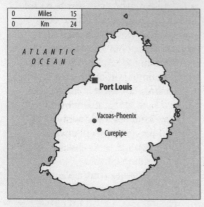

Land area: 2,000 sq. km.
Population: 1.2 million—urban 43%
Capital city: Port Louis, 144,000
People: Indo-Mauritian 68%, creole 27%, Sino-Mauritian 3%, Franco-Mauritian 2%
Language: Creole, French, English, Hindi, Urdu
Religion: Hindu 48%, Roman Catholic 24%, Muslim 17%, other 11%
Government: Republic
Life expectancy: 72 years
GDP per capita: $PPP 11,287
Currency: Mauritian rupee
Major exports: Textiles, garments, sugar

Mauritius, an island in the Indian ocean surrounded by coral reefs, is the site of an ancient volcano. The rainiest and most mountainous part of the island is in the south, falling to a plateau in the centre and then to a drier broad plain in the north.

The Mauritian population contains distinct and diverse groups. Two-thirds are of Indian origin— descendants of around half a million Indians who after 1850 were brought to work as indentured labourers in the sugar plantations—of these around one-quarter are Muslim and the rest Hindu. Then there are the creoles, descendants of slaves—the previous sugar workers. There are also smaller numbers of whites, the European colonizers, as well as Chinese and other Asian immigrants who came later as traders.

While there has been relatively little tension, the communities do remain fairly distinct. This is probably the only country where the end of Ramadan, All Saints' Day, and the Chinese New Year are all official holidays.

Each community uses its own language but three have become dominant: English is the official administrative language; French is the language of the major newspapers; creole the unofficial language used in the streets.

Communal divisions used also to correspond to a division of labour— with the Indians as farmers, creoles as artisans, Chinese as traders, and whites as large landowners. Today, these distinctions have been blurred. This is partly the result of increasing levels of education. But it also reflects a rapid modernization of the economy which has offered new opportunities to all. Indeed, Mauritius is one of Africa's rare success stories.

Standards of human development are high. Education is free to university level and health facilities are good.

Up to a few decades ago the major source of wealth was sugar— capitalizing on preferential access to the European market. Sugar still covers most of the arable land and is responsible for around 4% of GNP, but it now employs only around 7% of the workforce and future prospects are poor as a result of trade liberalization that will reduce preferential access to overseas markets. With little land on which to grow rice, most food has to be imported.

Fortunately, Mauritius has invested much of the earnings from sugar into creating a new manufacturing sector based in export-processing zones. Since the early 1980s, the country has attracted more than 500 companies, employing 13% of the workforce—two-thirds of them women. Around 80% of production is of garments—again taking advantage of preferential access to markets in Europe and North America. Companies were also attracted by Mauritius's stable, healthy, and well-educated society, good infrastructure, and cheap labour. Most investment has come from Hong Kong and from France and other European countries.

500 export-processing companies

However, the removal of trade barriers as a result of globalization will erode the Mauritian advantage. And there has been a steady rise in local wages: direct labour costs are now around 50% higher than in southern Africa, and one-third higher than in South-East Asia. Employment in garments has already started to fall. This will demand a move to higher-value products.

With this in mind the government wants to reinvent Mauritius as a 'cyber island' and has built an informatics park that can provide call-centre and other services to foreign companies.

The third main pillar of the economy is tourism, which accounts for around 8% of GDP and, directly and indirectly, employs more than 7% of the workforce. More than 700,000 visitors arrive each year, with France as the leading source. Most head for the beaches, but there is also eco-tourism. Mauritius sees itself as an 'exclusive' destination.

Surprisingly perhaps for a country with a booming private sector, politics in Mauritius has been dominated by political parties that are avowedly socialist. This has also involved informal power-sharing between Hindus, Muslims, and whites, though excluding the creoles.

In 1991 the country, which until then had the British monarch as head of state, introduced a new constitution and became a republic.

The 1995 election for the National Assembly was won by a coalition of the Labour Party (LP), led by Navin Rangoolam who became prime minister, and the Mouvement militant mauricien (MMM), though the MMM withdrew in 1997 and went into opposition.

The election in 2000, however, was won by the MMM, led by Paul Bérenger, and the Mouvement socialiste militant (MSM), led by Sir Anerood Jugnauth. Under a power-sharing agreement Sir Anerood initially became prime minister and was replaced by Bérenger in 2003. Sir Anerood then took over as president.

In the 2005 general elections, however, arguments within the MMM–MSM resulted in a defeat at the hands of the Alliance sociale coalition headed by the Labour Party whose leader, Navin Rangoolam, returned as prime minister.

Mayotte

Mayotte is a part of France and anxious to remain so

Mayotte is a volcanic island in the Indian Ocean between Madagascar and Mozambique. Its capital and port, Dzaoudzi, is not on the main island, but on a rock linked by a causeway to the nearby islet of Pamandzi.

The people, the Mahorais, are of Malagasy origin and the vast majority are Muslim. In addition there are minorities of Indians, Creoles, and Madagascans. There are also some 'M'Zoungous'—the local name for people who have come from metropolitan France. The official language is French though most people use Shimaore which is based on Swahili.

Here, French law and Muslim law work in tandem—the latter allowing polygamy, for example. Many young children also attend Koranic school before going to a French primary school later in the day. The population is very young (60% of people are under 20 years old) and continues to grow rapidly, a result both of natural increase and immigration from neighbouring Comoros.

Most people are farmers, and grow some food crops as well as cash crops such as vanilla, cloves, and trees for the perfume extract ylang-ylang. Their island has few natural resources, but their link with France entitles them to free education and health services, a guaranteed monthly minimum wage

Land area: 375 sq. km.
Population: 0.2 million—urban 41%
Capital city: Dzaoudzi, 15,000
People: Mahorais
Language: Shimaore, French
Religion: Muslim
Government: Territory of France
Life expectancy: 62 years
GDP per capita: $PPP 2,600
Currency: Euro
Major exports: Ylang-ylang, vanilla, copra

of $400, and French citizenship.

Mayotte was the fourth island of the Comoros when they were a French overseas territory. The split occurred in 1975 when the Comoros as a whole unilaterally declared its independence—despite protests from Mayotte. The Comoros continues to claim Mayotte.

France would be happy to relinquish an expensive colonial memento but will only do so if the local people want this. In 1976, a referendum in Mayotte produced a 99% vote in favour of retaining the link with France.

In a further referendum in 2000, 70% of people voted for greater autonomy while keeping their French status. France reluctantly gave them the status of 'departmental collectivity'—a compromise between a full department of France and an overseas territory. In 2003 the status was changed to 'overseas collectivity'. The Mahorais have a 19-member general council usually made up of four or five parties. The current president is Saïd Omar Ouali. They also elect members to the French National Assembly and to the Senate.

The Mahorais are unlikely to sever the link with France in the near future, especially given the general chaos and poverty in the Comoros.

Mexico

Mexico is no longer a one-party state—elections are now fiercely contested

Land area: 1,958,000 sq. km.
Population: 104 million—urban 76%
Capital city: Mexico City, 18.0 million
People: Mestizo 60%, Amerindian 30%, other 10%
Language: Spanish and some Indian
Religion: Roman Catholic 89%, other 11%
Government: Republic
Life expectancy: 75 years
GDP per capita: $PPP 9,168
Currency: Peso
Major exports: Cars, oil

Some have suggested that Mexico's most important geographical feature is the 2,000-mile border it shares with the USA. But Mexico has a striking diversity all of its own—from the snow-capped mountains of the Sierra Madre, to the deserts of Sonora, to the tropical rainforests of Yucatán. Most of the population is to be found, however, in the central region in the high plateau and the surrounding mountains.

Today's Mexicans are largely mestizo, a mixture of Indian and European ancestry. But about one-third are of purer Indian origin, in more than 60 ethnic groups concentrated in the south. They also tend to be the poorest. Overall, 10% of the population live below the national poverty line.

Education levels have improved in the past ten years and only 10% of adults are now illiterate. Even so,

one-third of adults have not completed primary education. Standards of health are also better, with a substantial reduction in infant mortality, though rates vary markedly between states.

Population growth and rural–urban migration have contributed to an explosive growth of cities, particularly of Mexico City, which with a population of 18 million is the second largest city in the world. It also has some of the most polluted air and is suffering from rising levels of crime.

Apart from migrating to cities, Mexicans have also been moving to the USA. By 2002, 9.5 million Mexicans were living in the USA, equivalent to 9% of Mexico's population, and sending $13 billion annually in remittances. Around 400,000 people emigrate annually.

A major stimulus for emigration has been the poor state of Mexican agriculture. Agriculture only contributes 4% of GDP but is responsible for 16% of employment. Only about one-fifth of the country is suitable for arable farming, and even then the soil is often thin and levels of technology are low. Over half of this land consists of 'ejidos', small collective farms generally growing maize and beans. From the mid-1980s, the government liberalized agriculture—reducing subsidies and allowing the ejidos to be sold and consolidated. But this has brought few benefits to small farmers who find it difficult to get bank loans. Following the 1994 North American Free-Trade Agreement (NAFTA), Mexican maize farmers also find it very hard to compete with imports from the USA. Since 1994, corn prices have fallen by 70% and one-third of corn is imported. The brightest agricultural development has been in the north, where irrigation has permitted a flourishing export trade in fruit and vegetables to the USA.

Aside from the land, one of Mexico's greatest natural assets is oil. Mexico is the world's fifth largest producer and the third largest source of crude oil to the USA. At present rates of extraction, reserves should last another 14 years. Another valuable asset is silver: Mexico has the world's largest silver production.

Most industrial employment is in manufacturing which since 1994 has been transformed by NAFTA. Many small companies collapsed faced with US competition while others gained. One of the most dynamic parts of the economy, which preceded NAFTA, has been the 'maquiladora', the 3,300 duty-free assembly plants strung along the border with the USA which employ one million people—10% of formal sector jobs. However, maquiladora output has fallen, particularly in garments when faced with lower wage competition from China and elsewhere.

3,300 assembly plants on the US border

Presiding over Mexico's development for 71 years was the Partido Revolucionario Institucional (PRI), which was in power continuously between 1929 and 2000. This one-party electoral dictatorship at times lapsed into brutal episodes of repression. But the monolith started to crumble in the 1990s. This was partly because in a liberalized economy the state had fewer opportunities for lavish patronage, but also because of more vigorous opposition, most dramatically from a rebel group, the Ejército Zapatista de Liberación Nacional, which in 1994 led an uprising in the poor southern state of Chiapas. More disruption comes from the drug mafias that control routes from South America into the USA.

The PRI was finally defeated in the July 2000 presidential election. The victor was Vicente Fox of the right-wing Partido Acción Nacional (PAN) whose strongest support is in the north and central states and among the middle classes. Fox, a former head of Coca-Cola in Mexico with limited political experience, had trouble getting support for economic reforms even within his own party, so ulimately had little impact.

The presidential election of 2006 was a closely fought contest between left and right. On the left was Andrés Manuel López Obrador, a populist former mayor of Mexico City, representing a coalition led by the centre-left Partido de la Revolución Democrática (PRD). On the right was the PAN candidate, Felipe Calderón —who won by 100,000 votes out of 42 million. The PRI candidate came a distant third. Calderón may soon be able to benefit from a PAN majority in congress but will probably form a coalition with the PRI.

Micronesia

Heavily dependent on US aid, Micronesia will need other sources of income

Land area: 1,000 sq. km.
Population: 0.1 million—urban 29%
Capital city: Palikir
People: Micronesian and Polynesian groups
Language: English, Chukese, Pohnpeian, Yapese, Kosraean
Religion: Roman Catholic 50%, Protestant 47%, other 3%
Government: Republic in free association with the USA
Life expectancy: 68 years
GDP per capita: $PPP 3,900
Currency: US dollar
Major exports: Copra, fish, pepper

The Federated States of Micronesia are Yap, Chuuk, Pohnpei, and Kosrae, and consist of more than 600 islands and coral atolls in the Pacific.

The people of the federation, though few in number, include a great variety of cultures and languages. Chuuk is the largest state, with around half the population, followed by Pohnpei.

Most productive activity centres on subsistence farming and fishing. The main crops include breadfruit, taro, and coconuts, and there is a small export income from copra, black pepper, and handicrafts. The most important fishing catch is tuna though much of this goes to foreign fleets.

The islanders' primary source of cash income, however, is the government which is the main employer, providing 40% of total employment, and its expenditure accounts for over 40% of GDP.

Micronesia relies on transfers of US aid through a 'Compact of Free Association'. Over the period 1986–2001 the country received $1.3 billion in US grants and in 2003 aid was worth $923 per person. This assistance is now being phased out. From 2004 the US government revised the compact and for the first three years of the subsequent 20-year period it will give $76 million in economic assistance grants, to be allocated among education, health, and other sectors. Another $16 million annually will go to a trust fund.

Tourism offers some potential, but so far has been confined to niche markets such as eco-tourism, visits to some ancient ruins, and shipwreck diving. However there are concerns about the impact of increased arrivals on the fragile environment.

There has also been some investment in manufacturing, notably in garment production, but expansion is hampered by poor infrastructure and the distance from markets.

Until 1986, the states were part of a UN Trust Territory. Then they became an independent federation, self-governing but in free association with the USA. Each of the four states has its own elected governor and legislative assembly. There are no political parties. In addition, there is an elected federal congress, which chooses the president. In 2003, it selected Joseph J. Urusemal.

In many respects Micronesia functions more as four separate states. There are however signs of greater co-operation on common issues such as the environment and tourism.

Moldova

Moldova has effectively been split in two by a long fight for secession

Land area: 34,000 sq. km.
Population: 4 million—urban 46%
Capital city: Chisinau, 717,000
People: Moldovan 78%, Ukrainian 8%, Russian 6%, Gagauz 4%, other 4%
Language: Moldovan, Russian, Gagauz
Religion: Eastern Orthodox
Government: Republic
Life expectancy: 68 years
GDP per capita: $PPP 1,510
Currency: Moldovan leu
Major exports: Food products, beverages, tobacco

Most of Moldova lies between two rivers that flow into the Black Sea: the Prut, which forms the western border with Romania; and the Dniester which flows roughly in parallel with the eastern border with Ukraine. The land is largely a hilly plain, though also cut through with many steep ravines. The soil—the rich, black 'chernozem'—is very fertile, and two-thirds of the country is forested.

Moldova's difficulties in building a new nation since 1991 reflect its fairly recent assembly into one territory. In 1940, the Soviet Union united the land from the Prut to the Dniester, Bessarabiya, which had been part of Romania, with the strip of land from the Dniester to the eastern border, Transdniestria. The two were merged as the Moldavian Soviet Socialist Republic.

After the Second World War, the Soviet Union made strong efforts to weaken the republic's Romanian past. This included changing the name of the language from Romanian to Moldovan, and switching from the Roman to the Cyrillic alphabet, while encouraging immigration from Russia and Ukraine. The break-up of the Soviet Union then provoked a crisis of national identity.

During the late 1980s, two separatist movements had appeared. And when the Moldavian Republic declared its independence in 1991, as 'Moldova', both groups feared that independence would lead to unification with Romania and both declared themselves independent. The smaller group were the Gagauz in the south-east who subsequently laid down their arms in exchange for a measure of autonomy.

The larger and more debilitating rebellion came from the Russians in Transdniestria, and in 1992 this escalated into a civil war, in which Russia backed the separatists while Romania backed Moldova. A peace treaty was signed in 1992 and has been enforced with Russian troops.

This has left the country effectively partitioned in two. Transdniestria, which has 11% of the territory, and 17% of the people, has its own government in Tiraspol headed by Igor Smirnov, who was returned as president for the third time in 2001. He is an unreconstructed communist whose heavy-handed and repressive rule is attempting to sustain the Soviet economic model.

Transdniestria has its own government

Moldova has the highest population density of the countries of the former Soviet Union, and it would be higher still if around 10% of the population did not work abroad, chiefly in Russia and Italy. It is also one of the poorest successor states and has the worst health standards.

Economically Moldova has suffered since a high proportion of heavy industrial activity is in Transdniestria. With what remains Moldova struggles with market reforms. In the 1990s the government privatized many industries by distributing vouchers and also made some cash sales. Most manufacturing consist of food processing and beverages, including a wine industry most of whose output goes to former communist countries.

With limited immediate prospects for industrial development, Moldova remains highly dependent on agriculture which accounts for around one-fifth of GDP and 40% of employment, and is the main source of export income. Following the break-up of collective farms, most production now comes from private farmers or household plots. The most important crops are cereals, sugar beet, tobacco, grapes, and other fruit. Although the land is fertile it has suffered from over-intensive cultivation and the heavy use of fertilizers, and production is vulnerable to the weather: harvests are regularly hit by droughts and floods.

In 1994, Moldovans voted for a new constitution to maintain the country's current borders but grant extensive autonomy to Gagauz and Transdniestria.

The 1996 presidential election was won by Petru Lucinschi, a leading official from the Soviet era. And the 1998 parliamentary elections made the Communist Party of Moldova (PCRM) the largest party, with 30% of the vote, but unable to form a government. This led to several years of instability and short-lived administrations.

The situation was only resolved after the 2001 legislative election gave the PCRM 70% of the seats. Meanwhile the constitution had been amended to allow the parliament to elect the president and they chose a former Soviet-era politician, Vladimir Voronin.

The PCRM government has not reversed market reforms but it has taken other controversial steps such as boosting the Russian language and playing down links with Romania. Nevertheless in 2005 the PCRM again won parliamentary elections and with the co-operation of the opposition re-elected Voronin for a second presidential term.

Moldova's future seems secure. The prospect of unification with Romania has receded and the leaders of Transdniestria accept that the best they are likely to achieve is a confederation of equal states. The chances of peace brightened in 2005 when Ukraine put forward a peace plan backed by the EU.

Mongolia

Long a Soviet satellite state, Mongolia is struggling to create a market economy

RUSSIA

Erdenet ● Darhan ● ● Choybalsan

Ulaanbaatar ■

CHINA

| 0 | Miles | 300 |
| 0 | Km | 480 |

Land area: 1,567,000 sq. km.
Population: 3 million—urban 57%
Capital city: Ulaanbaatar, 928,000
People: Mongol 95%, Kazakh 4%, other 1%
Language: Khalkha Mongol 90%, Russian or Turkic 10%
Religion: Tibetan Buddhist
Government: Republic
Life expectancy: 64 years
GDP per capita: $PPP 1,850
Currency: Togrog
Major exports: Fuels, minerals, metals, garments

Mongolia's territory, which is around the size of Western Europe, consists largely of a vast plateau between 900 and 1,500 metres above sea level. It has a harsh semi-arid climate, with bitterly cold winters, but has extensive pasture land. The greatest mountain ranges are to the north and west, while the land to the east and south stretches out into rolling steppes and the Gobi Desert.

Most Mongolians now live in cities—one-third in the capital, Ulaanbaatar. In ethnic terms the country is fairly homogenous, the largest minority being Kazakhs in the south. In the 1930s, a Stalinist government purged much of the country's Tibetan Buddhist heritage, though with a democratic government and freedom of religion Buddhism is now staging a revival.

In 1991, following seven decades of socialism and the sudden withdrawal of Soviet aid, the government embarked on a course of 'shock therapy', shutting some industries, privatizing others, and making severe cuts in social services. Despite more rapid economic growth in recent years, more than one-third of the population are still poor and one-fifth of the workforce are unemployed or underemployed, with the most severe problems for urban youth.

Education services also came under pressure and the government reduced the number of dormitory places that are essential for herder children, though in recent years it has increased expenditure on education. One of the most disturbing developments has been the emergence of thousands of street children in Ulaanbaatar. Because winter temperatures can drop to 30 degrees below zero the children have to live in the sewers.

Around 45% of the workforce work in agriculture. Nomadic Mongol tribes, with their circular felt tents or 'gers', and their huge herds of sheep, goats, horses, and yaks, offer a distinctive picture of Mongolian life. Their herds —29 million animals— had been collectivized during the communist period and then privatized again after 1990. The land remains in state ownership. Privatization revived production but it has also increased inequality and larger herds are harming the pastures—two-thirds of which are now degraded.

Mongolia's small farms can only produce around one-third of the country's cereal needs so the rest has to be imported.

Mongolia also has considerable mineral wealth—with rich deposits of coal, iron ore, copper, molybdenum, fluorspar, and tungsten. Fuels, minerals, and metals are responsible for more than two-thirds of exports. Revenues have fallen with the decline in world prices, but mineral extraction is one of the country's main priorities. Efforts have also been made to increase manufacturing production, and the next largest export sector is garments and textiles, including cashmere from goat hair.

Mongolia's main trade disadvantage is its isolation. Getting goods out to a seaport overland involves a three-day railway journey

Landlocked Mongolia's shipping fleet

through Russia or China. Surprisingly, however, Mongolia is now getting extra income through a shipping register managed by an agency in Singapore; the 'national' fleet now has 300 vessels.

Mongolia's economy shrank rapidly following the withdrawal of Soviet aid, which had been equivalent to 30% of GDP. It also lost supplies of cheap oil. Since then, the economy has recovered and for the past six years has seen steady growth.

Mongolia became the world's second communist state in 1924, ruled by the Mongolian People's Revolutionary Party (MPRP). For 70 years it was bankrolled by the Soviet Union as a buffer state against China. When communism collapsed in the Soviet Union it duly collapsed in Mongolia too, and in 1990 and 1992 Mongolia's constitution was amended to allow other parties to compete for seats in a single-chamber legislature, the Great Hural. In addition there was to be a directly elected president.

The MPRP, which by then was espousing economic liberalism, won the first two general elections. In 1996, however, it was defeated in the Great Hural election by the centre-right Democratic Union, consisting of the Mongolian National Democratic Party and the Mongolian Social Democratic Party. But matters were complicated later that year when an MPRP candidate, Natsagiin Bagabandi, was elected president.

In 1998 there was a political crisis when the MPRP started to boycott the Great Hural. Things seemed to settle down subsequently with a new administration. But rising poverty made the government increasingly unpopular. In the July 2000 parliamentary elections the MPRP was returned to power with a substantial majority, and Bagabandi was returned as president in May 2001.

At that point the prime minister was Namburiin Enkhbayar, a modernizer who dubbed himself the 'Tony Blair of the steppes'. The MPRP also gained the most seats in the 2004 elections to the Great Hural, though not an overall majority.

In May 2005, Enkhbayar was elected president, and in January 2006 the MPRP chose Miyeegombo Enkhbold as prime minister.

Montenegro

The newest member of the United Nations

Montenegro has a narrow coastal strip along the Adriatic, behind which rise the spectacular Dynaric Alps. The rest of the country consists of a rugged mountainous massif cut through by deep valleys and gorges.

Around half the population are ethnic Montenegrin (the country had been independent between the mid-1800s and 1918). There is also a substantial proportion of Serbs, particularly in areas bordering on Serbia. Although standards of health and education dipped following the break-up of former Yugoslavia they have now recovered somewhat.

Although not a very poor country, average salaries here are low, at only around $300, and 18% of the workforce is unemployed.

Only around 3% of workers work in agriculture, and because of the mountainous terrain they have been able to cultivate less than 10% of the country; most settled agriculture is largely in the Zeta valley and around Lake Scutari.

Around 30% work in industry of which the most important is aluminium. The country's largest enterprise is the aluminium smelter KAP, which has been bought by the Russian company Rusal.

Nowadays most workers, more than 60%, are employed in services. Many of these are linked to the tourist industry which takes advantage of

Land area: 14,026 sq. km.
Population: 0.6 million—urban 41%
Capital city: Podgorica, 118,000
People: Montenegrin 43%, Serbian 32%, Bosniak 8%, Albanian 5%, other 12%
Language: Serbian
Religion: Orthodox Christian, Muslim
Government: Republic
Life expectancy: 73 years
GDP per capita: $PPP 3,800
Currency: Euro
Major exports: Aluminium

beautiful beaches on the Adriatic coast as well as the mountains and lakes. However the industry will need considerable investment to bring it up to international standards.

Montenegro is the latest independent fragment of the disintegration of Yugoslavia. From 1992 the country had been yoked with republic Serbia as the rump Federal Republic of Yugoslavia. But this was always an uneasy alliance and its survival was cast into doubt from 1997 with the election of pro-western Milo Djukanovic, of the Democratic Party of Socialists, as president of Montenegro. After years of friction the two created a loose union in 2002 called Serbia and Montenegro.

However, Djukanovic had reserved the right to hold a referendum on independence. This was duly held in May 2006 when 55.5% of Montenegrins voted to break free, narrowly exceeding the 55% threshold that the EU had stipulated as necessary for the new state to be recognized.

Djukanovic who had been re-elected in 2003 seems likely to win again in autumn 2006, if only because the main function of the opposition parties had been to maintain the union with Serbia.

Morocco

Morocco now has a more liberal monarch but continues its illegal occupation of Western Sahara

Land area: 447,000 sq. km.
Population: 31 million—urban 57%
Capital city: Rabat, 668,000
People: Arab-Berber 99%, other 1%
Language: Arabic, Amazigh, French, Spanish
Religion: Muslim
Government: Constitutional monarchy
Life expectancy: 70 years
GDP per capita: $PPP 4,004
Currency: Dirham
Major exports: Phosphate rock, phosphoric acid, textiles

Morocco is the most mountainous country of North Africa. The two main chains are the Er Rif, along the Mediterranean coast, and the Atlas Mountains which dominate the country from north-east to south-west. Most people live in the lowlands that lie between these two chains and the Atlantic coast. Morocco also controls the desert area of Western Sahara to the south.

Morocco's people are largely Arabized Berbers, but around one-third are less-assimilated Berbers, most of whom live in the mountains and whose language, Amazigh, was given official recognition in 1994.

Morocco remains one of the poorest Arab countries. More than half the population are illiterate and more than 10% of children do not enrol in primary school. In the rural areas only half the population have access

to safe water. Even in urban areas there are high levels of poverty, and unemployment in 2004 was 11%. As a result, many Moroccans have chosen to emigrate clandestinely, making the perilous 15-kilometre sea-crossing to Spain. Around 1.7 million Moroccans now live overseas, chiefly in France and Spain; their remittances in 2004 were $4.2 billion—the second largest source of foreign exchange.

Agriculture employs about half the labour force. Most of the land is in the hands of smaller farmers growing cereals, potatoes, and other staples. Since they lack irrigation they have to rely on fairly erratic rainfall. The tenth of the arable land that is irrigated is mostly in larger farms which grow citrus fruits, grapes, and other export crops. Meanwhile, especially during drought years, much of the country's cereal needs have to be imported.

One of Morocco's main priorities is to improve the irrigation network—taking water from the areas of good rainfall to those regularly hit by drought. A new canal was opened in 1999 to take water from Guerdane in the east to the south of the country to irrigate citrus crops.

Many of the smaller farmers in the mountains also raise cattle, sheep, and goats. Morocco's fishing industry is a major supplier of sardines to the EU, though catches have fallen due to over-fishing.

Morocco's main industrial enterprises are state-owned and linked to phosphates. With Western Sahara, Morocco has three-quarters of the world's reserves, and is the third largest producer. The other major export industry is textiles, which expanded rapidly in the 1980s, though is now facing intense foreign competition. Many other smaller enterprises produce high-quality leather goods, rugs, and carpets.

Of the service industries, one of the most important is tourism. Around three million visitors arrive each year and provide employment to 6% of the labour force. Earnings have been falling, however, and many of Morocco's tourist facilities are in need of a facelift. The government intends to upgrade the resorts and develop new ones—five on the Atlantic coast and one on the Mediterranean.

Morocco is a constitutional monarchy. There is an elected legislative assembly. But it only has limited powers. The king acts like an executive president, appointing both the government and prime minister and presiding over the cabinet. From 1961 Morocco was ruled by King Hassan II—an autocratic ruler whose determination to retain tight control resulted in thousands of arbitrary arrests and disappearances.

Morocco's king really rules

On his death Hassan was succeeded by his son who became King Mohammad VI. King Mohammad promised to be somewhat different and did make some important changes. He released political prisoners, allowed exiles to return, and dismantled repressive security forces. He also supported women's rights by rewriting the 'moudawana', the personal and family law, to give women equal rights in marriage. But although he also says he will modernize the monarchy he is in no hurry to restrict his own powers.

Members of parliament can criticize government policy, but they cannot question the role of the monarchy. The 2002 elections to the House of Representatives, the country's first really free ballot, distributed the 325 seats among 22 parties. Of these, 94 went to socialist parties, 40 to the liberal right, and 48 to nationalists. But the most striking gains were for the moderate Islamist Parti de la justice et du développement. The cabinet, with Driss Jettou as prime minister, included representatives of the six largest parties.

Of more concern to the king are the radical groups such as Al Sunna wal Jamaa which try to enforce 'Islamic' behaviour, often with violence, and opposed the new moudawana. After 9/11, Morocco detained hundreds of their members. But Morocco too has experienced violence, including suicide bombings in Casablanca in 2003 that killed 44 people.

Another major issue is Western Sahara which Morocco seized in 1975—despite armed resistance and international protests. The UN has proposed a referendum on independence but Morocco has tried to bias this by encouraging settlement by Moroccans.

Mozambique

**Devastating floods set back
Mozambique's efforts to
capitalize on a period of peace
and stability**

Land area: 802,000 sq. km.
Population: 19 million—urban 36%
Capital city: Maputo, 1.7 million
People: Makua-Lomwe, Tsonga, Yao, Sena-Nyanja
Language: Portuguese, indigenous dialects
Religion: Catholic 24%, Muslim 18%, Zionist Christian 18%, other 18%, none 32%
Government: Republic
Life expectancy: 42 years
GDP per capita: $PPP 1,117
Currency: Metical
Major exports: Prawns, cotton, cashew nuts

Mozambique can be divided by
the Zambezi River into two main
geographical regions. The southern
part of the country is mostly flat
and low-lying, apart from highlands
on the eastern border. North of the
river the land rises to plateaux and
to mountains along the border with
Malawi.

The Zambezi also divides the
country's many different ethnic
groups. Most of the population is
concentrated in the better land north
of the river, particularly in the north-
east. Groups here include the Makua,
who are farmers, and the Muslim Yao.
South of the river, where people live
more along the coast, are groups such
as the Tsonga. None of their languages
is widely spoken around the country
so the official language is Portuguese.

Mozambique's civil war, which
ended in 1992, devastated much of

the country's social infrastructure,
destroying schools, hospitals, and
clinics—in what was already one of
the world's poorest countries. Since
then, the government has managed to
rebuild most of the schools. School
attendance has been rising and 55% of
children are now enrolled in primary
schools. Even so, adult literacy is still
only 47%. The health system also
needs more investment. The HIV/
AIDS infection rate in 2004 was 16%.

Most people are very poor.
Although the capital, Maputo, is
now a lot smarter, around 70% of the
population languish below the national
poverty line. Poverty is severe in
the rural areas and particularly in
the north. Those who have had
few employment opportunities at
home have traditionally migrated to
neighbouring South Africa which
currently has around one million
immigrant Mozambicans.

Around 80% of the workforce
make their living from agriculture.
Given good weather, Mozambique
should be self-sufficient in basic
foodstuffs such as maize and cassava.
The main cash crops are cotton and
especially cashew nuts. Production
of both food and cash crops has
increased, but farmers are still

hampered by poor infrastructure; many parts of the country are cut off during the rainy season. All land remains state-owned; farmers, even local ones, can only lease it.

Mozambique also exports prawns, along with some shellfish. The fishing industry was less affected than others by the war. Even so, the local fleet is relatively small and the government gains extra income by selling licences to boats from Europe, South Africa, and Japan.

Mozambique's industrial sector was badly damaged by the fighting. Since the mid-1990s, however, and following an extensive programme of privatization along with foreign investment, industrial output has revived. One of the country's largest enterprises is the Cahora Bassa hydroelectric dam on the Zambezi which exports electricity to South Africa and also supplies local demand, including the new $1.3 billion Mozal aluminium smelter, a joint venture led by BHP-Billeton, which has turned Mozambique into one of the world's largest aluminium producers. There are also plans to export natural gas to South Africa.

Much of this activity, however, has

Development is widening the north–south divide

been in the southern half of the country, which also tends to be economically more integrated with neighbouring countries, thus sharpening the north–south divide.

Mozambique remains heavily dependent on aid which accounts for around 13% of GNP—mostly in the form of grants. Mozambique, as one of the world's poorest countries, has also benefited from debt relief under

the Heavily Indebted Poor Countries initiative and debt servicing is now more manageable.

Since independence in 1975 the government has been in the hands of the Frente de Libertação de Moçambique (Frelimo). Frelimo's Marxist policies, which included extensive nationalization and state control, alarmed the then white governments of Rhodesia and South Africa, who armed and funded a guerrilla group, Resistência National de Moçambique (Renamo). Renamo rapidly distinguished itself as one of the world's most vicious guerrilla armies and Mozambique sank into a civil war that killed around 900,000 people.

The war ended with a stalemate and a peace accord in 1992, followed by very effective UN-organized programmes which demobilized the soldiers and repatriated more than one million refugees. In the elections in 1994, Frelimo, by then headed by Joaquím Chissano, won a surprisingly narrow victory over Renamo.

By the mid-1980s Frelimo had renounced Marxism and continued economic liberalization. In 1999, Chissano won another narrow victory in the presidential election and Frelimo gained a small majority in the legislative assembly.

After legislative and presidential elections in 2004, and despite widespread accusations of fraud, Frelimo were again returned to power, but with Armando Guebuza as president. Guebuza is less popular with the public than Chissano. He also has a less conciliatory style and takes an aggressive approach to the opposition and particularly Renamo.

Namibia

Namibia is one of the world's most unequal societies, and now faces an AIDS crisis

Land area: 824,000 sq. km.
Population: 2 million—urban 32%
Capital city: Windhoek, 234,000
People: Ovambo 50%, Kavangos 9%, Herero 7%, Damara 7%, mixed 7%, white 6%, Nama 5%, other 9%
Language: English, Afrikaans, German, Oshivambo, Herero, Nama
Religion: Christian 80%, indigenous religions 20%
Government: Republic
Life expectancy: 48 years
GDP per capita: $PPP 6,180
Currency: Namibia dollar
Major exports: Diamonds, uranium, gold

Namibia has three main geographical regions. Its long, distinctive Atlantic coastline consists of the dunes and rocks of the Namib desert. This rises to a broad plateau that includes mountains rising to 2,500 metres. To the east, the plateau descends to the sands of the Kalahari desert. The climate is hot and dry and the rainfall erratic—the only year-round rivers flow along the northern and southern borders.

Namibia is sparsely populated. Around half the population live in the far north. The other area of concentration is in and around the capital, Windhoek. Namibia has a number of different ethnic groups. The largest are the Ovambo, who are also politically dominant. Linguistically, the country is very diverse. Only a small minority speak English, which is the official language; most others use languages of the Ovambo or their own ethnic group, though many also understand Afrikaans or German.

Of the minority groups, the most discontented have been the 100,000 or so Lozi who live in the Caprivi strip, an odd sliver of land in the far north-east. In 1999 a Lozi uprising was brutally repressed by Namibian security forces, though since then the unrest seems to have subsided.

Another fundamental division is between the whites and the rest. Namibia's per capita income is fairly high but it has one of the world's most unequal distributions of income—the richest 10% of households get 65% of the national income. Efforts to overcome this imbalance have included heavy investment in education, which in 2005 took up 23% of the budget.

Health too is a major problem; most facilities are concentrated in the urban areas. Though Namibians suffer from diseases like malaria and tuberculosis, the leading cause of death is now HIV/AIDS—20% of the population are infected. In 2005, some 17,000 people died of the disease.

Agriculture represents only around 5% of GDP, but more than one-quarter of the population rely on farming for at least part of their livelihood. The staple crop is millet, which is grown by subsistence farmers in the north—though output is vulnerable to erratic rainfall and Namibia normally has to import almost half its cereal needs.

Commercial farming, which is mostly the prerogative of white settlers in the centre and south, is devoted primarily to livestock—raising cattle and sheep for export to South Africa and the EU. One of the most contentious political issues is land reform, since whites own around 40% of the land area. So far, the process has been slow and expensive because the constitution requires the government to provide full compensation for any land taken. As a result, the government has started appropriating land.

Whites own around 40% of farmland in Namibia

A larger component of Namibia's GDP comes from mining which contributes around 10% of GDP. Namibia has high-quality diamonds that are extracted by the Namdeb Diamond Corporation, a joint venture between the government and the South African De Beers Centenary company. An increasingly high proportion of these are now coming from offshore fields using sea-bed crawler mining equipment.

Another important mineral is uranium. Namibia's Rössing mine is the world's largest open-pit uranium mine. Namibia also produces gold, silver, and copper and in 2003 opened a new zinc mine and refinery.

Other industrial activity is more limited. Manufacturing has been concentrated in food processing, though the government has attracted other industries to work in an export-processing zone.

After agriculture, the main employer in Namibia is the government. In 2003, 78,000 people worked in the civil service.

Namibia was previously occupied by South Africa. But the South-West Africa People's Organization (SWAPO) fought a long and successful guerrilla war and since independence in 1988 it has retained political power.

SWAPO won the first election for the national assembly in 1989 and its leader Sam Nujoma was also elected president. The main opposition came from the Democratic Turnhalle Alliance—which had formed the government during the South African occupation.

SWAPO consolidated its position in the 1994 National Assembly and presidential elections. But Nujoma's rule grew increasingly autocratic—handing out most of its favours to the Ovambo people, while clamping down on the Lozi. He also changed the constitution to let him run for president a third time in 1999.

Nujoma won that election convincingly, though against a new and more credible opposition party, the Congress of Democrats, whose leader Ben Ulenga is a former SWAPO dissident.

For the 2004 presidential election the SWAPO candidate, who was duly elected, was the party's vice-president, Hifikepunye Pohamba. However since Nujoma remains party president, there are factional disputes within SWAPO. To boost his authority Pohamba may also stand for the party presidency.

Nepal

A poor and remote mountain kingdom afflicted by political turmoil and civil war

Land area: 141,000 sq. km.
Population: 26 million—urban 15%
Capital city: Kathmandu, 672,000
People: Chhettri 16%, Brahman-Hill 13%, Magar 7%, Tharu 7%, other 57%
Language: Nepali and 30 tribal languages
Religion: Hindu 81%, Buddhist 11%, Muslim 4%, Kirant 4%
Government: Constitutional monarchy
Life expectancy: 62 years
GDP per capita: $PPP 1,420
Currency: Nepalese rupee
Major exports: Carpets, garments

Nepal's territory can be divided roughly into three bands descending from north to south. The northernmost band includes the Great Himalayas dominated by Mount Everest at 8,848 metres. The central band has the lower Mahabharat range and a number of major river systems and valleys, including the densely-settled Kathmandu Valley. The band along the southern border with India has forested lower slopes that descend to the fertile Terai plain.

Nepal's people comprise more than 70 ethnic groups that can be considered in two sets. The first and smaller, the Tibeto-Nepalese, are the result of immigration from Tibet. They are found in the bleak high mountain areas as well as in the middle band, and are usually Buddhist. The larger set, of Indo-Aryan ancestry, have immigrated from India and elsewhere, and are largely Hindu and rigidly stratified into higher and lower castes.

In recent years, the Nepalese have become even poorer. Despite economic growth, 42% of the population live below the poverty line. Standards of education and health are very low: only 51% of the population are literate, 70% have no access to safe sanitation, and around half of children are malnourished.

The situation is particularly severe for women. While in most countries women's biological advantage enables them to live longer than men, in Nepal women's lifespan is somewhat shorter. This is the outcome of many kinds of discrimination, especially in health care, which results in high rates of maternal mortality—740 mothers die in childbirth per 100,000 live births.

Survival in Nepal depends primarily on agriculture, which generates about 40% of the country's GDP and involves two-thirds of the workforce—primarily cultivating rice, maize, and wheat, as well as raising livestock. But productivity is low and output is erratic. Cultivation is particularly arduous in the terraced farms of the hilly regions. Land holdings are small, irrigation is difficult, and the situation is being aggravated by soil erosion and deforestation. The position is somewhat better in the land of the Terai plain, which accounts for about half the cultivable land and where there are better prospects for irrigation. But land ownership here is highly concentrated so the benefits are unevenly spread. The government in 2001 announced fresh plans for land reform and there are officially new ceilings on landholdings, but so far nothing has happened.

Most of Nepal's industry is on a small scale for local consumption. One of the main export industries is carpet weaving, but carpet sales in Europe have been affected by accusations of the exploitation of child labour. Some Indian garment manufacturers have also established factories to take advantage of Nepal's quotas. The service sector has been growing too but this is primarily the result of government development expenditure, two-thirds of which is financed by foreign aid.

Another important source of income is tourism. However, the industry has been hard hit by concerns about law and order. In 2005 there were 277,000 arrivals. At the best of times it is difficult to travel around Nepal, and the vast majority of tourists get little further than Kathmandu.

Few tourists get much further than Kathmandu

Kathmandu. The government has been trying to encourage more arrivals, but even with peace it is doubtful that the infrastructure could cope with many more people.

Until 1990, Nepal was an absolute monarchy. Inspired by democratic changes elsewhere, many political groupings organized a series of protests that culminated in a mass march on the royal palace. Eventually, King Birendra gave way and in 1990 he introduced a new constitution based on multiparty democracy.

The first election in 1991 was won by the Nepal Congress Party, led by Girija Prasad Koirala, but this was to be the first in a rapid sequence of different coalition or minority administrations.

After 1996 the political situation was further destabilized by a Maoist insurgency demanding land reform and a republican state. This conflict has cost 13,000 lives and led to abuses of human rights on both sides.

In June 2001 the drama moved to the royal palace where Crown Prince Dipendra shot and killed King Birendra, after which Birendra's brother Gyanendra was crowned king.

In October 2002 Gyanendra dismissed the government but made little progress and in February 2005 declared a state of emergency and assumed absolute power. He proved an inept and arbitrary ruler and as the insurgency spread there seemed a real possibility that Nepal would degenerate into a failed state.

The political parties and the Maoists then formed an alliance against the King. A boycott of an election in February 2006 was followed by street demonstrations.

Eventually the army intervened and called on the political parties to form a new government. Girija Prasad Koirala, by now 84 years old, became prime minister for the fifth time. The new parliament has voted to reduce the monarchy to a ceremonial role. Relations with the rebels, however, remain fragile.

Netherlands

One of Europe's most tolerant, and most prosperous, societies

True to its name, most of the Netherlands is 'low lands': more than one-quarter of the country lies below sea level, protected from flooding by coastal dunes and by specially constructed dykes. For centuries, the Dutch have been reclaiming land from the sea and keeping it drained, using pumps that initially were powered by windmills. But even the higher and drier parts of the country seldom rise much above 60 metres.

This is one of Europe's most densely populated countries 456 per square kilometre; most people are concentrated in the west and centre. This makes land expensive and its use is controlled and regulated more closely than in most other countries.

The Dutch enjoy one of the world's most advanced systems of social welfare. They also have very liberal social attitudes: in 1994 this was the first country to permit euthanasia, and in 1998 the first to

Land area: 37,000 sq. km.
Population: 16 million—urban 66%
Capital city: Amsterdam, 0.7 million; jointly with The Hague, 0.5 million
People: Dutch 83%, other 17%
Language: Dutch
Religion: Roman Catholic 31%, Protestant 21%, Muslim 4%, other 4%, unaffiliated 40%
Government: Constitutional monarchy
Life expectancy: 78 years
GDP per capita: $PPP 29,371
Currency: Euro
Major exports: Manufactured goods, chemicals, food, agricultural products

register homosexual partnerships. The Dutch have also gone a long way towards decriminalizing the use of soft drugs—with marijuana readily available in 'coffee houses'.

Like other European countries, the Netherlands is facing a falling birth rate; its fertility rate is only 1.5 children per woman of childbearing age. This has been offset to some extent by immigration. The foreign born make up 4% of the population; previously they came from former colonies such as Indonesia or Suriname, though now the largest groups are from Turkey and Morocco.

Reclaiming land from the sea helped make the Netherlands one of Europe's leading agricultural nations, and it maintains that position today, even though only around 2% of the labour force work on the land. Thanks to very productive small farms, it has steadily increased agricultural output. The country is largely self-sufficient in food and also exports dairy products, meat, flowers, and bulbs.

Because it had few raw materials, the Netherlands did not develop as much heavy industry as other advanced economies. Instead it based its economy more on processing imported materials and on international trade. Thus Rotterdam is the world's largest port; it contributes around 10% of GDP, and is a transhipment point for imported oil that is destined for other European countries. Amsterdam's Schipol airport is Europe's second largest hub for air freight. The Netherlands' external orientation has given rise to important multinational companies such as Unilever, Royal Dutch/Shell, Philips, and Heineken. Even so, two-thirds of the workforce are employed in service industries.

One of the most important economic developments in the 20th century was the discovery in 1959 of huge deposits of natural gas in the north of the country. This rapidly turned the Netherlands into a major gas exporter. Unfortunately, this

Acquired and cured the 'Dutch disease'

also distorted other aspects of development, driving up the exchange rate and wages, and making industry less competitive—a phenomenon subsequently dubbed the 'Dutch disease'. The disease now seems to have been cured. Following decades of wage moderation, the Netherlands enjoyed rapid growth, low inflation, and low unemployment.

This success is the result of a unique social model that allows many different groups to participate in setting social and economic policy. Thus there is a Social Economic Council, with representation from employers and trade unions, that considers such issues as collective agreements and welfare provision and has helped keep wage claims within

manageable limits. However, the pact has come under strain: during a period of recession the government in 2004 started to make cuts in welfare.

Until recently, Dutch politics has been very consensual. The head of state is Queen Beatrix, but her role is largely ceremonial since almost all power rests in the parliament. In the past, political parties have been divided by ideology or by religion—Catholic or Protestant. But these distinctions have steadily been eroded. The religion-based parties merged in 1977 to form Christian Democratic Appeal (CDA) and ruled until the election of 1994, following which the centre-left Labour Party governed in a 'purple coalition' with the centre-right Liberals and the smaller libertarian party, D-66. This coalition, with Wim Kok as prime minister, retained power in the 1998 election.

The political landscape was reshaped, however, in 2001 by the rise of Pim Fortuyn who combined right-wing anti-immigration populism with social tolerance. Although he was murdered in 2002 his party, the Lijst Pim Fortuyn (LPF), was surprisingly successful in the general election that year and was briefly part of a right-wing coalition government.

This soon collapsed, however, and following an election in May 2003 in which the LPF's support was sharply reduced, a new centre-right coalition was formed, consisting of the CDA, D-66, and the right-wing People's Party for Freedom and Democracy (VVD), with the CDA's Jan Peter Balkenende as prime minister. When this collapsed in June 2006 he formed a temporary minority goverment to last until the November election.

Netherlands Antilles

A dispersed Dutch federation that is about to be disbanded

The Netherlands Antilles comprises five Caribbean islands. The northern group has the three smaller and greener islands of Sint Eustatius, Saba, and Sint Maarten (the southern part of St Martin). The southern group, off the coast of Venezuela, is formed by the two larger and drier islands of Curaçao and Bonaire.

The people are racially mixed—of combined African, European, and Amerindian ancestry. Those in the northern islands have a stronger black component; those in the south have a stronger Latin component. Though the official language is Dutch, on the northern islands the usual spoken language is English, while in the south it is the creole language Papiamento. Around three-quarters of the total population live on Curaçao. Standards of human development are quite high, boosted by aid from the Netherlands, but there have been high levels of unemployment—15% in 2004.

The islands have few natural resources and rely mostly on oil refining, tourism, and financial services. The proximity of Curaçao to Venezuela, and its location on important shipping lanes, made the island a major centre for oil storage and transhipment, cargo handling, and ship repair. This activity has been boosted by recent investment from Petróleos de Venezuela. While the oil

Land area:	800 sq. km.
Population:	0.2 million—urban 70%
Capital city:	Willemstad, 94,000
People:	Mixed black 85%, Carib Amerindian, white, and East Asian 15%
Language:	Papiamento, English, Dutch
Religion:	Roman Catholic, Protestant, Jewish, Seventh-Day Adventist
Government:	Dependency of the Netherlands
Life expectancy:	76 years
GDP per capita:	$PPP 16,000
Currency:	Netherlands Antilles guilder
Major exports:	Petroleum products

business has brought many economic benefits it has also caused some environmental damage.

The expansion of tourism helped to offer alternative employment, particularly on Sint Maarten. Bonaire is another tourist centre, with some spectacular scuba diving.

The islands are also an important stopping-off point for money. This is an offshore financial centre, with 42 registered banks which, the USA complains, are used for laundering drug money. Agriculture is very limited, employing only 3% of the workforce, though farmers grow oranges to make the liqueur Curaçao.

At present the islands are a federation with a Dutch-appointed governor, as well as an autonomous government for internal affairs. The legislature, the Staten, has representatives from each island.

Following referendums on all the islands their status is now to be changed. By July 2007 they hope to have a new constitution. Curaçao and Sint Maarten will loosen links with the Netherlands to become autonomous units, while the three smaller islands will become 'Kingdom Islands'—a status yet to be defined.

New Caledonia

Divided over the question of independence from France

Most of this Pacific country is formed by Grande Terre, a long island dominated by a mountain range that runs its entire length and descends steeply to hills and a narrow coastal plain. The country also includes many other islands, notably the Loyalty Islands and the Isle of Pines.

The original Melanesian inhabitants, known as the Kanaks, are now in a minority. France established New Caledonia as a country of settlement with immigrants from France, the 'Caldoches', and indentured workers from elsewhere in the Pacific and Asia.

New Caledonia's main resources are its mineral deposits. The country is the world's fourth largest producer of nickel and has around one-third of world reserves. This, combined with French aid of around $900 million per year, has given New Caledonians a relatively high standard of living, but low prices for nickel have reduced their income, increasing the need to develop other resources such as fishing and tourism to combat rising levels of unemployment.

New Caledonia is an overseas territory of France and sends deputies to the French parliament, as well as electing members to local assemblies. Since the 1970s, there has been a running conflict between the Kanaks, who favour independence, and the

---

Land area: 19,000 sq. km.
Population: 206,000—urban 62%
Capital city: Nouméa , 91,000
People: Melanesian 43%, European 37%, Wallisian 8%, Polynesian 4%, Indonesian 4%, Vietnamese 2%, other 2%
Language: French, varous Melanesian languages
Religion: Roman Catholic 60%, Protestant 30%, other 10%
Government: Overseas territory of France
Life expectancy: 73 years
GDP per capita: $PPP 11,400
Currency: CCP franc
Major exports: Ferronickel, nickel ore

Caldoches, who favour continuing as a French territory.

Although a referendum on independence was planned for 1998, this was ultimately considered too divisive, and instead the two sides reached the 'Nouméa Accords' through which France would allow considerable autonomy between 2000 and 2014, at which date there would finally be a referendum on independence.

Implementation of the accord has been slow and the outcome remains uncertain. The Kanaks have a higher birth rate than the Caldoches, and if migration from France declines they could be in the majority by 2014. On the other hand, the next generation of Kanaks will have had ten more years of French aid, so may have grown to depend on it.

The main pro-independence political group is the Front de libération nationale kanak socialiste (FLNKS). Anti-independence are the Avenir Ensemble and the Rassemblement pour la Calédonie dans la république (RPCR)—who formed a coalition after the 2004 election.

New Zealand

New Zealand now has one of the world's least regulated economies

Land area: 271,000 sq. km.
Population: 4 million—urban 86%
Capital city: Wellington, 340,000
People: European 70%, Maori 8%, Asian 6%, Pacific Islander 4%, other 12%
Language: English, Maori
Religion: Anglican 15%, Roman Catholic 13%, other Christian 26%, other 46%
Government: Constitutional monarchy
Life expectancy: 79 years
GDP per capita: $PPP 22,582
Currency: New Zealand dollar
Major exports: Wool, meat, dairy products, chemicals, forestry products

New Zealand consists of two main islands. The larger, South Island, is also the more mountainous, dominated by the snow-capped Southern Alps which run down its western half. In North Island the mountains are on the eastern side and are somewhat lower. Parts of New Zealand are of volcanic origin and there are a number of hot springs and geysers.

Three-quarters of the population live in the North Island. Most are of European, and particularly British, extraction—the product of more than a century of immigration that gave preference to 'traditional source countries', a policy that ended officially only in 1986. New Zealand has also been a country of net emigration, primarily to Australia. Worried about this brain drain, the government from 1991 made greater efforts to attract skilled immigrants and investors. This increased immigration, primarily from Asia, a shift that has created a political reaction.

The earlier arrivals came around the 14th century. These are the Maori who are now a minority, almost all of whom live on the North Island. 'Maori' is Maori for 'normal people', to distinguish themselves from the whites, the 'Pakeha'. The Maori name for New Zealand is Aotearoa, 'land of the long white cloud'. For a hundred years, the Maori were bitter that the 1840 Treaty of Waitangi that had guaranteed their land rights had not been honoured. They got some recognition in 1994 and 1995 in the form of compensation treaties with the government—and a formal apology from the British queen Elizabeth II.

The third largest group of New Zealanders come from various Polynesian islands, many of whom arrived from the 1960s onwards to meet the demand for unskilled labour.

Agriculture still employs around 10% of the population. New Zealand's soil is not particularly fertile, but the temperate climate and its grassy hills and meadows sheltered by the mountains make it ideal for pastoral farming. The vast sheep herd, the fourth largest in the world, enables New Zealand to be one of the world's leading producers of lamb, mutton, and wool. It also exports beef and dairy products. The main arable crop is barley for animal feed. Other crops include kiwifruit and apples for export, along with the grapes that have enabled New Zealand to become a leading wine producer.

Manufacturing industry is dominated by the processing of meat and dairy products for export, but New Zealand is now also a major producer of pulp and paper.

New Zealand has the service industries common to most industrial countries. But its tourist industry is increasingly important. More than two million visitors come each year to see the country's attractive scenery—some of which posed as 'Middle Earth' in the movie trilogy, *The Lord of the Rings*.

New Zealand poses as Middle Earth

New Zealand is a constitutional monarchy, headed by the British queen. Because the population is of overwhelmingly British origin there seems to be little pressure for change to a republic.

Until the mid-1980s, New Zealand had a highly regulated welfare state. This changed dramatically from 1984 following an electoral victory by the Labour Party which embarked on a radical programme of economic liberalization, removing agricultural subsidies and import controls, and in its next term from 1987 privatizing many public enterprises.

By the end of the second term, splits in the Labour Party contributed to its downfall. In the 1990 elections the right-wing National Party returned to power and, with Jim Bolger as prime minister, continued and extended Labour's liberalization programme while cutting many welfare benefits. In 1993, the National Party was narrowly re-elected.

Meanwhile a new right-wing nationalist party, New Zealand First, was formed to campaign against immigration, particularly from Asia.

The 1996 election was the first under a new system of proportional representation and resulted in a coalition government that paired the National Party with New Zealand First. This soon started to unravel. Bolger was ousted as National Party leader and Jenny Shipley became New Zealand's first woman prime minister.

The election in 1999 brought Labour, led by Helen Clark, back to power in a coalition with the centre-left Alliance Party. After years of neo-liberal economic policies, Labour shifted back somewhat to the left, increasing spending on health and education and taking initiatives for Maoris and Pacific Islanders. However the Alliance party fell apart and its founder Jim Anderton created what is now the Progressive Party.

Clark called a snap election in July 2002, and although Labour came out well ahead, she still had to form a coalition with the Progressive Party.

The 2005 election was a close thing. The National Party, led by Don Brash, campaigned against extra privileges for Maoris. Of the 121 total seats 50 went to Labour, 48 to the National Party, and 7 to New Zealand First. Clark decided to form a minority administration, with New Zealand First as an unlikely coalition partner.

Nicaragua

Nicaraguans of left and right now seem more prepared to work together

Land area: *130,000 sq. km.*
Population: *5 million—urban 57%*
Capital city: *Managua, 820,000*
People: *Mestizo 69%, white 17%, black 9%, Amerindian 5%*
Language: *Spanish, English, Amerindian*
Religion: *Roman Catholic*
Government: *Republic*
Life expectancy: *70 years*
GDP per capita: *$PPP 3,262*
Currency: *Córdoba*
Major exports: *Coffee, seafood, meat, sugar*

Nicaragua is the largest country in Central America—and the least densely populated. The emptiest area consists of tropical forests in the east that stretch down to the marshy area called the Miskito Coast, named after the Miskito Indians who inhabited it when British buccaneers arrived in the 17th century. Even today, this region, which is also home to most of Nicaragua's black English-speaking population, remains relatively isolated from the rest of the country.

The majority of the population, mestizo and white, are to be found in the plains on the Pacific coast or in the central volcanic highlands, though in the north even these highlands are sparsely populated.

Nicaragua is one of the world's more disaster-prone countries, vulnerable to earthquakes and flooding as well as to hurricanes. In October 1998 Hurricane Mitch killed 3,000 people and caused $1.5 billion-worth of damage.

Standards of human development are low. Basic education levels were boosted by a literacy drive by the Sandinistas in the 1980s, but since then progress has been slow and 24% of children still do not enter primary school. The Sandinistas also started mass health campaigns. Now, however, spending on health is inadequate and health standards, particularly in the rural areas, are among the lowest in the Americas.

This is also one of the poorest countries in the region, with more than 48% of the population below the poverty line. Meanwhile inequality is increasing: the richest 20% of the population get almost half the income.

Around 40% of the workforce make their living from agriculture. Many grow basic food crops in the central regions and along the Pacific coast, while in the uplands the main crop is coffee. Agricultural yields for food-crops are very low. This is partly because more than one-third of landholders lack secure title over their land so are reluctant to invest in irrigation or other improvements. Coffee output on the other hand has increased, but farmers have suffered from low world prices.

Nicaragua also has some light industry. The most dynamic production is in the 15 'maquila' or free-trade zones. In 2004 these employed 66,000 people making a range of products including garments, shoes, and electrical equipment. But jobs elsewhere are hard to find. In 2004 around 45% of the workforce were unemployed or underemployed; the situation is particularly bad on the Atlantic coast. Almost two-thirds of all jobs are in the informal sector.

Nicaragua found itself the focus of world attention after the 1979 overthrow of corrupt dictator Anastasio Somoza and the arrival of the government of the Frente Sandinista de Liberación National (FSLN). The 'Sandinistas' had strong socialist principles,

Reagan versus the Sandinistas and distributed land and property to the peasants. But they faced resolute opposition from the Reagan administration in the USA which accused it of fostering communism and for most of the 1980s funded 40,000 'contra' rebels operating from bases in Honduras.

The debilitating civil war that ensued, combined with poor economic management, brought the country to its knees.

The FSLN survived but in 1990 they were voted out. As well as protesting against poverty, Nicaraguans had become disillusioned with a party that seemed to have become less democratic. They opted instead for a right-wing coalition led by Violeta Chamorro. She set the country out on a new course, reducing public expenditure, and privatizing state enterprises. She also adopted a conciliatory line with the FSLN, accepting their land reforms.

The 1997 election was a violent affair which again saw a victory for the right wing, this time a coalition of parties led by Arnoldo Alemán Lacayo of the Partido Liberal Constitutionalista (PLC) which won both the presidential election and a majority in the Legislative Assembly. The FSLN was, and continues to be, led by one of its original personalities, Daniel Ortega.

Alemán was a populist and conservative president, though was prepared to deal with the Sandinistas. The 1997 Property Law agreement, for example, protected beneficiaries of the Sandinista land redistribution.

Alemán and Ortega moved even closer together early in 2000 when the government made constitutional changes that entrenched the position of the PLC and the FSLN. Many important institutions, such as the Supreme Court, would henceforth have their members nominated by the two parties.

Alemán could not stand for re-election in 2001. The victorious PLC candidate this time was Enrique Bolaños Geyer. Bolaños, however, soon turned on Alemán, eventually stripping him of his immunity as a deputy and having him prosecuted for fraud and money-laundering for which he received a 20-year jail sentence. This cost Bolaños the support of the PLC in the National Assembly so he has had to rely on support from the FSLN.

The elections due in November 2006 will be closely fought, between Ortega, inevitably, for the FSLN and Eduardo Montealegre for the PLC.

Niger

Niger's underground wealth has been of little benefit to most of its people

Land area: 1,267,000 sq. km.
Population: 13 million—urban 22%
Capital city: Niamey, 675,000
People: Hausa 56%, Djerma 22%, Fulani 9%, Tuareg 8%, Beri Beri 4%, other 1%
Language: French, Hausa, Djerma
Religion: Muslim 80%, indigenous beliefs and Christian 20%
Government: Republic
Life expectancy: 44 years
GDP per capita: $PPP 800
Currency: CFA franc
Major exports: Uranium, livestock, cotton

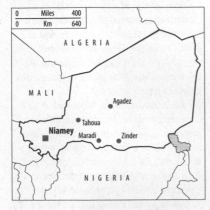

Niger has three main geographical zones. The zone to the north, approximately half the country, falls within the Sahara Desert. This arid territory includes the Air Mountains with sandy desert on either side. The central zone falls within the Sahel, with thin soil, scrubland, and sparse vegetation. The most fertile zone is the south, which benefits from annual rains, as well as, in the far south-west, flooding from the Niger River which flows south into Nigeria.

Most people live in the fertile south where sedentary farming groups include the Hausa and Djerma. The nomadic herding communities, including the Fulani, who raise cattle, and the Tuareg, who also have extensive herds of goats and camels, are likely to be found in the Sahel area, and sometimes in the desert to the north. But creeping desertification and over-use of the land by a growing population are steadily undermining the nomadic lifestyle—and many nomads are settling as farmers or moving to the cities. The annual population growth rate over the period 2001 to 2015 is expected to be 3.6%.

Levels of human development are desperately low. Niger is one of the few countries that even in peacetime has seen no improvement in social indicators over the past decade. In 2004, Niger came last but one in UNDP's human development index, a consequence not just of low income, with more than 60% of the population below the poverty line, but also of dismal standards of health and education. Around 15% of children die before their fifth birthday and 40% are malnourished. Only 38% of children enrol in primary school and around 70% work.

Most people make their living from agriculture which accounts for around 40% of GNP and employs 80% of the labour force. Their food crops include millet, sorghum, rice, and cassava, and their cash crops include cotton and groundnuts. Most of this cultivation is in the south-west of the country and since it is rain-fed it remains very vulnerable to the often erratic climate, which in recent years has led to severe shortages. In a good year the country is more or less self-sufficient in food. Even then, during the dry season many Nigeriens migrate to neighbouring countries in search of work. The main agricultural export is livestock, much of which walks to Nigeria to escape taxation.

Niger's main export and an important source of government revenue has been uranium ore. Niger has extensive deposits of high-grade uranium ore in the desert to the east of the Air Mountains. Niger is the world's fourth largest producer, sending a lot of its output to nuclear-power reactors in France. The mines are run by two independent companies though the government has shareholdings in both.

Niger's uranium fuels French reactors

In the 1970s uranium provided a much-needed boost to national income, but since the late 1980s the uranium price has collapsed and revenues have fallen drastically.

This also led the country deep into debt. Debt service payments were reduced, however, in 2004 as a result of the Heavily Indebted Poor Countries initiative which cancelled about half the country's debt.

Niger's democratic performance since independence in 1960 has not been impressive. Brief periods of multiparty governance have been interspersed with bouts of military rule. There have also been armed rebellions by the Tuareg who, with the support of Libya, have at times pressed for independence though they signed a ceasefire in 1995.

The last-but-one coup was in 1996, when Colonel Ibrahim Baré

Maïnassara seized power and arrested both the president and the prime minister and banned political activity. Later that year he did, however, introduce a new constitution that put power much more firmly in the hands of the president. Unsurprisingly, Maïnassara presented himself for election and took most of the vote in an election boycotted by the opposition. In 1997 he launched his own party, the Rassemblement pour la démocratie et le progrès.

Maïnassara's rule became steadily more repressive and any group protesting against government policy was subject to harassment or arrest. In April 1999, however, he was shot dead by soldiers in what the prime minister called an 'unfortunate accident', but is assumed to have been a palace coup.

At any rate, the military seized power until fresh elections were held in November 1999 when Mamadou Tandja, a retired army colonel, won the presidential election and his party, the Mouvement national pour la societé de développement (MNSD), won 38 of the 83 seats in the National Assembly while its coalition partner the Convention démocratique et sociale (CDS) took 17. They were later joined by the Alliance nigérienne pour la démocratie et le progrès.

Tandja and the MNSD were returned to power in the elections in December 2004. Again the CDS, led by Mahamane Ousmane, offered support. With four other parties, the two parties have 88 seats in the 113-seat parliament.

In 2005 the government denied it was facing a food crisis and suppressed international reporting.

Nigeria

Nigeria's democratic government is struggling to reverse decades of economic and social decline

Land area: *924,000 sq. km.*
Population: *126 million—urban 47%*
Capital city: *Abuja, 2.0 million*
People: *Hausa-Fulani 29%, Yoruba 21%, Igbo 18%, Ijaw 10%, Kanuri 4%, other 18%*
Language: *English, Hausa, Yoruba, Igbo, Fulani*
Religion: *Muslim 50%, Christian 40%, indigenous beliefs 10%*
Government: *Republic*
Life expectancy: *43 years*
GDP per capita: *$PPP 1,050*
Currency: *Naira*
Major exports: *Oil, cocoa, rubber, cotton*

estimate for 2003 was 124 million. As with much else, this is a highly political issue since it has a critical bearing on the distribution of funds between the federal government and the states. The population is likely to grow at an average of 2.5% per year over the next decade or so—and could reach 240 million by 2025.

Nigeria can be divided into four main regions. The humid coastal belt includes extensive swamplands and lagoons that can extend 15 kilometres inland, and further still in the area around the Niger River delta. Further inland the swamps give way first to hilly tropical rainforests and then the land rises to several plateaux—the Jos plateau in the centre and the Biu plateau in the north-east. Further north, these descend to savannah grasslands and eventually to semi-desert areas. The principal feature of the east-central border area with Cameroon is the Adamawa plateau, which includes the highest point in the country, Mount Dimlang.

Nigeria has Africa's largest population, though the actual size is a matter of some dispute. The government has claimed that the 2001 total was 119 million, though the UN

Nigeria is a federation of 36 states with more than 400 ethnic groups, though more than half belong to the three main groups: the Hausa-Fulani, who are Muslims and live in the north; the Yoruba, who are followers of both Christian and Islamic faiths and live in the south-west; and the Igbo, many of whom are Christians and live in the south-east. They largely live together peacefully, but there are at times outbreaks of violence. In 1999, for example, there were clashes between Yoruba and Hausa and hundreds of people died in various outbreaks of ethnic violence. Even more alarming were the deaths in early 2000 of hundreds of Christians and Muslims in disputes over the latter's determination to impose Shariah law in northern states in which Muslims are in the majority.

Despite its potential wealth,

Nigeria has been slipping backwards. More than half the population live below the poverty line. The education system is in poor shape and the literacy rate is only 67%. Those who can afford private education

Nigerian children packed into crowded classrooms

make progress but most children are packed into crowded and dilapidated classrooms. The public health system too is increasingly over-burdened. Maternal mortality is three times the average for developing countries. HIV/AIDS is also taking a heavy toll: in 2005 the official estimate was that 4% of adults were HIV-positive, though the real figure could be higher.

Despite rapid urbanization, most Nigerians still earn their living from agriculture, which makes up around 40% of GDP. The vast majority are subsistence farmers who have small plots and use primitive tools. In the south they tend to grow root crops like yams and cassava, while in the north the main crops are sorghum, millet, and maize.

With little investment, and over-exploitation of the land, output has often lagged behind population growth. As a result Nigeria, which used to export food in large quantities, is now a major importer. Small farmers also grow some cash crops, but of these only cocoa now provides any significant export income—and output has halved in the last 30 years.

Agriculture, as with much else in Nigeria, has been pushed into the background by the oil industry. Oil was discovered in 1956 and output grew rapidly in the 1960s and 1970s, helped by the high quality

and low sulphur content and also by the location—most of the fields are onshore in the Niger delta and convenient for export. One-third goes to the USA. In 2004, proven reserves were around 31 billion barrels—sufficient for another 35 years. Nigeria also has some of the world's largest reserves of natural gas.

In 2004, oil accounted for 98% of export earnings and around two-thirds of federal government revenue. So when the oil price falls Nigeria is in serious trouble, and since the mid-1980s Nigerians have watched their infrastructure decay and their public services decline.

The government-owned Nigerian National Petroleum Company has joint venture agreements with most of the oil companies, including Elf, Agip, Mobil, and Chevron. But around half the oil output comes from one company, Royal Dutch/Shell, which has come under heavy local criticism. People in the Niger delta have seen little benefit from oil production that has often caused environmental damage.

The Ogoni people in particular have been demanding greater control over the oil, and in 1995 Ken Saro-Wiwa and eight other Ogoni were executed. Shell was accused of

Shell accused of collusion in human rights abuses

collusion with the government and although it denies this, the company has now accepted greater responsibility for community development in the area. In 1999, when it announced an $8 billion oil investment plan, it also said it would spend around $40 million per year in the 1,500 communities where it works, though that has stopped neither the hostage-taking nor the inter-ethnic violence.

Most of the hundreds of billions of dollars received in oil wealth since the early 1970s have been squandered, either in wasteful development projects, or in graft or theft. Corruption started at the top. Military leaders simply stole much of the money. But scarcely any service

Oil wealth squandered

is available without the payment of what is known as 'chop'.

One of the most ironic effects of corruption is that Nigeria has suffered from chronic shortages of petrol. Local refineries have been deliberately run down so that government officials can get their chop from the sale of lucrative licences to import petrol, much of which finishes up on the black market.

From the early 1970s, the government invested some of the oil money in nationalized heavy industry—particularly petrochemicals, steel, and fertilizers—though such industries have been poorly managed and Nigeria still depends on imports. Most productive industrial activity is based on small-scale manufacturing, and a high proportion of the urban populations struggle in the informal sector.

Nigeria's political history since independence in 1960 has been fraught with ethnic conflicts and demands for greater independence for the states, which wanted to seize greater shares of the oil revenue. In 1967 the military governor of the Eastern Region attempted to secede, provoking a civil war that lasted two-and-a-half years and killed one million

people. Most of the years since then have involved a series of coups with brief interludes of civilian rule.

The most recent military regime, that of General Sani Abacha starting from 1993, was by general consent the most brutal and rapacious. Abacha had come to power following the annulment of a presidential election that gave victory to a Yoruba, Chief Abiola. Abacha subsequently imprisoned Abiola, launched ferocious attacks on opposition figures, and hanged Saro-Wiwa— drawing widespread international condemnation. Abacha had planned to restore 'civilian' rule, with himself as president. But to widespread relief he died in 1998—as did Abiola in prison (of a heart attack).

This led to a presidential election in 1999 and a victory for Olesegun Obasanjo, and his People's Democratic Party (PDP). Although a former military ruler, he did relinquish the office voluntarily in 1979.

Obasanjo came to power with the support of the northern elite, but tried to be even-handed in distributing state appointments across regional and ethnic lines. Many other aspects of his government have been disappointing. He has made little progress in tackling corruption. Nor has there been much reduction in ethnic violence, with at least 10,000 deaths since 1999. Human rights abuses have continued.

Obasanjo and the PDP were re-elected in May 2003, defeating Mohammadu Buhari, a former military ruler and leader of the All Nigeria People's Party (ANPP)— though amidst allegations of fraud. He may try to change the constitution to allow himself to run again in 2007.

Norway

Norway is one of the world's richest countries, but is trying to wean itself from over-reliance on oil

Land area: 324,000 sq. km.
Population: 5 million—urban 79%
Capital city: Oslo, 529,000
People: Norwegian, Sami
Language: Norwegian (Bokmål, Nynorsk)
Religion: Church of Norway 86%, other 14%
Government: Constitutional monarchy
Life expectancy: 79 years
GDP per capita: $PPP 37,670
Currency: Norwegian krone
Major exports: Oil and gas, metals, machinery

Norway is a mountainous country with a distinctively complex coastline. Much of the landscape consists of rolling plateaux, called 'vidder', interspersed with high peaks. To the west the mountains descend steeply to the coast which, during the Ice Age, was deeply cut by glaciers creating long, narrow inlets, the 'fjords', beyond which lie hundreds of islands. Though one-third of the country is above the Arctic Circle, the warm waters of the gulf stream usually keep the fjords above freezing point.

Norway is ethnically fairly homogenous. Almost everyone speaks the same language, though this exists in two mutually intelligible forms. The older of the two, Bokmål, is more widely used—in the national newspapers and in most schools—and is the product of previous centuries of Danish occupation of Norway. 'New Norwegian', Nynorsk, is both newer and older—the result of efforts to revive earlier dialects of Norse.

The oldest minority are the country's original inhabitants, the Sami, who herd their reindeer in the far north. But from the 1980s Norway started to receive significant flows of asylum seekers, first from Asia and Africa and later from former Yugoslavia. By 2005, 8% of the population were foreign born. In 2005, Norway also occupied first place in the UNDP human development index—a reflection of the country's wealth and its comprehensive system of welfare.

Norway's economic prospects were transformed in 1962 by the discovery of the Ekofisk oilfield in the North Sea. This and numerous other fields discovered subsequently have made Norway the world's third largest oil exporter. Oil reserves should last for another 50 years and gas reserves for another 100 years.

Oil and gas have fuelled Norway's economic growth and provided more than half of export income, and in 2004 one-quarter of government revenue. Unlike many other countries, Norway has been setting much of this windfall aside into a Petroleum Fund, which by 2005 had reached

$165 billion. One disadvantage of the oil bonanza, however, is that it has stifled investment in other activities. Manufacturing, for example, accounts for only 18% of GDP. Meanwhile, some older industries have been stagnating.

Norway's dramatic, serrated coastline has ensured a close relationship with the sea, and the country has established a strong shipping industry that in 2004 still gave it around 10% of the world's commercial fleet—including one-quarter of all cruise vessels and 20% of gas and chemical tankers.

Agriculture has long been in decline, though the government has subsidized it heavily and agriculture still employs 4% of the labour force. Fishing too, for herring, cod, and mackerel, has also declined as a result of over-fishing.

Despite weakness outside the oil sector Norway's unemployment rate in 2006 was only around 4%.

Norway is a constitutional monarchy, currently ruled by King Harald, who acceded to the throne in 1991. For most of the 20th century Norway was governed by the social-democratic Labour Party. One of the leading figures in the 1980s and 1990s was Gro Harlem Brundtland, who was a Labour prime minister in the periods 1986–89 and 1990–99. Brundtland applied for Norwegian membership of the EU and achieved significant concessions. But the electorate did not share her enthusiasm and in a referendum in 1994 declined to join—52% voted against. Brundtland resigned in 1996. She had already

Norwegians reject EU membership

been a major international figure, having chaired the World Commission on Environment and Development, and later served as head of the World Health Organization.

Brundtland was replaced as Labour leader by Thorbjorn Jagland. Although Labour received 35% of the vote in the 1997 election, this was less than what Jagland had declared he needed to govern and he gave way to a minority centrist government led by Lutheran priest Kjell Magne Bondevik of the Christian Democrat Party (KrF) in coalition with the rural Centre Party and the free-market Liberals. Bondevik took the unusual step in late 1998 of disappearing from view as a result of 'a depressive reaction to overwork'—an act which, if anything, seemed to enhance his popularity.

In February 2000 Labour replaced Jagland as their leader with Jens Stoltenberg. He set his sights on the government and in March brought it down through a no-confidence vote, after which the king asked him to form a new coalition government.

At the next general election in September 2001 there was no clear mandate and after extensive negotiations Bondevik and the KrF returned at the head of a 'co-operation government' that included the Conservatives and the Liberals.

People were becoming increasingly concerned however that Norway's wealth was not being translated into even better public services—with staff shortages in hospitals and other services. As a result Labour and Jens Stoltenberg returned to office in 2005 with a comfortable majority, forming a coalition with the Socialist Left Party and the Centre Party.

Oman

Oman is less well endowed with oil than its neighbours, and is keen to diversify

Land area: 212,000 sq. km.
Population: 3 million—urban 78%
Capital city: Muscat, 665,000
People: Arab, Baluchi, South Asian
Language: Arabic, English, Baluchi, Urdu, Indian dialects
Religion: Ibadhi Muslim 75%, Sunni Muslim, Shia Muslim, Hindu, or other 25%
Government: Sultanate
Life expectancy: 74 years
GDP per capita: $PPP 13,584
Currency: Omani rial
Major exports: Oil, animal products, textiles

Oman has three main geographical areas. The Al-Batinah plain extends for about 270 kilometres along the northern coast but only about 10 kilometres inland. From here, the land rises to the Al-Hajar mountain range, whose highest peak is around 3,000 metres. Beyond the mountains to the south, the rest of the country, around three-quarters of the land area, is largely desert until it reaches the mountains of Dhofar in the far south-west.

Oman's people are culturally diverse. Most Omani citizens are Arab and Muslim, though around three-quarters belong to the Ibadhi sect, an early breakaway from the Shia. The rest are Sunni or Shia. But this is not a strict Islamic state.

One-third of the population are immigrants, and make up a high proportion of the labour force—77% of the private sector and 18% of the public sector. Most come from the Indian subcontinent, though there are also expatriate Europeans. Immigrants do all the lowliest jobs—two-thirds earn less than the minimum wage.

From around 1970, thanks to oil income, Oman's people made swift improvements in human development—some of the most rapid progress ever recorded. Between 1970 and 2003 life expectancy increased from 40 to 74 years, and over the same period infant mortality fell from 126 per thousand live births to 10. Education too expanded rapidly, though it has not produced an optimum balance of skills.

Most Omani graduates expect managerial or professional jobs, preferably with the government, and are less willing to acquire vocational skills. But with the population of Omanis growing at around 3% per year there are not enough of these jobs to go around.

Oil has fuelled much of Oman's progress, accounting for 25% of GDP, but it is a dwindling asset and on present trends will last less than 20 years. Since it must be extracted from over 100 small and complex fields in the interior of the country and in Dhofar, production costs are higher than in neighbouring countries. There are other reserves but these would probably be too expensive to extract. Most production is in the hands of one company: Petroleum Development Oman.

The other main source of income is agriculture. Though accounting for less than 2% of GDP, it employs around one-third of the workforce. On the northern Al-Batinah plain their main crop is dates, while on a smaller scale farmers on the southern coastal plain grow coconuts and bananas. There is also some subsistence agriculture in the mountains as well as livestock-rearing in the interior. But Omani agriculture is running short of water. Most has to be pumped from underground and the advent of diesel pumps has encouraged over-extraction. Coastal areas also suffer from intrusion of saline water.

Farmers are running out of water

Oman's first attempts to diversify involved creating new industrial estates for manufacturing. These factories are now working, producing textiles and other goods for export. But they remain over-reliant on foreign workers.

Better prospects seem likely to emerge from recent discoveries of natural gas which have encouraged three major industrial projects. The most important is a large liquefaction plant to enable gas exports—primarily to South Korea, Japan, and India. The other possible uses are more problematic. Plans to develop an ammonia-urea fertilizer plant have been set back by low world prices for fertilizers. And plans to use the gas to fuel aluminium smelting have suffered a number of setbacks.

As well as diversification Oman is also pursuing 'Omanization' of private industry, setting quotas for different industries and establishing fines and other penalties for enterprises that fall short. But with so few trained or willing recruits most companies will be hard pressed to meet their targets.

Presiding over these developments is an absolute monarch. Since the 18th century Oman has been controlled by the Al bu Said tribe, and since 1970 the sultan has been Qaboos bin Said. In his early years he was faced with a communist insurrection in Dhofar. He finally crushed this in 1975 with the help of the USA, which still has a military presence. Since then opposition has been much more limited and the sultan has managed to satisfy the various tribes, regions, and interest groups.

He has also been edging towards democracy. In 1997 the sultan promulgated a new 'Basic Law' (based on Islamic law) that serves as a constitution and establishes some limited democratic rights. In addition, he has set up two consultative bodies. The first is the Majlis al-Shura whose 83 elected members were in 2003 for the first time elected by universal suffrage. The second is the State Council whose members are appointed from tribal leaders and former government officials.

The Basic Law also establishes that the sultan's successor must come from the Al-Turki branch of his family. It says too that if the family cannot agree on a suitable successor within three days of the sultan's death, they will have to accept his choice—which will be left in a letter.

Pakistan

Asia's great underachiever. Democracy has yet to take hold and human development is stifled by semi-feudal landlords and corruption

---

Land area: 796,000 sq. km.
Population: 152 million—urban 34%
Capital city: Islamabad, 901,000
People: Punjabi, Sindhi, Pathan, Baluch, Muhajir
Language: Punjabi 48%, Sindhi 12%, Siraiki 10%, Pashtu 8%, Urdu 8%, other 14%
Religion: Sunni Muslim 77%, Shia Muslim 20%, other 3%
Government: Republic
Life expectancy: 63 years
GDP per capita: $PPP 2,097
Currency: Pakistani rupee
Major exports: Cotton yarn, garments, cotton textiles

Pakistan has four main geographical regions. First, in the far north is the Hindu Kush and Pakistan's section of the Himalayas, including K2, the world's second highest mountain. Second, in the west and south-west, are a number of other mountain ranges and the Baluchistan plateau. Third, to the east are the desert areas which join with the Thar Desert across the border in India. Fourth, in the centre of the country is the rich agricultural plain of the Indus River, which traverses the entire length of Pakistan, emerging from the foothills of the Himalayas and flowing down to the Arabian Sea.

Pakistan is also divided politically into four provinces: Baluchistan in the west; North-West Frontier Province, on the border with Afghanistan; Sindh in the south; and Punjab in the heartland of the Indus Valley.

The population of Pakistan has been formed by waves of migration from many different directions and is a mixture of many influences with relatively few ethnic divisions. The Punjabis account for around two-thirds of the population, and the Sindhi for 13%. The other main groups are the Baluchs and the Pathans in North-West Frontier. Added to these are the Muhajir ('refugees'), descendants of the eight million Muslims who fled from India in 1947 at the time of partition—most of whom are in the urban areas of Sindh and particularly Karachi.

Faced with limited opportunities at home, millions of Pakistanis have headed overseas, chiefly to the Gulf states. They send home remittances worth around $3 billion per year.

Almost all Pakistanis are Muslims. Pakistan was created out of the partition of British India specifically to provide a state for the Muslim community. Even so, it was not initially envisaged as an Islamic state. Only in recent years has Islam become a dominant political force, having frequently been exploited by authoritarian governments seeking to rally support. There have been frequent clashes between Sunni and Shia Muslims.

Economically, Pakistan appears at times to have been fairly successful But it has failed to translate economic growth into real improvements in human well-being. Around 28% of the population live below the poverty line. Education levels are low: one-third of children do not enrol for primary education and over half of the adult population are illiterate. There are similar problems in health. Health services are poor and two-fifths of the population do not have safe sanitation.

Particularly disturbing is the slow advance of women. Thus while boys on average have 2.9 years of schooling, girls have only 0.7 years, and the literacy rate for women is only half that of men. **Pakistani women have been making very slow progress** Women also have fewer employment opportunities— particularly in the socially more conservative provinces of Baluchistan and North-West Frontier.

Women's generally low status is also reflected in poor standards of reproductive health: a consequence of official neglect, religious intransigence and considerable violence against women. Contraceptive use is low, so the birth rate is high and thousands of women die each year in childbirth. This also contributes to Pakistan's high annual population growth rate of 1.9%. By 2050 this will be the world's third most populous nation—after China and India.

Agriculture still employs around 40% of the workforce and makes up one-quarter of GDP. But more than 40% of the arable land is controlled by rural élites in farms of 50 acres or more where it is often used poorly— growing the most profitable crops like cotton but using inefficient farming methods that have made the land waterlogged or saline. Since 1959, there have been half-hearted efforts at land reform, but little has happened since the main landowners are also usually politicians. Pakistan used to be self-sufficient in wheat and rice but production has not kept pace with population growth and millions of tons of wheat have to be imported.

Industry has been fairly slow to develop. Pakistan is one of the world's major cotton producers but still exports most of its output as yarn rather than cloth. Immediately after independence, manufacturing was aimed at import-substitution.

More recently there have been efforts to open up the economy— removing some bureaucratic restrictions and lowering tariffs. But manufacturing industry is still unimpressive—dominated by low-tech enterprises. Industrialists claim they cannot invest because unlike the politically powerful landowners they have to pay so much tax.

Pakistan has been equally slow to develop its mineral deposits. The country does have some oil and gas. But oil production is only sufficient for about 20% of the country's needs, while reserves are falling. And exploitation of gas in remote areas of Baluchistan is hindered by opposition from belligerent tribesmen. Pakistan also has large deposits of rock salt, limestone, and other minerals but these have been little exploited.

Pakistan has frequently been

embroiled in serious disputes with its neighbours, particularly India. In 1998, Pakistan responded to Indian nuclear tests with six tests of its own (Pakistan is the world's only nuclear-armed Islamic state). The most dangerous issue is Kashmir, a territory whose sovereignty Pakistan disputes with India. For many years, this has involved just sporadic gunfire, but it took an alarming turn from 1999 when Pakistan encouraged rebel fighters to cross into Indian-held territory. Early in 2004, relations seemed to improve, however—a thaw that included an historic series of Pakistan–Indian cricket matches.

The only nuclear-armed Islamic state

Democracy in Pakistan has at best been fragile. For more than half its history since independence in 1947 the country has been under military rule. From 1978, the military ruler was General Zia ul-Haq, who had been steering Pakistan towards becoming a full Islamic state.

Following Zia's death in 1988, Pakistan then had a series of short-lived administrations that terminated with their dismissal by the president. Thus the 1988 election resulted in a victory for Benazir Bhutto and the Pakistan People's Party (PPP), the party founded by her father—a former prime minister whom Zia had executed. Her administration was regarded as incompetent and corrupt and she was dismissed in 1990.

The ensuing election resulted in a victory for a coalition headed by Nawaz Sharif, a Punjabi businessman whose party was the Pakistan Muslim League (PML). He too soon fell foul

of the president and was dismissed in 1993. Bhutto then returned only to be dismissed again in 1996. The 1997 election saw Sharif back but his downfall came in October 1999; after he tried to fire the army chief, General Pervez Musharraf, he was deposed by him in a bloodless coup.

Initially the international community condemned the coup, but following Musharraf's support for the US-led war in Afghanistan and the struggle against al-Qaida (support which is domestically unpopular) this criticism has subsided. Indeed he has since benefited from considerable economic aid. Even the revelation in 2004 that Pakistani scientists had been passing nuclear weapons technology to Iran, North Korea, and Libya produced only a mild response. And after having been expelled from the Commonwealth, Pakistan was re-admitted in May 2004.

Musharraf, who has survived several assassination attempts, has developed a hybrid form of military and quasi-democratic rule. He remains head of the army but in 2002 he held a referendum, of doubtful validity, to have himself appointed as president for five years. He also held dubious general and provincial elections.

In January 2005, Musharraf said that, contrary to a previous promise, he would not relinquish his post as head of the army and become a civilian leader—citing the need for 'unity of command'.

Meanwhile the opposition remains weak. Benazir Bhutto and Nawaz Sharif are still in exile, though they have brought the PPP and the PML together as the Alliance for the Restoration of Democracy.

Palestine

The chances of a peaceful settlement with Israel look more remote than ever

| 0 | Miles | 30 |
| 0 | Km | 48 |

Mediterranean Sea

Nabulus

JORDAN

Jericho

Jerusalem ■

Gaza

ISRAEL

Land area: 6,000 sq. km.
Population: 7 million— urban 71%
Capital city: East Jerusalem, 388,000
People: Palestinian Arab and other 83%, Jewish 17%
Language: Arabic, Hebrew, English
Religion: Muslim 75%, Jewish 17%, Christian and other 8%
Government: Republic
Life expectancy: 73 years
GDP per capita: $PPP 1,100
Currency: Israeli shekel, Jordanian dinar
Major exports: Industrial goods, food

Palestine comprises two territories linked by land corridors through Israel. The larger of the two is the West Bank, most of which consists of hills of up to 900 metres that descend eastwards from the border with Israel down to the Jordan river and the Dead Sea. The land in the north-west is relatively well watered, but rainfall is much sparser in the south. The remainder of Palestine is the Gaza Strip, which makes up only 6% of the territory—a flat and sandy region on the Mediterranean coast between Israel and Egypt.

The people are predominantly Arab, but the territory is dotted with Israeli settlements. In 1997, the population was 1.9 million in the West Bank and one million in the more densely settled Gaza Strip. Of these, the UN classifies 1.5 million as refugees, who live in 19 camps in the West Bank and eight in Gaza. There

are also 170,000 Jewish settlers. The population is young—47% are under 15 years.

Thanks to international aid, standards of education and health are relatively high, though services have been disrupted by violence and unrest. Nevertheless around 50% of people live on less than $2 per day.

Many Palestinians have sought work and refuge overseas. In 1998 there were around 4.2 million Palestinians abroad, chiefly in Jordan (2.3 million), Israel (1.0 million), Lebanon (0.4 million), and Syria (0.4 million), as well as in other Arab countries, in Europe, and in the Americas.

Until fairly recently, both the West Bank and Gaza were predominantly agricultural, and agriculture still accounts for around 8% of GDP. Most of this is on small family farms, chiefly growing olives, citrus fruits, and vegetables in sufficient quantities for export, as well as raising livestock, but farmers' incomes have been dramatically reduced by Israel's closing of the border.

Industry remains largely undeveloped, accounting for 15% of GDP in 1999. This consists mostly of small enterprises engaged in food processing, textiles, clothing, leather, and metal working. But political uncertainty has hampered investment so productivity is low.

Israel's invasion of 1967 effectively incorporated what is now Palestine into Israel—and made it dependent on Israel for most of its markets; 85% of exports still go there. Just as important, Israel became a major source of work: by 1992 one-third of the workforce were commuting to Israel. At times, however, Israel has effectively closed the border to workers, most recently from 2000. This has had a devastating effect on the economy. By 2004 average unemployment was 29%.

Israel has crippled the West Bank's economy

Were it not for international aid Palestine would be even worse off. In 2004, the budget assumed that aid from bilateral and multilateral donors was $1.7 billion—most of which was in the form of emergency rather than development funds and an increasing proportion of which has come from the Arab League.

Until the Israeli invasion of 1967, Jordan controlled the West Bank and continued to claim it until 1987, when it relinquished it in favour of the Palestine Liberation Organization (PLO). The PLO had been formed in 1964 and was recognized by the Arab League as the representative of the Palestinians.

From the outset, the PLO had been engaged in armed struggle and in 1987 it embarked on a more general uprising called the 'intifada'. Subsequently, however, it shifted towards political activity and negotiations took a dramatic turn

in 1993 when secret talks in Oslo produced a peace agreement. This involved mutual recognition between Israel and the PLO, as well as a staged Israeli withdrawal.

As a result, by 1996 much of Palestine was being administered by a Palestinian National Authority run by the PLO. Elections in 1996 gave 87% of the vote to PLO chairman Yasir Arafat as president and also gave a majority to Arafat's Fatah faction of the PLO in the Legislative Council. Arafat's administration was autocratic, with no real opposition and tainted by corruption.

A provocative visit in 2000 by Israeli leader Ariel Sharon to the Temple Mount in Jerusalem triggered a second intifada and the creation of a special Fatah unit, the al-Aqsa Martyrs Brigades which, together with other militant groups such as Hamas and Islamic Jihad, carried out suicide bombings in Israel. Israel responded by bombing the PLO headquarters and building a 'security fence' that cuts deep into Palestinian territory.

Arafat died in 2004 and presidential elections in January 2005 resulted in a victory for former prime minister Mahmoud Abbas of Fatah (Abu Mazen) who negotiated an end to the intifada. Another hopeful development was Israel's withdrawal of settlers from the Gaza strip.

But the victory for Hamas in the January 2006 parliamentary election created an even more complex situation. Hamas does not recognize Israel, and many Western governments classify it as a terrorist group and have withdrawn aid. Hamas also has bitter disputes with Fatah and Mahmoud Abbas.

Panama

Panama has a new master plan to develop and widen its greatest asset

	Miles	400
0		
0	Km	640

Caribbean Sea

Colón
Panama
David

PACIFIC OCEAN

Land area: 76,000 sq. km.
Population: 3 million—urban 57%
Capital city: Panama, 1.3 million
People: Mestizo 70%, West Indian 14%, white 10%, Amerindian 6%
Language: Spanish, English
Religion: Roman Catholic 85%, Protestant 15%
Government: Republic
Life expectancy: 75 years
GDP per capita: $PPP 6,854
Currency: Balboa (US dollar)
Major exports: Bananas, seafood

Panama does have mountain ranges, including an extinct volcano, and a temperate zone above 2,000 feet. But 85% of the country is lower and flat, and one-third is forested. Its most significant geographical feature, however, is that at its narrowest it is only 80 kilometres wide—an ideal location for a canal to link the eastern and western coasts of the USA.

The USA has long been the dominant influence here. In 1903, it leased from Colombia the strip of land on either side of a canal that had been started and abandoned by a French company. When the Colombian senate refused to ratify the agreement the USA organized a rebellion that resulted in the secession of Panama from Colombia. A treaty with the new country gave the USA rights to the land in perpetuity and the canal was completed in 1914.

Panama's location as a crossroads

for communications has given it a very diverse population, most of whom would be considered mestizo. The largest single group are the black descendants of the West Indians who were brought in to build first a railway and then the canal itself. But there are also significant Amerindian communities, including the Guaymí who live in the west of the country.

Compared with their neighbours, Panamanians are better off, with good standards of health and education, though there are often striking contrasts between people linked with the canal and the modern sector and those in the rural areas. Unemployment in 2006 stood at 9%.

Most Panamanians occupy the land on either side of the canal—although population pressure is forcing people to clear more of the rainforests. Around one-quarter are engaged in agriculture, growing subsistence crops on small farms or producing for export—chiefly bananas and sugar—or fishing for shrimps.

In the past most other employment has been derived from canal services. The canal itself has a labour force of around 9,000. Some 14,000 vessels go through the canal each year for which they pay up to $110,000 per transit creating annual revenues of around $980 million.

But Panama has also diversified into other industries. At the Caribbean end of the canal, Colón has the second largest free-trade zone in the world, though it employs only 1% of the national workforce. On the other side of the country, Panama has also become a major financial centre with more than 70 offshore banks. In addition, Panama's shipping registry gives it the largest 'fleet' in the world with 10,500 registered vessels.

Panama's turbulent political history has been shaped by its relationship with the USA. The US bases in the Canal Zone were home to the 'Southern Command'. But years of pressure and occasional riots provoked a nationalist backlash and Panamanians demanded sovereignty over the canal. In 1978, the military government of General Omar Torrijos, which had introduced popular measures such as agricultural reform, also negotiated a new treaty with US president Jimmy Carter, who agreed a handover for the end of 1999.

This did not signal an immediate end to US involvement in Panamanian political life. The most dramatic intervention in recent years was in

The USA invades to overthrow Noriega

1989 when the USA invaded to overthrow the government of General Manuel Noriega—later tried and jailed for 40 years in Miami for corruption and drug smuggling. Several hundred people were killed.

The 1994 presidential election was won by Ernesto Pérez Balladares of the populist Partido Revolucionario Democrático. But he was not to supervise the handover of the canal. Although his party won a large majority in the 1999 legislative elections, he was defeated in the presidential election by Mireya Moscosa of the conservative-populist Arnulfista party, who became the country's first woman president.

The handover of the canal went ahead as planned. But the Stars and Stripes was lowered in a rather grudging manner, as no senior US official chose to attend. Former president Carter showed up to complete the process he had started.

As well as providing more direct income for the government, the handover of the canal has also created a property bonanza as ex-US buildings are converted to other uses. But the poor have gained little advantage from this and many are doubtful about the benefits of US withdrawal if it just lines the pockets of the rich.

Moscosa had also achieved control of the Legislative Assembly at the head of a coalition, the Unión por Panamá (UPP). The coalition has wavered but she largely managed to keep her majority. However, high levels of unemployment and concerns about extensive corruption have reduced public support for her party.

The May 2004 presidential elections resulted in a victory for Martín Torrijos of the Partido Revolucionario Democrático (PRD). He is the illegitimate son of the former president, Omar Torrijos. He is keen on achieving a free-trade deal with the United States but one of the most ambitious undertakings is a $5-billion development plan for the canal that has to be approved in a referendum.

Papua New Guinea

More than 800 groups live in the rainforests—above some of the world's richest deposits of minerals

Land area: *463,000 sq. km.*
Population: *6 million—urban 13%*
Capital city: *Port Moresby, 254,000*
People: *Melanesian, Polynesian*
Language: *Pidgin English, English, and more than 700 indigenous languages*
Religion: *Christian 66%, indigenous 34%*
Government: *Constitutional monarchy*
Life expectancy: *55 years*
GDP per capita: *$PPP 2,619*
Currency: *Kina*
Major exports: *Gold, copper ore, oil, logs, palm oil, coffee, cocoa*

Papua New Guinea consists of the eastern half of the island of New Guinea, along with around 600 other islands. The mainland, with around 85% of the territory, includes extensive swampy plains to the north and south and a rugged chain of central highlands. But there are a number of other substantial islands, mostly of volcanic origin, including New Britain, New Ireland, and Bougainville. More than 80% of the country is covered with dense tropical rainforest.

Communications are difficult and much internal travel is only feasible by air, which makes it hard to help isolated communities at times of natural disaster, as when a tidal wave swept into the north-west coast of the mainland in July 1998, killing up to 10,000 people.

Papua New Guinea has one of the world's most complex societies, with 800 or more ethnic groups or languages. The lingua franca is Tok Pisin—Pidgin English. For most Papua New Guineans, the primary allegiance is to their tribe or clan linked by a common language, called in Pidgin a wontok ('one talk').

Papua New Guineans have a rich culture, but low levels of human development. Around one-third of children under five years old are underweight, and only two-fifths have access to safe water. Nevertheless, almost everyone benefits from basic health services, whether provided by the government, the churches, or non-governmental organizations. And since people have access to commonly held land few are likely to starve unless there is a severe drought.

The country is also making some progress in education: primary school enrolment is now 73% and the literacy rate is 57%, though secondary education is still weak.

More than 85% of the population depend on agriculture. They grow mainly subsistence food crops such as sweet potatoes, bananas, sugar cane, and maize. But smallholders also grow cash crops for export including coffee, cocoa, and copra. In addition, there are large estates producing rubber, palm oil, and tea. Forestry is another source of export income. But commercial agricultural development generally has been slow—hampered by difficult communications.

Modernization is drawing more people to the towns and particularly to the capital, Port Moresby, where 40% of people live in shanty towns with no sanitation or basic services.

Port Moresby has very high levels of crime Here and in other urban areas there has been considerable civil unrest, along with high levels of crime— both of which have served to slow social and economic development.

Papua New Guinea's greatest cash wealth lies underground in extensive mineral deposits that generate more than 80% of exports. The earliest to be exploited was gold in the 19th century, then copper, notably on the island of Bougainville in the Panguna mine. The only copper mine now, however, is the Ok Tedi mine near the western border, owned by an Australian company, BHP. Since 1989 a series of major gold mines have also opened up. More recently, there have also been discoveries of oil and gas. Gas could be particularly important, with an undersea pipeline to Australia.

Distributing the country's mineral wealth has long been a vexed issue—complicated by communal land ownership. There is also severe environmental damage. In 1999, BHP admitted that the pollution from the Ok Tedi mine was so severe that the only solution will be to close down by 2010—an alarming economic prospect for Papua New Guinea.

A perennial problem has been the threat of secession by Bougainville. Local people felt they were getting little benefit from the Panguna copper mine. An armed insurrection in 1989 closed the mine permanently and the next ten years of fighting cost around 15,000 lives. Following mediation, from Australia and particularly from New Zealand, the fighting ended in January 1999 and in March 2000, under the 'Loloate Understanding', a Bougainville Interim Provincial government was established to offer autonomy. Later there could be a referendum on independence.

Papua New Guinea's political structure matches the complexity of its society. This is a constitutional monarchy, headed by the British monarch. But since independence from Australia in 1975 its single-chamber parliamentary system has yet to develop a stable party basis. Voters choose MPs largely on personality or tribe. Party allegiances are weak and, once elected, MPs feel free to 'cross the floor'.

All governments have been unstable coalitions. The 1997 election produced a four-party coalition which chose as prime minister the somewhat erratic Bill Skate. He resigned in 1999 to be replaced by Sir Mekere Morauta of the People's Democratic Movement.

The elections in 2002 were fairly chaotic with more than 3,000 candidates. The party emerging with the largest number of seats (19 out of 109) was Sir Michael Somare of the National Alliance who formed a 13-party coalition government. Known as 'the Chief', it looks like he will stay in place until elections in 2007.

Paraguay

Paraguay produces more electricity per person than any other country, but remains very poor

Land area: *407,000 sq. km.*
Population: *6 million—urban 57%*
Capital city: *Asunción , 0.5 million*
People: *Mestizo 95%, other 5%*
Language: *Guaraní, Spanish*
Religion: *Catholic 90%, Protestant 10%*
Government: *Republic*
Life expectancy: *71 years*
GDP per capita: *$PPP 4,684*
Currency: *Guaraní*
Major exports: *Electricity, soya, timber, cotton*

The Paraguay River, which flows through the country from north to south, divides Paraguay into two very different geographical zones. To the west, making up three-fifths of the country is the flat, featureless scrubland of the Gran Chaco, which extends into Bolivia, Argentina, and Brazil. To the east, the remaining two-fifths, where most of the population live, is more fertile and humid and is an extension of the Brazilian Highlands. Much of the eastern and southern border is formed by another major river, the Paraná.

The majority of Paraguayans are mestizo, a mixture of European and Guaraní Indian. But the Guaraní influence remains strong. For about one-third of the population Guaraní is the first language, and most people understand it as well as Spanish. More recent immigrants have been Brazilian farmers who have settled on land in the east. There are also small numbers of Indians—in 17 ethnic groups who live both in the Chaco and the east.

Paraguayans are among the poorest people in South America: one-fifth of them live in poverty, and only half have electricity. Around one-fifth are infected with the debilitating insect-transmitted Chagas' disease.

Agriculture remains an important source of income, accounting for one-quarter of GDP and employing 40% of the workforce. But land is very unevenly distributed, with the top 1% of landowners controlling around two-thirds of the land. The main export crop used to be cotton, which is grown primarily by 200,000 or so peasant farmers on small plots of land, but of late they have been trying to diversify into other crops like wheat, rice, and sunflowers. Nowadays, the major export crop is soya, which is typically grown on the more efficient farms. Both cotton and soya, however, remain vulnerable to the vagaries of the weather and international prices.

Livestock is also important, with cattle-raising in the Chaco and in the south. Again, ownership is highly concentrated—1% of producers own around three-fifths of the total herd.

Paraguay has little manufacturing

industry. Most is in the hands of small firms producing for local consumption. Foreign investors have shown little interest—deterred by the poor state of infrastructure and the low levels of education and skill.

Although only around 7% of the workforce is unemployed, around 24% more are thought to be underemployed.

One important and expanding industry is electricity production. Paraguay's rivers have enormous hydroelectric potential. The world's largest single source of hydroelectricity is the Itaipú dam on the Paraná River. This project, most of whose funding came from Brazil, began production in 1984. Since Paraguay's share of the output is more than 20 times its total consumption, it sells most of the electricity to Brazil where the dam provides one-quarter of the country's electricity.

A second, more controversial joint enterprise is with Argentina further down the Paraná at Yaciretá. This started production early in 1998,

Protests over a new dam

and provides 40% of Argentina's electricity, but has been the subject of international protests against environmental damage and the displacement of 50,000 local people.

Paraguay's political system has yet to develop any coherent shape following the end of the dictatorship of General Alfredo Stroessner. He seized power in 1954, along with the leadership of Paraguay's dominant Partido Colorado (PC), and his dictatorship continued until he was ousted in a coup in 1989 by General Andrés Rodríguez.

Rodríguez subsequently won the 1989 election as the Colorado candidate.

The Colorado candidate for the 1993 presidential election a businessman, Juan Carlos Wasnoy. Wasnoy tried to liberalize the economy further but faced resistance both from conservative elements and an emerging peasant movement that demanded land reform. Wasnoy also crossed swords with General Lino Oviedo, forcing him to resign.

In 1996, Wasnoy imprisoned Oviedo for an attempted coup. But Oviedo did not stay in jail long. Raul Cubas Grau won the 1998 election and promptly pardoned him.

Grau was opposed in this by his vice-president, Luis Maria Argaña, who threatened to impeach him. Matters came to a head in March 1999 when Argaña was murdered and Grau resigned to avoid impeachment. Oviedo fled to Argentina. One of Argaña's supporters, senate president Luis Gonzalez Macchi, was sworn in as president.

For the elections in 2003, Oviedo formed a new party from exile in Brazil, but was refused permission to return to Paraguay. The PC presidential candidate was Nicanor Duarte Frutos who duly won while the PC also won a majority in the national assembly. In power since 1946, the PC is the world's longest serving political party. Having achieved a new agreement with the IMF, Duarte Frutos now has to struggle with economic reform.

The main opposition throughout this period has been the largely rural-based Partido Liberal Radical Auténtico which has been unable to extend its support.

Peru

Peru has returned to democracy but now needs to find ways to cut poverty

Land area: 1,285,000 sq. km.	
Population: 27 million—urban 74%	
Capital city: Lima, 7.8 million	
People: Amerindian 45%, mestizo 37%, white 15%, other 3%	
Language: Spanish, Quechua, Aymara	
Religion: Roman Catholic 81%, other 19%	
Government: Republic	
Life expectancy: 70 years	
GDP per capita: $PPP 5,260	
Currency: Nuevo sol	
Major exports: Copper, fish products, gold	

Peru's striking terrain falls into three well-defined zones. The Pacific coastal region is a long strip of desert, broken only by 60 or so rivers crossing to the sea. Further inland rise the dramatic Andean mountains, whose plateaux and valleys nurtured the Inca civilization. Then to the east, the mountains descend more gently through densely forested foothills to the vast sparsely populated tropical rainforest of the Amazon Basin.

The Quechua and Aymara Indian descendants of the Incas still make up close to half the population. Millions of them are to be found in peasant farming communities in the highlands. But many others, driven by violence or poverty, have now made their way to the towns and cities, particularly to those on the coast.

The warmer tropical land to the east, though it makes up more than 60% of the country has only around 5% of the population, mostly native Indians living in scattered settlements.

Peru remains very poor. Based on the national poverty line, more than half the population are classified as poor and one-quarter as extremely poor. In recent years, the government has increased spending on basic social services but investment still lags far behind what is required. There is also a wide gap between the urban and rural areas: child malnutrition and under-five mortality are twice as high in rural areas as in the towns and cities. But even in the sprawling capital Lima half the population struggle to make a living in the informal sector.

Agriculture employs around one-third of the workforce. The most productive areas are around the rivers in the coastal areas where farmers grow cotton, sugar, and food for the cities. Agriculture in the highlands, the altiplano, remains fairly primitive. Peasant farmers grow maize and potatoes and raise alpacas and llamas. However, the most lucrative crop in the highlands is still coca—worth perhaps $1 billion per year though production has been falling.

On the coast one of Peru's most important activities is fishing. Peru is one of the world's largest fishing nations, catching around 10 million tons per year, mostly sardines and anchovies that are processed locally into fishmeal for export.

Peru's most important official foreign exchange earner, however, is still mining. Copper, gold, zinc, and lead provide more than half of export earnings. Privatization of many of the mines since 1994 has encouraged new foreign owners to step up investment. Peru also has reserves of oil, both in the jungle regions and offshore, as well as large natural gas fields.

Tourism to the spectacular Inca sites has also provided important income, and has revived now that guerilla warfare has subsided, with around 1.3 million visitors per year.

In the 1970s Peru had a series of military or autocratic governments. But even the restoration of democracy in the 1980s did little to improve

Fighting Sendero Luminoso
the lot of the poor. On one side the populist government of Alan García had by the late 1980s contributed to rampant inflation. On the other side, two guerrilla movements had emerged: the Cuban-inspired Movimiento Revolucionario Tupac Amaru (MRTA); and the Maoist Sendero Luminoso. The ensuing struggles killed 28,000 people and forced more than 700,000 to flee.

It was in this chaotic environment that Peruvians in 1990 elected as president Alberto Fujimori, a former university rector and descendant of Japanese immigrants. Fujimori proved an effective antidote to many of these problems—though at a democratic cost. In 1992, in an army-backed 'autocoup', he closed the Congress and suspended the judiciary. In 1995,

however, with a new constitution he won a free election.

On the military front, his successes included the 1992 arrest of Sendero Luminoso leader Abimael Guzmán. Though not eliminated, the guerrilla threat has subsided. Fujimori embarked on free-market reforms and opened Peru up to foreign owners.

For the election in 2000, despite widespread manipulation, Fujimori narrowly failed to win the first round against the candidate of the Perú Posible party, Alejandro Toledo —an economist of Amerindian descent. Toledo withdrew from the second round, accusing Fujimori of fraud. Fujimori duly claimed victory, to widespread international condemnation.

His downfall came in September 2000 when his security chief, Vladimiro Montesinos, was videotaped bribing television stations to support Fujimori's re-election campaign. Fujimori fled to Japan from where he resigned.

Fresh elections in April 2001 gave Toledo victory over former president Alan García of the Alianza Popular Revolucionaria American (Apra). Toledo struggled to live up to his election promises to increase employment and improve public services. His government also faced several scandals.

The 2006 elections concluded with a run-off between the centre-left former-president, Alan García, and the nationalist Ollanta Humala of the recently formed Partido Nacionalista Peruano. García won and returned to office but this time promised more prudent economic management while aiming to reduce poverty.

Philippines

The Philippines has lagged behind the 'tigers' of South-east Asia—always on the brink of economic take-off

Land area: 300,000 sq. km.
Population: 80 million—urban 61%
Capital city: Manila, 9.9 million
People: Christian Malay 92%, Muslim Malay 4%, Chinese 2%, other 2%
Language: Filipino (Tagalog), English
Religion: Roman Catholic 81%, Protestant 9%, Muslim 5%, other 5%
Government: Republic
Life expectancy: 70 years
GDP per capita: $PPP 4,320
Currency: Peso
Major exports: Electronics, machinery and transport, garments

The Philippines is an archipelago of more than 7,000 islands, though many are tiny and less than half are named. The two largest are Luzon in the north and Mindanao in the south. In between lie the group of islands known collectively as the Visayas. Most of the islands are mountainous, with the ranges running from north to south, and they include 20 active volcanoes. The Philippines is environmentally rich and diverse but since the 1970s much of the forest land has been cleared and tropical forests now cover only around one-fifth of the territory.

Filipinos are ethnically relatively homogenous, though there are a number of small indigenous groups. The main distinction is between Christians, who make up the majority of the population, and Muslims, most of whom are to be found on Mindanao. The country had two major periods of colonization, first by Spain and later by the USA, and it bears strong traces of both. Though the main language is Filipino, many people also speak English, which is the principal language for higher education. There are also smaller ethnic groups on different islands.

Many Filipinos are well educated, and the literacy rate is high, but nutrition and health standards are less impressive. Around one-third of children are malnourished and health services are very unequally distributed: half the doctors work in the National Capital Region.

Around 37% of people are living below the poverty line—of whom most are in the rural areas. Emigration is one solution to poverty, and there are thought to be around seven million Filipinos abroad, with two million in the USA alone. One-third of Filipino children live in households where at least one parent has gone overseas. Their remittances, which in 2005 amounted to $10.6 billion, not only sustain families, they also prop up the economy.

The largest source of employment at home remains agriculture, which employs more than 40% of the population. But the Philippines is a very unequal society, especially in landholding. In 1988, 2% of landowners had 36% of the land. The main food crop is rice, most of which is grown in Luzon. The second largest crop is coconuts, with half the production in Mindanao. Sugar used to be a major cash crop, particularly on the island of Negros, but following a collapse in the world price output has declined steeply. Fishing is another source of livelihood, though over-fishing by commercial fleets has been hitting the catches of inshore subsistence fishing communities.

The Philippines was slower than many other South-East Asian countries to modernize its industries. Like land, industry is often in the hands of powerful families. But its educated, and often English-speaking, population is proving increasingly attractive to foreign investors. Many companies have now established **Assets controlled by rich families** factories in dozens of export processing zones, particularly for assembly of electronic products, which now account for around two-thirds of exports. The Philippines also has some mineral potential: though it has little oil, it does have deposits of copper, nickel, gold, and silver.

The Philippines has to deal with two main guerrilla movements. The first is the communist New People's Army (NPA) which was very active in the 1980s. Since then there have been a number of ceasefires, but in any case support for the NPA has dwindled and it does not have much of a future. Of greater concern are various Islamic groups struggling for the independence of Mindanao. Here too there have been ceasefires, most recently with the Moro Islamic Liberation Front. But other more radical and violent groups, such as Abu Sayyaf, have emerged which have links with al-Qaeda.

Politics in the Philippines tends to be based more on personality than ideology or principle. This was particularly evident during the authoritarian rule of Ferdinand Marcos from 1965 to 1984. Although the economy grew during this period, he and his high-profile wife, Imelda, siphoned off much of the wealth. Marcos was eventually ousted by a 'people-power' revolution and Corazon Aquino, wife of a murdered opposition leader, was elected president in 1987. She was succeeded in 1992 by former army leader Fidel Ramos, who embarked upon a period of economic reform.

In 1998, Ramos was succeeded as president by a former movie actor, Joseph Estrada. By early 2000, however, his government had sunk into cronyism and corruption. People-power surfaced yet again and eventually he was forced to resign.

In January 2001 he was replaced by his vice-president, Gloria Macapagal Arroyo. She tried to continue with Ramos's economic reforms with little success. In the May 2004 elections she ran against another populist film star, Fernando Poe Junior. After a long dispute she was finally declared the winner. Arroyo's government has been fragile. She had to survive impeachment in 2005, for allegedly trying to rig the election, and a supposed coup attempt in 2006.

Poland

Poland has been economically one of the most successful countries in Eastern Europe, but politically very volatile

Land area: 313,000 sq. km.
Population: 39 million—urban 62%
Capital city: Warsaw, 1.7 million
People: Polish
Language: Polish
Religion: Roman Catholic 90%, other 10%
Government: Republic
Life expectancy: 74 years
GDP per capita: $PPP 11,379
Currency: Zloty
Major exports: Manufactures, machinery, mineral fuels

Poland has swamps and sand dunes along its northern coast. But the main part of the country comprises the central plains and lowlands; 'Poland' derives from a Slavic word for 'plain'. The mountainous one-third of the country includes the Carpathians along the southern border with Slovakia.

Poland's population has become ethnically more homogenous—largely as a result of the deaths and expulsions of Jews and other minorities during the Second World War. At the same time, many ethnic Poles have also emigrated, a trend which continued in the early 1990s as many people moved to Germany and elsewhere in search of work. Around one-third of ethnic Poles are thought to live abroad. An important unifying factor has been the Catholic Church, which played an important part in the struggle against communism—and also provided the previous Pope, Karol Wojtyla, John Paul II. However, the Church's political influence has diminished and it now restricts itself more to social teaching.

Poles have not just become wealthier, they have also become healthier. Since the early 1990s, life expectancy has been increasing and infant mortality falling. This was helped by better food and by cleaner air, though Poles still suffer from high levels of smoking and alcohol abuse. Education levels had also been high during the communist period, though low salaries have dissuaded potential teachers, particularly in the rural areas, and standards have fallen.

Poland is considered a successful example of economic 'shock therapy'. From 1989, the government rapidly liberalized prices, imposed wage controls, and reduced subsidies to state-owned enterprises while encouraging private businesses. This helped to control inflation but ushered in a deep recession with high unemployment. From around 1992, however, the economy started to recover and by the mid-1990s Poland had embarked on rapid economic growth. Although the rate has slowed in recent years, by 2005 it was up to 3%. Manufacturing led the way, fuelled by foreign investors who had been encouraged by a liberal economic environment, good skills, and low wages. Some of the older industries have faded but there has been strong growth in lighter industry, notably food and drink and some heavier manufacturing such as automobiles. Between 1989 and 2004, the private sector's share of GDP rose from 18% to 75%. This has still not been enough to reduce unemployment which in 2005 was 17%.

Poland is one of the world's largest coal producers. But in recent years output has been falling while wage costs have risen and the industry is losing money.

Agriculture remains important. Although it accounted for only 5% of GDP in 2004 it employed 16% of the labour force. Polish farmers had managed to resist collectivization during the communist period. All

Poland has two million small farms

the 1.8 million farms are privately owned but most are small—averaging only 8.4 hectares—and very labour intensive.

In May 2004 Poland joined the EU where it has established itself as a tough negotiator. Poland's farmers will benefit from the EU's common agricultural policy but will only gradually get the same levels of subsidy as those in other states.

Poland's political transformation out of communism was famously the work of the Solidarity trade union based in the Gdansk shipyard. This momentum also swept Solidarity into political power and its leader, Lech Walesa, was elected president in 1990. Solidarity achieved many economic

reforms but later fragmented into numerous centre-right groups.

In 1993, after a series of short-lived Solidarity governments, parliamentary elections resulted in a victory for the former communists of the Left Democratic Alliance (SLD) and in 1995 its founder, Aleksander Kwasniewski, was elected president. Even so, the SLD continued Solidarity's economic reforms. In May 1997, a national referendum narrowly approved a new constitution that reduced the powers of the presidency.

Nevertheless Solidarity had been regrouping and under the leadership of Marian Krzaklewski came back together as a 37-party coalition, Solidarity Electoral Action (AWS). As a result, after the 1997 parliamentary election, AWS was able to form a government in coalition with another group, Freedom Union (UW).

In the November 2000 presidential election Kwasniewski defeated both Krzaklewski, Solidarity's candidate, and also Lech Walesa who subsequently withdrew from politics.

For the parliamentary elections in 2001 SLD, headed by Leszek Miller, returned to government. When Miller resigned in 2004 he was replaced by Marek Belka.

In the 2005 election, however, it was the conservative Law and Justice (PiS) party which came out ahead, subsequently forming a coalition with two smaller parties, and with Kazimierz Marcinkiewicz as prime minister. In addition, the PiS candidate Lech Kaczynski won the presidential election, and in July 2006 his twin brother Jaroslaw took over as prime minister. This has not proved a very impressive pairing.

Portugal

Political stability and EU membership have stimulated development

Land area: 92,000 sq. km.
Population: 10 million—urban 55%
Capital city: Lisbon, 2.0 million
People: Portuguese
Language: Portuguese
Religion: Roman Catholic 94%, other 6%
Government: Republic
Life expectancy: 77 years
GDP per capita: $PPP 18,126
Currency: Euro
Major exports: Textiles, garments, cars, footwear, wood products

Portugal's major river, the Tagus, flows across the country from east to west, reaching the Atlantic ocean at the capital, Lisbon. To the north of the Tagus, the land is mountainous and the climate cooler and wetter. The southern part of the country, the Alentejo ('beyond the Tagus') consists of rolling and generally arid lowlands. Portugal also includes the Madeira Islands and the Azores.

Portugal is one of the least industrialized countries in Western Europe and two-thirds of the population live in rural areas, with greater concentrations in the north.

By the standards of most European countries, the population is fairly homogenous, with relatively small numbers of immigrants from its former African colonies. Indeed until the 1980s Portugal was one of Europe's leading sources of emigrants who were drawn away by higher

wages in France and Germany. Some 4.5 million Portuguese still live overseas. However, many former emigrants have been returning and immigration has been rising: in 2000 2% of the population were foreign born.

Though unemployment is fairly low, at 8% in 2006, wages have also been low, though have recently started rising. Portugal certainly benefited from membership of the EU. Between 1986 when it joined and 1996, per capita income grew from 54% to 74% of the EU average, but has remained at that level ever since.

Portugal's slow development has been linked to the relatively poor state of agriculture, which employs one-fifth of the workforce. Most farms, particularly in the north, are small and relatively unproductive. Thin soil and unreliable rainfall, combined with low levels of investment, lead to yields one-third lower than the EU average. More than half the country's food has to be imported.

Nevertheless, Portugal does have agriculturally based exports, notably tomato paste, port, and other wines. Forestry also generates important exports: more than one-third of the country is forested, much of this with cork-oak trees. Portugal has one-third of the world's cork trees. The country's long coastline also supports a significant fishing industry, largely catching cod for local consumption and sardines for export, though more recently over-fishing has caused catches to fall steeply.

Manufacturing in Portugal is still fairly low-tech—much of it concentrated in textiles, clothing, and footwear—but the government has gone out of its way to attract higher-tech investment. One of the most striking examples was the record $1-billion subsidy for the joint Ford–Volkswagen Auto Europa car factory just outside Lisbon which is responsible for around one-tenth of merchandise exports. Portugal also has some mineral resources, particularly in the Alentejo which has the Neves Corvo copper mine as well as centres of marble production.

Portugal offers $1 billion for a car plant

Within the service sector, one of the most important industries is tourism. Portugal attracts more than 14 million visitors each year, mostly from Europe's chillier countries, heading for Madeira and the southern Algarve coastline. Tourism employs around 10% of the workforce and makes a major contribution to the balance of payments.

The years after 1976, which marked the end of the 36-year dictatorship of Antonio Salazar and the introduction of a democratic constitution, were a period of rapid political change. In the first 15 years of democracy, there were 12 changes of government. In the mid-1980s, however, the situation settled down. In 1985 the largely ceremonial post of

president was taken by Mario Soares, formerly prime minister and leader of the centre-left Socialist Party (PS). And in 1987 the Social Democratic Party (PSD) achieved a parliamentary majority and Anibal Cabaço Silva became prime minister. This pairing, which was to last until 1995, ushered in a period of relative stability—due to Portugal's 1986 entry into the EU.

Discontent with the PSD led to their defeat in the 1995 parliamentary elections, following which the PS took over again, this time with the urbane Antonio Guterres as prime minister. In 1996 Cabaço Silva ran for president but was defeated by the socialist candidate Jorge Sampaio.

The socialist administration, which moved towards the centre, benefited from several years of rapid economic growth. With unemployment below 5%, and inflation under control, Guterres remained popular and in the 1999 parliamentary elections the socialists substantially increased their representation. By 2001, however, the economic situation had deteriorated and Guterres resigned, prompting a general election in March 2002.

This time the PSD, now led by José Manuel Durão Barroso, came out ahead and formed a coalition with the small Popular Party. In July 2004 Barroso stepped down to become president of the European Commission, making way for Pedro Santana Lopes, who proved so inept that the president dissolved parliament.

In the ensuing elections in February 2005 José Sócrates, the new leader of the socialists, won an historic outright majority and pressed ahead with economic reform.

Puerto Rico

Puerto Rico has decided not to become a US state, yet. Now it wants greater economic independence

Land area: 9,000 sq. km.
Population: 4 million—urban 71%
Capital city: San Juan, 428,000
People: White 81%, black 8%, other 11%
Language: Spanish, English
Religion: Roman Catholic 85%, other 15%
Government: Commonwealth (US)
Life expectancy: 78 years
GDP per capita: $PPP 18,600
Currency: US dollar
Major exports: Chemicals, manufactures, food

Puerto Rico is largely mountainous. The Central Cordillera range extends east–west along most of the length of the island, descending steeply to the narrow southern coastal plain, and more gently to the broader plains of the northern coast. There is also a north–south distinction in climate: humid and tropical in the north; drier in the south.

The population is racially fairly homogenous, and there is little overt discrimination though, as elsewhere in Latin America, the people in positions of power tend to be those with lighter skin colour. Puerto Ricans are US citizens, but their standard of living is far lower than that in the USA: the per capita income is around half that of the poorest US state, Mississippi; the unemployment rate is around twice as high as in the USA; and half the population live below the US poverty line. Not surprisingly, many people driven by poverty and unemployment have headed for the USA, which now has around 3.4 million people of Puerto Rican descent, most of whom are in New York.

Almost everyone speaks Spanish, which is the main language of instruction in schools. English is also an official language, but far fewer speak it fluently and many people want to resist any further imposition.

Puerto Rico's main economic activity is manufacturing, which accounts for more than 40% of GDP. Many US firms have been attracted to Puerto Rico by low wages. Although Puerto Rico is subject to US minimum wage legislation, wages tend to be one-quarter less than in the USA. In the past there have also been significant tax concessions. These are now being phased out though local politicians are lobbying for them to be retained.

Initially this drew in firms engaged in labour-intensive industries like garments manufacture. But in recent decades companies have invested in higher levels of technology—around two-thirds of current manufacturing is in chemicals, and another quarter is in metal products and machinery. Most of this output is for export, primarily to the USA.

Though some garment production remains, and employs around 30,000 people, many labour-intensive industries, faced with increases in the US minimum wage, have migrated in search of cheaper labour elsewhere—a process that accelerated after the signing of the North American Free-Trade Agreement which heightened competition from Mexico.

Little is left of Puerto Rican agriculture, which now contributes less than 1% of GDP.

Another important source of employment and income is tourism. San Juan, the capital, is a favourite port of call for cruise liners. Around 1.4 million cruise visitors disembark each year along with five million stopover tourists, primarily from the USA.

The USA acquired Puerto Rico in 1899 at the end of the Spanish American War.

Puerto Rico's strange status

Since then, its status has remained unresolved. The US Supreme Court has asserted that Puerto Rico is an 'unincorporated territory of the United States'—a possession of, but not a part of, the USA. A new constitution in 1952 established a system for local government, making Puerto Rico a 'Commonwealth'.

So although Puerto Ricans are US citizens, they cannot vote in presidential elections and they do not pay federal income tax. They elect their own governor and Congress which have autonomy in many areas such as tax, education, and criminal justice. The US government deals with defence, monetary issues, and trade and also extends some federal social

programmes to the island. Around 30% of government expenditure comes from federal grants.

There have been a number of attempts to change the relationship, since Puerto Rico's position is anomalous in international law; unless it is a US state or an independent country it is effectively a colony. On this view, 'commonwealth' is a transitional status and Puerto Ricans should periodically vote to select their ultimate status—to become independent, or become a US state. Non-binding plebiscites in 1993 and 1998, with more than 70% turnouts, voted narrowly in favour of retaining commonwealth status.

The argument for statehood is chiefly financial. As part of the USA, the island would get full representation in Washington; people would pay federal income tax but they would also qualify for more federal funds. The argument against is chiefly cultural—Puerto Rico would lose its distinctive identity.

Given the significance of this issue, politics in Puerto Rico tends to be organized around it. The two main parties are the Partido Nuevo Progresista (PNP), which is pro-statehood, and the Partido Popular Democrático (PDP), which is pro-commonwealth. Since 1968 power has tended to alternate between these two parties.

In the elections in 2000, for example, after eight years of PNP rule the PDP returned with Sila María Calderón elected as Puerto Rico's first woman governor. She was replaced after the 2004 election by another PDP figure, Aníbal Acevedo Vilá, though the PNP controls the legislature.

Qatar

With a huge new gas field, Qatar could become the world's richest nation

Though it has some wells and sand depressions that provide water for limited agriculture, the peninsula of Qatar is mostly a flat and arid desert.

Native Qataris are in the minority—only around one-third of the population. Though they are entitled to free education, this is not compulsory and 11% of adults are illiterate. The health service is also free, though some low charges now have been introduced.

Most of the manual work is done by an immigrant workforce, drawn from the Indian subcontinent and Iran.

Oil was discovered in 1939 and has fuelled this small state's rapid development, providing two-thirds of government revenue. The country has also made efforts to diversify. Qatar has had a steel industry since 1978 and many foreign companies have invested in the production of fertilizers and other petro-chemicals.

Even so, the future lies with gas. Qatar's Dukhan field has been exploited since 1980, but most attention is now focused on the North field, which is the world's largest gas field not associated with oil—with reserves that could last 300 years. To pay for the investment in gas Qatar has been over-pumping oil at a rate that could exhaust reserves in about 20 years, as well as borrowing from

abroad. But this investment should pay off over the next few decades, probably doubling the country's income.

Qatar is ruled by the emir, Sheikh Hamad al-Thani, an absolute monarch who took over in 1995 when he ousted his father in a bloodless coup. As crown prince, Sheikh Hamad had been working to open up Qatar economically and politically, and when he became emir he immediately ended press censorship and established al-Jazira as one of the region's most outspoken TV stations—a key source of information during the war in Iraq. Qatar also has close relations with the US and provided military facilities during the 2003 war.

In 2003 the Sheikh introduced the country's first constitution, agreed in a referendum in 2004, which sets up a new 45-member parliament, the Shura Council, two-thirds of which will be elected and which will have legislative powers—although the emir will retain a power of veto. The first elections for the council take place in 2007.

This sets an unnerving precedent for other, and not very democratic, Gulf countries though Qatar has a reputation as a regional maverick.

Land area: 11,000 sq. km.
Population: 0.7 million—urban 92%
Capital city: Doha, 340,000
People: Arab 40%, Pakistani 18%, Indian 18%, Iranian 10%, other 14%
Language: Arabic, English
Religion: Muslim 95%, other 5%
Government: Absolute monarchy
Life expectancy: 73 years
GDP per capita: $PPP 27,400
Currency: Qatari riyal
Major exports: Petroleum products, fertilizers, steel

Réunion

A French dependency in the Indian ocean with high levels of unemployment

Réunion is a volcanic island with rugged mountains surrounded by basins and plateaux that lead down to the tropical coastal lowlands. The island is often exposed to violent cyclones, and there is still a live volcano.

Most people are of African or creole (mixed) descent and are far poorer than the French and Asian minorities, who are generally the richest. The population is growing fast and unemployment remains high: 33% in 2003. This has at times led to violent protests as people have demanded social security payments equivalent to those in France. Many have also emigrated to France in search of work.

The island was colonized as a potential sugar producer, and sugar still dominates the economy, providing 80% of exports as well as the raw material for rum and molasses. Other crops include vanilla beans and geraniums, which are grown for perfume essences. Another source of income is tourism, from 400,000 visitors per year, of whom 80% come from France. But the country remains heavily dependent on French aid.

Réunion is the largest overseas department of France and sends five deputies and three senators to the French National Assembly. Locally, it is administered by a French-appointed

Land area: 2,500 sq. km.
Population: 755,000—urban 68%
Capital city: Saint-Denis, 122,000
People: African, French, Malagasy, Asian
Language: French, creole
Religion: Roman Catholic 86%, Hindu, Muslim, or Buddhist 14%
Government: Republic
Life expectancy: 75 years
GDP per capita: $PPP 5,600
Currency: Euro
Major exports: Sugar, rum, molasses, perfume essence

prefect and has a general council and a regional council.

Political parties on the island include one of the main French parties, the right-wing Union pour la démocratie française (UDF). This also included the Rassemblement Pour la République until 2002 when it merged with part of the UDF and another party to form Union pour un Mouvement Populaire (UMP).

But two local parties have also held sway. One is the Parti communiste réunionnais (PCR); the other had a more distinctive origin, arising out of a pirate television station. The authorities had tried to suppress a pirate radio station Télé-Free-DOM but in the 1992 regional council election the most seats went to candidates supporting the station's right to broadcast. In a new election in 1993 Free-DOM won again and the station-owner's wife Marguerite Sudre was elected president of the regional council.

Since then, more conventional parties have been in charge. Following the elections of April 2004 the UMP emerged as the largest party and Nassimah Dindar-Mangrolia, a Muslim, became president of the council.

Romania

Romania has avoided ethnic disaster, but has suffered from political confusion and an increase in poverty

Land area: *238,000 sq. km.*
Population: *22 million—urban 55%*
Capital city: *Bucharest, 1.9 million*
People: *Romanian 90%, Hungarian 7%, Roma 2%, other 1%*
Language: *Romanian, Hungarian*
Religion: *Romanian Orthodox 87%, Roman Catholic 6%, Protestant 7%*
Government: *Republic*
Life expectancy: *71 years*
GDP per capita: *$PPP 7,277*
Currency: *Leu*
Major exports: *Textiles, footwear, basic metals*

Western Romania is dominated by the Carpathian Mountains which take up around one-third of the land area and run from the northern border with Ukraine down the centre of the country before veering west and exiting south-west into Serbia. East of the mountains are broad plains that extend to the south before reaching the coast of the Black Sea.

Most people are ethnically Romanian but there are significant minorities that have given rise to ethnic tensions. The largest minority are the two million Hungarians who are concentrated in the north-west of the country and have been pressing for greater rights. These efforts have born fruit in recent years with greater language rights and the creation of a Hungarian university. The other important minority are the Roma, or gypsies. The official figure is around

500,000 but there are probably nearer two million. They tend to live apart from other Romanians and have been subject to widespread discrimination and abuse. Many have fled to other countries, notably Germany. Around 1.8 million Romanians work abroad, especially in Italy, Spain, Germany, Israel, and Hungary.

Romania is one of the poorest countries in Eastern Europe. In 2003, 25% of the population lived in poverty—with particularly high poverty rates in the north-east. Health standards are desperately low: the rates for infant and maternal mortality are among the highest in Europe.

Moreover, population is falling. Communist Romania had been determined to boost the population and had prohibited abortion and contraception. As a result, many unwanted children finished up in grim orphanages that became symbols of the failures of communism. Now, as a result of economic uncertainty and family planning, the birth rate is very low. Combined with a rising death rate and emigration, this has led to a steady fall in population—by 5% between 1992 and 2002.

One-third of the workforce are

employed in agriculture. Around 96% of agricultural land is now in private hands, but this is mostly in the form of small inefficient plots in which there has been very little investment.

Communist governments had given a high priority to heavy industry, particularly engineering and chemical, but most of the equipment is now obsolete. Romania's manufacturing potential now lies in more labour-intensive light industries, such as textiles and footwear. There has been some foreign investment in the car and pharmaceutical industries, for example, but manufacturing overall has been slow to take off.

Romania's economic reforms have been sluggish—resisted by the bureaucracy, the trade unions, and by politicians milking the system. State enterprises continue to play a major role in utilities, banking, and manufacturing.

Romania places great store by the potential benefits of EU membership. The EU has said Romania might be able to join by 2007—and a public clock in Bucharest counts down the days. But there are still many significant obstacles to accession including corruption and restrictions on judicial and press freedom.

Countdown to the EU

Romania had one of the most traumatic transitions from communism. From 1965 it had largely been the personal fiefdom of a brutal dictator, Nicolae Ceausescu, who was finally overthrown and executed in 1989. But the government remained in the hands of former communists, and their new party, the National Salvation Front (NSF), won the 1990 parliamentary and presidential elections. A new constitution approved in 1991 was followed by further elections in 1992 which again chose Ion Iliescu as the president—who has a strong executive role. By this time, however, the NSF had split and subsequently it merged with several other parties. In 1993 it was renamed the Party of Social Democracy in Romania (PSDR). However the government was essentially in thrall to the old politicians and bureaucrats as well as to nationalist forces.

This era ended in 1996 with the election of an anti-communist government, led by a centre-right coalition: Democratic Convention (DC). Emil Constantinescu of the DC also defeated Iliescu in the presidential election and appointed Victor Ciorbia as prime minister. The DC governed in conjunction with the third largest party, the Social Democrat Union (SDU). The DC government attempted to restructure the economy but living standards continued to fall, provoking widespread social unrest.

In the elections in 2000 the most seats went to the PSDR whose candidate, Ion Iliescu, was also elected president. The PSDR subsequently merged with a smaller party as the Social Democratic Party and formed a minority government.

The 2004 election was also indecisive but finally resulted in a coalition led by the National Liberal Party (NLP) and the Democratic Party (DP), with Calin Popescu Tariceanu of the NLP as prime minister. Former DP leader, Traian Basescu, was elected president. The coalition has proved fractious, with difference on economic policy, and the prime minister complains of presidential interference.

Russia

Russia's president is determined to modernize this vast country but is also crushing its nascent democracy

Land area: 17,075,000 sq. km.	
Population: 147 million—urban 73%	
Capital city: Moscow, 10.1 million	
People: Russian 80%, Tartar 4%, Ukrainian 2%, Chuvash 1%, other 13%	
Language: Russian	
Religion: Russian Orthodox, Muslim	
Government: Republic	
Life expectancy: 65 years	
GDP per capita: $PPP 9,230	
Currency: Rouble	
Major exports: Fuels, metals, machinery	

Russia, which is officially the Russian Federation, is by far the world's largest country, spanning two continents—more than 10,000 kilometres across. Russia's territory is huge and diverse but can be considered as a series of regions west to east. First, to the west are the rolling plains and uplands of the European plain, through which flow major rivers including the Don and the Volga. These plains terminate in the east at the Ural mountains, which, running north to south, form a dividing line between Europe and Asia.

Beyond the Urals lies the vastness of the West Siberian plain, a mostly featureless and often marshy landscape that stretches east to the Yenisey River. This leads on to the Central Siberian Plateau and then to Russia's Far East, which ends in a number of mountain ranges before reaching the Pacific coast. Russia's climate is often harsh, with bitterly cold winters and short summers, especially in the north and east.

The majority of people, more than four-fifths, are ethnic Russians, but the country also has over 70 other nationalities, of which the largest are Tartars, Ukrainians, and Chuvash. The ethnic spectrum corresponds to some extent to the country's complex administrative structure. Russia has 89 different administrative units, including 21 'minority' republics, such as Tatarstan, Chechnya, and Karelia. However, these account for only around 20% of the total population, and even they have majority Russian populations.

Russia's people have endured a series of shocks since the demise of communism. Living standards have fallen steeply. In 2004, 18% lived below the official poverty line. There has also been a steep increase in inequality. In the Soviet era, the top 10% of earners got four times as much as the bottom 10%; by 2001 they got 14 times as much.

One of the most serious effects has been a decline in health standards. Between 1989 and 2002, Russia's death rate rose steeply and life expectancy for men dropped from 65 years to 59, though for women the drop was less steep—74 to 72. This has been ascribed to a combination **Bribes for health and education** of stresses and uncertainties that have contributed to heart problems, strokes, and alcoholism. The decline in the health system will also have played a part. Russia still has enough doctors, but they often work in poorly equipped hospitals and anyone who needs urgent treatment has to bribe the staff. Around one million Russians are thought to be HIV-positive.

The education system too has come under strain. School attendance is compulsory and free, but standards have fallen because of low investment. Education expenditure is however to increase, from 4% of the budget in 2004 to 12% in 2006. Universities survive largely on fees and bribes.

In the first half of the 20th century Russia, as the core of the Soviet Union, had made striking economic progress—particularly in heavy industry. But by the 1980s its centrally planned economy was unable to produce the goods evident in the rest of the world. Following the collapse of the Soviet Union, Russia from 1992 undertook radical reforms—privatizing industries, liberalizing prices, and reducing tariffs.

The first phase of privatization involved selling off small enterprises, such as shops or small restaurants, primarily to their workers. More difficult was the disposal of large-scale enterprises. Initially this was done in 1992 by giving every Russian citizen a voucher to buy shares. After 1995, the government started selling enterprises for cash, often at low prices, to large industrial groups, typically those with good political contacts. This created vast fortunes for a new group of 'oligarchs' who now control much of the country's industry and its financial sector.

Russian manufacturing was antiquated and once exposed to international competition went into steep decline. Between 1990 and 1998, output halved before starting to grow again. Compared with other industrial countries, where small and medium enterprises account for around half of GDP, in Russia they still contribute less than 15%. Around two-thirds of Russia's official GDP is controlled by 20 huge conglomerates. However, much commercial activity, perhaps 40%, probably takes place unrecorded in the informal sector.

One of the largest industrial sectors, and the one with the greatest potential, remains energy. Russia has about 5% of global oil reserves and is now the world's second largest exporter. Privatization and subsequent consolidation created major oil **Vast mineral reserves** companies—though the government now wants to regain control: one of the largest, Yukos, is effectively now state-owned following the imprisonment of its head, Mikhail Khodorkovsky.

Russia also has around one-third of global gas reserves, most of which are in the hands of the state controlled by Gazprom, Russia's largest company.

Russia is also a major coal producer and has large deposits of many other minerals, including diamonds, nickel, and platinum.

Agriculture, which still employs

around 13% of the workforce, has also suffered a sharp decline and now accounts for only 5% of GDP. Between 1992 and 1998 the grain harvest fell by half—though it has since recovered slightly. Most of the land is cultivated by partnerships or by companies created by reorganizing former state or collective farms—which still rely heavily on state subsidies. Only since 2002 has it been possible to sell agricultural land.

Although Russia's economy has started to grow again the country still lags way behind most countries in Europe, though the recent boost in global oil prices is changing the picture.

Privatization and the accumulation of all this new wealth have been accompanied by a proliferation of organized crime. Commerce is now dominated by the 'mafia' who can also rely on police co-operation.

The new era in Russia can be traced from 1985 when Mikhael Gorbachev started to open up the Soviet Union. This unleashed tensions between conservative and reformist forces that culminated in August 1991 in a failed coup by hard-line conservatives which triggered the break-up of the Soviet Union.

Until the end of 1999, political life in Russia was dominated by Boris Yeltsin who had been elected president of the Russian Federation in 1991. Yeltsin strengthened his position in 1993 after winning an overwhelming victory in a referendum that gave the president strong executive powers and allowed him to take many actions without endorsement by the state parliament, whose lower house is the Duma. The upper house is the Federation Council, consisting of regional governors and heads of regional assemblies.

Yeltsin's grip was constantly shaken by a number of rebellions, the most serious of which was in the southern republic of Chechnya where he unleashed a fierce civil war that continues to this day. Yeltsin was re-elected in 1996 and appointed and sacked a sequence of prime ministers, the final one being, in 1999, a former KGB colonel, Vladimir Putin. In December 1999 Putin and his supporters won a striking victory in the Duma elections.

Brutal war in Chechnya

Yeltsin resigned on 1 January, 2000—handing over presidential power to Putin. Portraying himself as a strong war leader, Putin easily won the March 2000 presidential election and, following the parliamentary elections in 2003, parties supporting Putin then dominated the Duma.

Putin has consolidated his position as an autocratic leader, determined to modernize the economy but at the same time crushing Russia's nascent democracy. He has repressed the media, jailing the owner of the only national independent TV station. He has also manipulated elections in Chechnya and elsewhere and harassed or expelled human rights groups. Meanwhile he has attacked any oligarchs, such as the head of Yukos, who might have served as an alternative focus of political power.

It came as no surprise that in March 2004 with the overwhelming and uncritical support of Russia's state-owned media Putin was re-elected president with 70% of the vote.

Rwanda

Rwanda has emerged from the period of genocide but now has an autocratic government

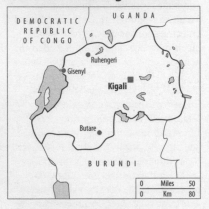

DEMOCRATIC
REPUBLIC
OF CONGO

UGANDA

Ruhengeri

Gisenyl

Kigali

Butare

BURUNDI

| 0 | Miles | 50 |
| 0 | Km | 80 |

Land area: *26,000 sq. km.*
Population: *9 million—urban 19%*
Capital city: *Kigali, 750,000*
People: *Hutu 84%, Tutsi 15%, Twa 1%*
Language: *Kinyarwanda, French, English, Kiswahili*
Religion: *Roman Catholic 57%, Protestant 26%, Muslim 5%, indigenous beliefs and other 12%*
Government: *Republic*
Life expectancy: *44 years*
GDP per capita: *$PPP 1,268*
Currency: *Rwandan franc*
Major exports: *Coffee, tea, hides*

From Lake Kivu on the western border, Rwanda's territory rises to a steep mountain range, with the volcanic Virunga Mountains in the north. To the east of these is the broad hilly plateau that covers most of the country before descending to the marshy land along the border with Tanzania.

The people of Rwanda are sharply divided between two main groups: the Hutu and the Tutsi, though the demarcation is more political and economic than cultural, since they speak the same language and follow the same religions.

Traditionally, the minority Tutsi, who were primarily cattle herders, had formed the ruling élite; the majority Hutu worked for the Tutsi while raising their own crops. Conflict between them culminated in genocide in 1994. Since then, a traumatized population has staged a remarkable

recovery. The economy is back to pre-war levels and with the help of international aid Rwanda has rebuilt its health and education systems.

Around 90% of Rwandans live in rural areas and depend on subsistence agriculture. Population density is high and most households survive on less than one hectare. For decades, their intensive cultivation of steep slopes has eroded the soil and the violence of 1994 resulted in part from competition for increasingly scarce land.

Farmers primarily grow staple crops like plantains, sweet potatoes, cassava, and maize. Food production has recovered from the war but has not kept pace with population growth. Livestock numbers are back to pre-war levels.

Rwandan farmers also grow cash crops, notably high-quality coffee in the north-west. And there is also a good export market for locally grown tea. Most industrial activity is devoted to processing coffee, tea, and other crops, along with production in a few factories manufacturing textiles and other simple consumer goods.

Rwanda also has some mineral resources including columbo-tantalite (coltan) which is used for making mobile phones.

For the longer term, high-value tourism also has potential.

The seeds for modern conflict were sown by the Belgian colonists who ruled through their favoured group, the Tutsi. As independence approached, the Hutu, anticipating majority rule, had become more assertive. In 1959, following a number of violent incidents the Tutsi king fled into exile along with hundreds of thousands of refugees.

Rwanda achieved independence in 1962—with a Hutu government led by the first president, Grégoire Kayibanda. The ethnic strife continued. The exiled Tutsi made several coup attempts and thousands of people died in reprisals.

Early indications of conflict

Kayibanda's government ended in 1973 when he was overthrown in a coup by the Hutu army chief, General Juvénal Habyarimana. More violence followed, with massacres of both Hutu and Tutsi. By 1990 more than 600,000 Tutsi had fled into exile. In a one-party state, Habyarimana was subsequently elected at the head of the Mouvement révolutionaire national pour le développement (MRND)—and re-elected in 1989.

Violence surged again in 1990 when the exiled Tutsi united as the Rwandan Patriotic Front (RPF) headed by Paul Kagame, and launched more offensives. Meanwhile, under donor pressure, the Hutu government had allowed other political parties to organize, and in 1991 Habyarimana agreed to a transitional government which, in 1993, made a peace agreement with the RPF at Arusha in Tanzania, though many in the MRND opposed this and started training Hutu militias: the 'Interahamwe'.

In 1994, President Habyarimana died in a plane crash. MRND dissidents took over and together with the Rwandan army they launched a genocidal massacre, mostly of the Tutsi. In response, the RPF intensified its campaign and by July 1994 had taken over the country. This ended the genocide, but by then up to 800,000 people had died. The RPF also pursued the Hutu into what is now Congo, provoking another horrific civil war in that country.

In principle the RPF was ruling as part of a coalition and until April 2000 the president was Pasteur Bizimunguy, an ethnic Hutu. When he resigned, however, he was replaced by Tutsi vice-president Paul Kagame, who had really been running the country. Kagame's position and that of the RPF was confirmed in presidential and legislative elections in 2003.

The RPF has remained firmly in control. It claims that it is not a tribal party: indeed it has taken great pains to remove the significance of ethnic identity having abolished the collection of any statistics based on ethnic distinctions. Returning Hutu rebels go through re-education camps where they are encouraged to be patriotic citizens, and all young Rwandans are obliged to imbibe similar non-tribal values in 'solidarity camps'.

This preoccupation has, however, created an authoritarian regime and a climate of silent resentment. There is no free press, any potential opposition groupings are harassed or jailed, and a number of opposition figures have disappeared.

St Lucia

A Caribbean island nation building a diverse economy

St Lucia is a volcanic island in the Caribbean, and subterranean activity surfaces via a number of hot springs. The mountain range that runs from north to south still has some dense forest cover though most of the land lower down has been cleared for agriculture.

Most St Lucians are black and English-speaking. But there is also a French influence: the majority are Roman Catholic and many also speak a French creole language.

Around 40% of the labour force work in agriculture on small farms that take advantage of generally fertile soil. They raise subsistence crops such as breadfruit and cassava. In addition, many have traditionally grown bananas which are the country's leading export.

But in recent years prices for bananas have been low. As a result, almost half of St Lucia's banana farmers are thought to have quit since the 1990s and their numbers are still falling. The situation will get worse when the island loses its preferential access to the EU.

The other main source of income is tourism. The island has a number of beach resorts that provide all-inclusive holidays. More than a quarter of a million visitors stay each year, and there are a similar number of short visits by cruise-ship passengers. Most tourists come from Europe and

---

Land area: 620 sq. km.
Population: 0.2 million—urban 31%
Capital city: Castries
People: Black 90%, mixed 6%, East Indian 3%, white 1%
Language: English, French creole
Religion: Roman Catholic 68%, other 32%
Government: Constitutional monarchy
Life expectancy: 72 years
GDP per capita: $PPP 5,709
Currency: East Caribbean dollar
Major exports: Bananas, clothing, cocoa, vegetables, fruits, coconut oil

---

the USA. Poorer St Lucians argue that relatively little of the income percolates down to them.

Manufacturing output is more diverse than in other Caribbean islands. As well as processing agricultural goods, local factories produce clothing and assemble electronic items.

St Lucia is a parliamentary democracy with the British monarch as its head of state, represented locally by a governor-general. The first elections after independence in 1979 were won by the St Lucia Labour Party (SLP), which established links with Cuba and North Korea. In 1982, power passed to the more conservative United Workers' Party (UWP), which opened up the economy and embarked on a series of IMF structural adjustment programmes.

These measures failed to reduce persistently high unemployment—of 30% or more—and this, combined with allegations of corruption, contributed to the UWP's downfall.

In the 1997 elections, the SLP won convincingly. Dr Kenny Anthony took over as prime minister and, having shifted the SLP towards the centre, won again in 2001. He is expected to stand for a third term in 2006.

St Vincent & the Grenadines

Rural poverty alongside hedonistic luxury

The country comprises one main island, St Vincent, which has around 90% of the territory and population, and the Grenadines, a group of 30 or more smaller islands, or cays, to the south. St Vincent is rugged and mountainous, with an active volcano, Soufrière, which last erupted in 1979.

Much of St Vincent is covered with dense tropical rainforests. The Grenadines have white sandy beaches and coral reefs.

There are still a few of the original Carib Amerindians, but most people on St Vincent, who are concentrated on the coast, are black or mulatto. They depend for survival primarily on agriculture, growing subsistence crops such as sweet potatoes and plantains.

The main cash crop, and the core of the country's economy, is bananas. These are grown both on small farms and large estates. New irrigation projects have boosted production and quality. But the loss of preferential access to the EU will strike a severe blow, and add to unemployment, which is already around 25%.

Efforts to diversify have had limited success, though marijuana could now be the second-largest export. US marines arrived in 1998 to destroy an estimated $1 million-worth of the crop, prompting local farmers to demand compensation.

The Grenadines are famous for

Land area: 390 sq. km.
Population: 0.1 million—urban 58%
Capital city: Kingstown, 16,000
People: Black 66%, mixed 19%, East Indian 6%, Carib Amerindian 2%, other 7%
Language: English, French creole
Religion: Protestant, Roman Catholic
Government: Constitutional monarchy
Life expectancy: 71 years
GDP per capita: $PPP 6,123
Currency: East Caribbean dollar
Major exports: Bananas

secluded luxury tourism on islands like Mustique, and are occasionally home to celebrities, but the islands also attract more than 70,000 other tourists each year, as well as many visitors from cruise liners.

The Grenadines have a drugs trade too, though in this case the drug is cocaine: the Grenadines are said to tranship 10% of Colombia's cocaine traffic to the USA.

A small offshore financial sector provides another source of income—sustained by strict legislation to preserve secrecy. This has caused the anti-money-laundering Financial Action Task Force to list it as a 'non-co-operative' jurisdiction.

The country is a constitutional monarchy whose head of state is the British monarch, represented by a governor-general. Government is based on a two-chamber legislature, with a 15-member house of assembly and a six-member appointed senate.

Between 1984 and 2001, the government was in the hands of Sir James Mitchell and the New Democratic Party. The two main opposition parties had merged in 1994 as the United Labour Party. This finally achieved power in 2001 with Ralph Gonsalves as prime minister and was returned to power in 2005.

Samoa

Samoa is forging closer links with its American neighbour

Samoa, which in 1998 changed its name from Western Samoa, consists of two principal Pacific islands: Upolu and Savai'i—volcanic islands surrounded by coral reefs. Less than one-quarter of the territory is covered by tropical forests. Samoa lies just east of the international date line, so had the distinction of being the last country to enter the 21st century.

Two-thirds of Samoans live on Upolu. Standards of human development are fairly high. The literacy rate is close to 100% for both men and women, and the leading diseases are those associated with a rich lifestyle, including hypertension, diabetes, and coronary heart disease. Although fertility rates are high, population growth has been slowed by emigration.

Two-thirds of the workforce are engaged in agriculture, mostly working on communally owned land. As well as growing subsistence crops, such as breadfruit and taro, they also grow the cash crops that produce the country's major exports—including cocoa and coconuts.

Samoa also has substantial forest reserves, though these are being depleted by illegal logging and land clearance. Both agriculture and forestry are often affected by the cyclones that regularly hit the islands. Fishing has also expanded in

Land area: 3,000 sq. km.
Population: 0.2 million—urban 22%
Capital city: Apia, 34,000
People: Samoan
Language: Samoan
Religion: Christian
Government: Constitutional monarchy
Life expectancy: 70 years
GDP per capita: $PPP 5,854
Currency: Tala
Major exports: Fish, copra, car parts

recent years, primarily for tuna, and accounted for 40% of export earnings in 2004.

Manufacturing primarily involves agricultural processing and garments, though the Japanese company Yazaki also exports car parts to Australia.

Samoa has been strengthening its links with neighbouring American Samoa—though the country's change of name to Samoa caused some friction. Co-operation helps promote tourism; the islands between them welcome more than 90,000 visitors each year.

Closer integration should also help boost trade; Samoa sends tuna to be canned in American Samoa, along with large quantities of beer.

Since independence from New Zealand in 1962, Samoa has been a constitutional monarchy, based on traditional systems of authority. The head of state is the high chief, who in 2006 was Malietoa Tanumafili II. When he dies his successor will be elected by the legislature, the Fono.

Since 1982 the government has been formed by the conservative Human Rights Protection Party. Their latest victory in elections to the Fono was in 2006 when Tuila'epa Sa'ilele Malielegaoi returned as prime minister. The main opposition is the Samoa National Development Party.

São Tomé and Príncipe

Bankrupted by political rows and now cursed by oil

Land area: 1,000 sq. km.
Population: 0.1 million—urban 38%
Capital city: São Tomé , 54,000
People: Mestiço, African
Language: Portuguese, creole languages
Religion: Roman Catholic, Protestant
Government: Republic
Life expectancy: 63 years
GDP per capita: $PPP 1,200
Currency: Dobra
Major exports: Cocoa

This small country consists of two main volcanic islands as well as a number of smaller islets in the Gulf of Guinea off the coast of West Africa. São Tomé has around 85% of the land area and almost all the people.

The islands were largely uninhabited until the Portuguese established plantations here—producing first sugar, then coffee and now cocoa. Today, most people are of mixed race, though of distinctly defined origins.

The forros, the ruling élite, are of Portuguese-African origin. The angolares are descendants of slaves from Angola. The serviçais are descendants of former contract labourers from Angola, Cape Verde, and Mozambique. Half the population still live in the rural areas, mostly on plantations. Although education and health standards are respectable by African standards, unemployment is very high and poverty has been increasing—to over 50% by 2001.

Since independence in 1975 production of the main crop, cocoa, has dropped by more than half, though still accounts for 85% of exports. There is also limited production of coffee. The second largest source of foreign exchange is licences from foreign vessels, mostly from the EU, fishing for tuna.

Prospects are likely to be transformed however by the discovery of offshore oil, production of which should start in 2010. Given the current widespread corruption and the dismal example of nearby Nigeria, the arrival of new funds that will swamp the existing budget is not likely to improve the quality of governance.

Since independence, the government has largely been controlled by the Movimento de Libertação de São Tomé e Príncipe (now the MLSTP-PSD). Initially this was a Marxist party and until a new constitution in 1990 it was the only legal one. Today the political scene has less to do with ideology than with personal power struggles and party infighting. Between 1991 and 1999, there were seven governments.

The presidential election in 2001 was won by Fradique de Menezes who had been supported by the main opposition group, the Moviemento Democrático das Forças da Mudança (MDFM). The elections in 2002 were indecisive, dividing the seats equally between the MLSTP-PSD and the MDFM. Then after a short-lived coup in July 2003 a government of national unity was formed, bringing in also Acção Democrático Independente (ADI). This coalition collapsed in 2004 to be replaced by one combining just the MLSTP-PSD and ADI.

Saudi Arabia

**Even Saudi Arabia has
to tighten its belt when
expenditure outpaces income**

Land area: 2,150,000 sq. km.
Population: 23 million—urban 88%
Capital city: Riyadh, 2.8 million
People: Arab 90%, Afro-Asian 10%
Language: Arabic
Religion: Muslim
Government: Monarchy
Life expectancy: 72 years
GDP per capita: $PPP 13,226
Currency: Riyal
Major exports: Oil

Along the Red Sea, Saudi Arabia has a narrow coastal plain. From here the land rises sharply to highlands that vary from 1,500 metres in the north to 3,000 metres in the south. Beyond these western highlands is a vast plateau that descends gently to the eastern coast along the Persian Gulf. In the centre of this plateau is the rocky expanse of the Najd, around which circles an arc of desert, which includes in the south the world's largest area of sand, the Rub' al-Khali—the 'empty quarter'. Saudi Arabia has no rivers and must take all its water from underground sources.

The native population is almost entirely Arab, though along the Red Sea coast there is also a black population. This is the original home of Islam. Most people are Sunni Muslim with a strict interpretation known as Wahhabism, but there are also up to two million Shia Muslims

who live largely in the east and have suffered repression. Many live abroad in exile. The population is growing fast—3.3% per year. Saudi Arabian citizens generally have benefited from the oil wealth. All receive free education and health care, and enjoy a range of subsidies. Those less wealthy are also entitled to a plot of land and a loan to build a house.

Women have traditionally been kept behind closed doors, and are still not allowed to drive, but their education levels are rising and many are now entering businesses. They can also take full advantage of the internet—two-thirds of Saudi users are women.

Although only one-quarter of the resident population consists of foreigners they make up 70% of the labour force. This includes immigrant workers from other Arab countries and from South and South-East Asia, who do the more menial jobs, as well as Europeans and Americans who do more specialized or technical work.

Nevertheless the boom years now seem to be over. Since the population has been growing faster than the economy Saudi Arabia's per capita income has been falling—it has more than halved in the past 20 years. As a result, Saudi nationals are having to accept more routine jobs such as security guards and taxi drivers.

Most economic activity revolves around oil, which is responsible for more than one-third of GDP, three-quarters of government revenue, and 85% of exports. Saudi Arabia has the world's largest oil reserves—one-quarter of the global total. This is extracted from huge oilfields—most of which are in the east, though there have also been discoveries elsewhere. At current rates of extraction, reserves should last 80 years or more.

The business is almost entirely in the hands of the government through Saudi Aramco. As well as having local refineries, the company also owns extensive refining and marketing operations in other countries.

Saudi Arabia has other important natural resources. Thus it has 4% of global gas reserves and is well endowed with many other minerals, including gold, iron, copper, and phosphates. Having concentrated on oil, it has barely exploited these but is planning to do so in the future.

The government has been making efforts to diversify and to encourage and build industries that would offer more employment—including petrochemicals, fertilizers, and steel. It has also been making greater efforts to attract foreign investment.

Despite its lack of water, and its underdeveloped agriculture, the country in the 1980s aimed to be self-sufficient in food, even in wheat. This meant extracting huge quantities of water from underground aquifers at rates that would exhaust the resource in around 30 years. Wheat production costs are four times the world price. In

Exhausting underground water

recent years, however, the government has been withdrawing subsidies and encouraging farmers to grow vegetables.

Another source of income is tourism. The annual 'haj' pilgrimage to Mecca draws in around three million foreigners a year, many of whom also take advantage of the opportunity to stock up on consumer goods. The government is now more interested in promoting tourism and has set up a tourist board.

Since the 1930s Saudi Arabia has been governed by the Al-Saud family, which in its various branches has 20,000 members, many of whom pursue a profligate lifestyle. The current leader is 83-year old King Abdullah bin Abdel-Aziz al-Saud who succeeded his brother, King Fahd, in 2005. Sultan bin Abdel-Aziz al-Saud was appointed as crown prince.

Successive kings have usually governed with some degree of consultation and in 1992 Fahd introduced a 'Basic Law' which serves as a kind of constitution. There is also the Majlis al-shura, a consultative council with 120 appointed representatives, chiefly from former officials and tribal leaders.

Since May 2003 Saudi Arabia has suffered as series of terrorist attacks. Osama bin Laden, who comes from a rich Saudi family, and al-Qaeda are bitterly opposed to Saudi Arabia's close relationship with the USA.

There appear to be chinks of a democratic opening. In 2005, for the first time there were open elections for seats on town councils. Women were excluded, though one candidate ran on a platform demanding that women be allowed to vote, and drive.

Senegal

One of Africa's most stable democracies, though it still faces armed insurgents

Land area: *197,000 sq. km.*
Population: *11 million—urban 50%*
Capital city: *Dakar, 2.4 million*
People: *Wolof 43%, Peuhl 24%,*
Serer 15%, Diola 4%, Mandinka 3%,
Soninke 1%, other 10%
Language: *French, Wolof, Pulaar, Diola,*
Mandingo
Religion: *Muslim 94%,*
Christian 5%, indigenous beliefs 1%
Government: *Republic*
Life expectancy: *56 years*
GDP per capita: *$PPP 1,648*
Currency: *CFA franc*
Major exports: *Fish, chemicals, groundnuts,*
cotton, phosphates

Senegal's northern border is formed by the Sénégal River whose valley provides a fertile strip of land. This soon gives way in the south and west to the flat, dry savannah of the Sahel, suitable only for raising livestock. The centre of the country consists largely of grasslands that are used more intensively for agriculture. Further south, beyond the 'finger' of the Gambia, a country that Senegal completely encloses, the Casamance region is greener, with more dependable rainfall and some tropical forests.

Senegal has more than a dozen ethnic groups of whom the largest are the Wolof, who are found particularly in the west. They make up only around two-fifths of the population and their language is also spoken by many others—though the official language remains French. Another large group, found in the west, are the Serer. Senegal also has many Peuhl (or Fulani). They are found throughout the country, and although traditionally they raise livestock many are now settled farmers. One of the politically most powerful groups is the Islamic Sufi sect, the Mourides.

The most significant dissident group are the Diola people in the Casamance region in the south who feel they have been exploited by northerners. Their armed organization, the Mouvement des forces démocratiques de la Casamance (MFDC), signed a peace agreement in 2004 though this could be undermined by splits within the MFDC.

Senegal is more developed than its immediate neighbours and Dakar, formerly the capital of French West Africa, is one of the region's more cosmopolitan cities, but the Senegalese are still very poor. More than one-third of the population are below the poverty line, only one-third are literate, and average life expectancy is 56 years. Health services are weak and concentrated in the cities.

Most people still depend directly or indirectly on agriculture, typically in small farms using basic production methods. The main food crops are sorghum and millet, grown in the north and centre, and rice in the Sénégal river valley and in the Casamance region in the south. Even so, Senegal still needs to import three-quarters of its rice consumption.

Most farmers also grow cash crops, of which the most important traditionally has been groundnuts. Groundnuts were also the main export earner but harvests over the last decade have been affected by erratic rainfall and more recently by privatization of the marketing system. Another important cash crop is cotton, much of which is bought by the local textile industry. Livestock production too has been affected by drought and the country also has to import meat.

The largest source of foreign exchange is fish. Senegal's Atlantic

Catching too much tuna

coastal waters are a rich source particularly of tuna. Most is caught from small boats by fishermen, who make up 15% of the workforce, but there is also industrial fishing. Tuna is canned and exported to the EU. The industry has expanded recently, making fish the largest source of export income, but over-fishing and increasing international competition have been reducing Senegal's tuna sales.

Senegal is one of the more industrialized countries in West Africa though productivity is not very high. Most manufacturing involves processing local raw materials. Thus, in addition to the fish canneries, there are groundnut-crushing mills, and four cotton-ginning plants, as well as textiles factories.

There is also a significant chemicals industry, based on the deposits of phosphates in the west of the country. This produces phosphoric acid and fertilizer for export.

Senegal has a reputation for being more democratic than most African countries, though in fact for the first four decades after independence in 1960 it was ruled by the same party, now called the Parti socialiste (PS). The PS tended to win because of its power of patronage and the whole system had become corrupt.

From 1981 to 2000 the president was Abdou Diouf. He and the PS won elections in 1983 and 1988. The main opposition party, the Parti démocratique sénégalais (PDS), and its leader Abdoulaye Wade alleged that the elections were fraudulent. In 1991, under donor pressure, Diouf brought Wade into the government.

Change came at last in February 2000. Wade had been gathering greater support from the younger urban population, and from the Mourides, and won the presidential election with 60% of the vote. To his credit, Diouf accepted defeat with good grace and stepped aside—one of the few elected African presidents to have done so.

Wade, who finally arrived in office at the age of 74, strengthened his position in 2001 when the PDS won the most seats in the legislative elections and headed a coalition: Convergence des actions autour du Président en perspective du 21ème siècle (CAP-21). Idrissa Seck was later appointed as prime minister but, seen as a rival to Wade, he was removed in 2004, tried for corruption, and replaced by Macky Sall.

Serbia

The steadily shrinking remnant of former Yugoslavia is likely to become even smaller

Land area: *88,361 sq. km.*
Population: *9 million*
Capital city: *Belgrade, 1.6 million*
People: *Serb 66%, Albanian 17%, Hungarian 4%, other 13%*
Language: *Serbian, Albanian*
Religion: *Serbian Orthodox, Muslim, Roman Catholic*
Government: *Republic*
Life expectancy: *74 years*
GDP per capita: *$PPP 4,400*
Currency: *Yugoslav dinar*
Major exports: *Manufactures, food, livestock*

Serbia was one of the core republics of the former Federal Republic of Yugoslavia. It has two provinces. The first in the north is Vojvodina, which has two-thirds of Serbia's territory, and consists of low-lying fertile plains, through which flows the River Danube, and which rise to forested mountains in the centre and south. Further south is the second province, Kosovo, which borders on Albania and Macedonia and consists of two inter-mountain basins.

The country is ethnically diverse, but the three main groups have been concentrated geographically—and even more so following the Kosovo War of 1999.

Two-thirds are Serbians who are Eastern Orthodox Christians and are now almost entirely in Serbia. The next largest are the ethnic Albanians, most of whom are Sunni Muslims who now form almost all the population of Kosovo. The third main group are Hungarians, who are in Vojvodina.

Most people are well educated and minority groups are entitled to education in their own language. However the education system suffered as a result of the wars and international isolation.

The health system too deteriorated after 1991. Much of the hospital equipment is antiquated and health expenditure is low.

Before the wars of the 1990s, Serbia formed the core of Yugoslavia and had a thriving economy based on diverse manufacturing industries—including chemicals, vehicles, furniture, and food processing. The NATO bombing campaign of 1999 destroyed much of this and the recovery has been slow. The chemicals industry has made some progress but others such as computers, telecommunications equipment, and textiles are still struggling.

Agriculture was also doing fairly well and Serbia has regularly had a food surplus. The chief crops are wheat and maize, along with a wide range of fruit and vegetables. But a lack of investment is holding back development and yields are around half of what they could be.

Throughout the country, warfare and economic mismanagement have savagely cut living standards. In 2006, inflation was around 15% and one-third of the workforce were still officially unemployed, though many of these actually work in the black economy.

The country's political history from the mid-1980s to 2000 was dominated by Slobodan Milosevic—a former

Dominated by Slobodan Milosevic

communist leader who in 1989 was elected president of Serbia. He and the Serbs had resisted the break-up of Yugoslavia. By 1992, Serbia and Montenegro were the only republics remaining and declared themselves to be the Federal Republic of Yugoslavia.

Milosevic was re-elected president of Serbia in 1992 and continued to support the Serbs in Bosnia who wanted to create a 'Greater Serbia'. In 1995, however, he took part in the negotiations in Dayton in the USA which ended the Bosnian War. Having had two terms as Serbian president, Milosevic stood instead for the presidency of the federation, to which he was elected in 1997.

Meanwhile, matters were coming to a head in Kosovo. During the communist era, both Kosovo and Vojvodina had enjoyed considerable autonomy, but to suppress the Albanian community Milosevic had suspended Kosovo's autonomy and imposed rule from Belgrade. Opposition to this turned violent in 1997–98 when a guerrilla movement, the Kosovo Liberation Army, stepped up its attacks on Serbian police.

Milosevic responded ferociously. After a failed peace conference, NATO started to bomb Serbia, while Serbia drove hundreds of thousands of Albanians out of Kosovo. The war ended after 72 days in June 1999 when Milosevic suddenly accepted a peace plan. Most of the Albanians returned, but then it was the Serbians' turn to flee—180,000 to Serbia, 30,000 to Montenegro.

The political situation changed dramatically following the presidential election for the federation in October 2000 when Milosevic was defeated by Vojislav Kostunica of the Democratic Party of Serbia (DSS), standing as the candidate of a coalition: the Democratic Opposition of Serbia (DOS). The DOS also won the general election in Serbia in December 2000. In June 2001 the government delivered Milosevic to a war crimes tribunal in The Hague.

In February 2003 under a new constitution the federal republic of Yugoslavia was replaced by a looser union: Serbia and Montenegro, with Svetozar Marovic as president. In Serbia, following the legislative election in December 2003, Kostunica and the DSS formed a minority coalition government that included G17 Plus and the Serbian Renewal Movement–New Serbia alliance. After several failures to get sufficient turnout, Boris Tadic of the DSS was elected president in July 2004.

Montenegro had meanwhile distanced itself from Serbia and, following a referendum in May 2006, declared its independence.

Next on the secession agenda is the future of Kosovo, whose status is now the subject of UN-supervised talks.

Sierra Leone

Sierra Leone was devastated by years of warfare. After maintaining a fragile peace, it has now started to rebuild

Land area: 72,000 sq. km.
Population: 5 million—urban 39%
Capital city: Freetown, 470,000
People: Mende 30%, Temne 30%, and many smaller groups
Language: Mende, Temne, Krio, English
Religion: Muslim 60%, indigenous beliefs 30%, Christian 10%
Government: Republic
Life expectancy: 41 years
GDP per capita: $PPP 548
Currency: Leone
Major exports: Rutile, diamonds, bauxite, cocoa

Sierra Leone takes its name ('lion mountain') from the distinctive shape of the mountainous peninsula that is the site of the capital, Freetown, and north of which there is a fine natural harbour. Most of the coastal area however is swampy for about 60 kilometres inland. Beyond this for around 100 kilometres is a broad plain of grasslands and woods, rising to a plateau broken by mountains that covers the eastern half of the country.

Sierra Leone's largest ethnic groups are the Temne and the Mende, each of which incorporates many subgroups. There is also a small population of Krios—descendants of liberated slaves who settled here from the end of the 18th century. Though few in number, they have had a disproportionate influence in the professions and the civil service.

The level of human development,

already low, sank further in recent years when the country was torn apart by violence. Sierra Leone's people have one of the world's shortest life expectancies—41 years. Around one-fifth of children die before their fifth birthday and around 2% of all births result in the death of the mother. HIV/AIDS is also a growing threat.

The war destroyed many of the health facilities. With donor support the government is now reconstructing and resupplying them.

Many schools were also destroyed, though parents are now rebuilding and employing teachers and around half have reopened. Less than one-third of the population are literate.

During the fighting many people fled to the relative safety of the cities, where they added to already critical problems of overcrowding, poor sanitation, and poverty.

Around two-thirds of people make their living from agriculture, principally growing rice along with cash crops such as food and coffee. Indeed, Sierra Leone was a major rice exporter. For most of the 1990s its people had to rely largely on food aid. Now that the fighting is over, most farmers have returned to the land, and output has recovered strongly and should continue to do so.

Sierra Leone should have been able to invest in much higher standards of human development, given its rich mineral resources. In the south it has extensive deposits of rutile (titanium ore) and bauxite (aluminium ore) that in the past provided the bulk of the country's export earnings. The fighting dramatically cut production, but operations have now resumed.

Sierra Leone's other main mineral, diamonds, was less affected by the war. Because the diamonds are alluvial, larger mining companies can use mechanical systems but individuals can also dig and pan. The main problem is that most diamonds are smuggled out of the country.

Surveys have also indicated offshore oil and gas deposits though so far there have been no significant discoveries.

Since independence in 1961, Sierra Leone has suffered from economic mismanagement and corruption on a

Corruption on a grand scale

grand scale. For two decades after 1970 this was effectively a one-party state. That government was overthrown partly as a result of the spillover from the fighting in neighbouring Liberia when in 1991 a local rebel group, the Revolutionary United Front (RUF), with Liberian help, launched an uprising in the south-east.

Then in 1992, a group of army officers who had been fighting with the rebels staged a coup. This intensified the ferocious civil war that killed around 15,000 people, and drove millions of people from their homes. The RUF were particularly brutal, recruiting thousands of child soldiers.

It was not until 1996, following a peace accord signed in Abidjan, that Sierra Leoneans had the opportunity to vote in an election. They chose a former UN bureaucrat, Ahmed Tejan Kabbah, of the Sierra Leone People's Party (SLPP), as president. Hopes of a lasting peace were dashed, however, in 1997 when the same army officers staged another coup and invited the RUF, led by Foday Sankoh, to share power, until the Economic Community of West African States, and its peacekeeping force ECOMOG, largely with Nigerian troops, helped to restore Kabbah as president.

The fighting continued until a UN-sponsored peace deal in July 1999. This kept Kabbah as president and put Sankoh in charge of diamond mining. In fact the RUF did not disarm and in April 2000, when Nigerian peacekeepers handed over to UN troops, the RUF captured hundreds of UN soldiers and started to advance towards Freetown.

This alarmed the international community, and particularly the former colonial power, the UK, which sent in troops and managed to protect Freetown and restore order. In May 2000 Sankoh was arrested.

The UN was then able to adopt a more positive peace-building role. By the end of 2001 the majority of the RUF fighters had been demobilized.

In May 2002 the SLPP gained an absolute majority in the legislative elections and Kabbah was returned as president with 70% of the vote. Since then the country has remained stable despite the withdrawal in 2005 of the UN peacekeeping force.

Singapore

An autocratic city-state that has become a leading manufacturing and financial centre

MALAYSIA

Singapore

Singapore Strait

0	Miles	10
0	Km	16

Land area: 643 sq. km.
Population: 4 million—urban 100%
Capital city: Singapore, 3.6 million
People: Chinese 77%, Malay 14%, Indian 8%, other 1%
Language: Chinese, Malay, Tamil, English
Religion: Buddhist, Muslim, Christian, Hindu
Government: Republic
Life expectancy: 79 years
GDP per capita: $PPP 24,481
Currency: Singapore dollar
Major exports: Petroleum products, machinery, chemicals

Singapore lies at the tip of the Malay peninsula, occupying the island of Singapore, along with 60 other adjacent islets. The city covers the southern part of the main island, but has extended itself further so that there is now little distinction between the city and the countryside. There is a small, highly productive agricultural sector but this is essentially an urban environment.

Singapore has a diverse immigrant-based population. The majority are of Chinese origin, though since they come from a range of provinces they often speak different dialects. The Indian population is also disparate, including Tamils, Malayalis, and Sikhs. The Malay population is linguistically more uniform. But most Singaporeans are at least bilingual and also speak English, the language of government and business. Racial

tensions, although muted, persist.

Singapore's rapid development has also created labour shortages—at both the bottom and the top of the market. More than one-quarter of the resident population are migrants. This includes 140,000 foreign maids as well as 300,000 foreign professionals. The government has a well-organized system for controlling this workforce, involving special levies on employers.

Singapore has now attained first-world status. Standards of education and health are good. The education system is highly competitive, with a strong emphasis on the literacy and numeracy requirements of a modern economy—though there are worries that regimented instruction is producing workers who are not sufficiently creative.

Most of these workers are employed in manufacturing and services. Singapore has for a century or more been an important regional centre for trade and business, and since the early 1960s the government has relentlessly propelled the economy along a path of export-led growth.

Manufacturing accounts for one quarter of GDP, around half of which comprises electronics: Singapore was the world's leading producer of computer disk drives but is now moving more into integrated circuits.

Since the early 1970s, Singapore has become one of the world's largest oil-refining centres—based on imported crude oil—and is a significant producer of chemicals and pharmaceuticals. It is also a centre for high-quality printing and publishing. Most of this activity is in the hands of 5,000 or so international companies along with a few large local enterprises.

A further one-quarter of GDP comprises finance and business services. Singapore's strong financial services sector combines a protected local banking system, dominated by four domestic banks, with a more open offshore banking sector. All are closely regulated.

Singapore's development has demanded a hectic process of construction. Until the mid-1980s this largely meant razing slums and replacing them with soulless apartment blocks and skyscrapers. More recently there have been attempts to preserve older buildings. High urban density has also created serious environmental problems, particularly from the automobile. In response, the government applies a sophisticated electronically monitored system of road pricing, along with a good network of public transport. One persistent problem is a shortage of water, which has to be piped from Malaysia.

Hectic construction of soulless apartments

This frantic expansion has been directed by an autocratic government. For two years following independence in 1963, Singapore was part of the Malaysian Federation, but it broke away in 1965 to become an independent country.

From the outset, Singapore has been governed by the multiracial People's Action Party (PAP), which for three decades was presided over by Lee Kuan Yew, who became a major international figure.

The PAP has shown little tolerance for opposition. A regular tactic is to sue political opponents for defamation, bankrupting them so they are ineligible to stand for parliament.

The government also maintains wide powers to detain people arbitrarily and to restrict their travel, freedom of speech, and rights to free association. It has a similarly fierce attitude towards the media, intimidating local journalists into self-censorship, and keeping a close watch on the foreign media. The autocratic impulse extends into the social sphere, with laws, for example, controlling chewing gum, and fines for not flushing public toilets.

In 1990, Lee stepped down in favour of Goh Chok Tong. Then in August 2004 he in turn was replaced by Lee Kuan Yew's son, Lee Hsien Loong, usually called 'BG' because he used to be a brigadier-general.

The May 2006 election followed the usual pattern. The PAP sued the Singapore Democratic Party for allegedly impuning its honesty, and poured money into the only two opposition-held seats. But it did not have it all its own way. The three opposition parties fielded candidates in half the constituencies reducing the PAP's share of the vote to 67%, though still won only two seats.

Slovakia

The poorer part of former Czechoslovakia has been reforming rapidly

Land area: 49,000 sq. km.		

Land area: 49,000 sq. km.
Population: 5 million—urban 58%
Capital city: Bratislava, 425,000
People: Slovak 86%, Hungarian 11%, Roma 2%, other 1%
Language: Slovak, Hungarian
Religion: Roman Catholic 69%, Protestant 10%, Orthodox 4%, other 17%
Government: Republic
Life expectancy: 74 years
GDP per capita: $PPP 13,494
Currency: Slovak crown
Major exports: Manufactured goods, machinery

Slovakia is dominated by the Carpathian Mountains. Three ranges run east to west and cover the northern half of the country. The highest point is in the Tatra Mountains—the Gerlachovský Peak at 2,655 metres. Most of the river valleys run north–south, so east–west communications are difficult. The main lowland areas are along the border with Hungary—one in the south-west including the capital Bratislava, and one in the east which includes the second city, Košice.

The majority of people are Slovaks but there are also important minorities. The largest group are ethnic Hungarians along the southern borders who since independence have been subject to increasing discrimination—especially against the use of their language. There is also a sizeable Roma, or gypsy, population, officially 2% but probably nearer 5%, many of whom are concentrated in the Tatra Mountains and live in constant fear of racist attack. Thousands have fled overseas as refugees, particularly to the UK.

Health standards and life expectancy are lower than in the more developed European countries. Though curative health services are extensive, less effort has gone into preventive health. As a result, men in particular have high death rates in middle age—probably the result of smoking, poor diet, and exposure to pollution. Education levels, however, have been high, comparable with those in Western Europe.

Industry now accounts for only one-quarter of GDP since production has fallen steeply, particularly in armaments. During the Czechoslovakia period, Slovakia also transformed imported raw materials into intermediate products such as steel and chemicals to be sent to more sophisticated manufacturing plants in the Czech lands. This pattern has continued, though at lower levels of production. More recent economic reforms have encouraged foreign investors. Volkswagen is now the largest private employer. Other major investors include US Steel, Whirlpool, and Peugeot-Citroën.

One serious consequence of the previous pattern of industrialization has been pollution: many factories have been located in valleys that trap the smoke generated by burning brown coal.

Employment in agriculture fell steeply in the 1990s and by 2005 employed only 4% of the labour force. Nevertheless, production has been maintained and the country is largely self-sufficient in food.

People who lost their jobs in industry and agriculture moved to services which by 2004 accounted for over half of employment. Many also work in the informal sector which is thought to account for 12% of GDP.

Slovakia's peaceful transition to a democratic free-market system was part of Czechoslovakia's 'velvet revolution' in 1990. But differences soon emerged between the two parts of the country. Slovakians thought that economic liberalization was going too fast for their less-advanced economy and wanted greater autonomy. Matters came to a head after the elections of 1992. In Slovakia the leading party was the Movement for a Democratic Slovakia (HZDS), led by a hard-line nationalist, Vladimir Meciar. This consisted predominantly of nationalists and former communists. Meciar negotiated with the Czech prime minister the terms for the 'velvet divorce'—which was not subject to a referendum since the split actually had little public support in either part of the country.

On 1 January 1993 Slovakia became an independent republic with

Slovakia breaks with the Czechs

Meciar as prime minister and another HZDS politician, Michel Kovác, as president (elected by parliament). Meciar's heavy-handed authoritarian instincts soon came to the fore as he concentrated power in his own hands, stifling opposition and curbing the press. He temporarily lost power in 1994 but was soon back and intensified his repression.

But by the elections of 1998 Meciar's grip was slipping, and he lost power to a four-party centre-right coalition led by the Slovak Democratic and Christian Union with Mikulas Dzurinda as prime minister.

In May 1999, Slovaks had their first direct election for president and chose Rudolf Schuster, who defeated Meciar.

Dzurinda set the country back on a more liberal and democratic path. He restored Hungarian language rights, for example, and started to rebuild the country's democratic institutions. In 2002 he was re-elected at the head of a slightly different centre-right coalition which enabled him to extend economic reform. For this turnaround Slovakia was rewarded with membership of both NATO and the EU, joining the latter in 2004.

In 2004, Meciar just lost the presidential election to his former right-hand man Ivan Gasparovic.

Then, in a surprising turnaround, the voters in 2006 rejected their relatively successful government in favour of a coalition led by an anti-reform populist, Robert Fico of Smer–SD. He is allied with Meciar's LS–HZDS, and Jan Slota of the racist, anti-Hungarian, Slovak National Party—though neither of these two have cabinet posts.

Slovenia

The richest country of ex-communist Europe and on course to join the EU

Land area: 20,000 sq. km.
Population: 2 million—urban 51%
Capital city: Ljubljana , 0.3 million
People: Slovene 83%, Croat 2%, Bosniak 1%, other 14%
Language: Slovenian 91%, Serbo-Croatian 6%, other 3%
Religion: Roman Catholic 58%, other 42%
Government: Republic
Life expectancy: 76 years
GDP per capita: $PPP 19,150
Currency: Tolar
Major exports: Manufactured goods, machinery, chemicals

Slovenia is predominantly mountainous. The highest areas are the Julian and Karavanken Alps on the north-western borders with Austria and Italy. These descend to a sub-alpine region on the edge of which is the Ljubljana basin. Slovenia has a short strip of coastline, which includes attractive beaches as well as the major port of Koper.

Slovenia's people, who are ethnically very homogenous, form an Alpine nation that has more in common with Germans, Austrians, and Italians than with the other former-Yugoslav republics to the south. Although health standards are generally good, education levels are below most EU countries.

The standard of living is higher than in other countries in Central Europe. Indeed Slovenes in general seem so contented that it is difficult to get them to move anywhere else:

foreign companies have problems persuading employees to work abroad. Even so, unemployment remains relatively high at around 10%.

Like other richer European countries, Slovenia is concerned about the ageing of the population since it offers generous pension benefits.

Slovenia only had about one-tenth of former Yugoslavia's population, but was responsible for around 20% of the output and was effectively subsidizing the poorer republics. The years immediately following independence were difficult as Slovenia lost markets in the other republics, but the economy soon revived as exporters found new buyers in the West.

Major manufacturing industries include metal products, furniture, paper, footwear, and textiles as well as pharmaceuticals and electronic appliances. Since the domestic market is very small most of the output has to be exported. About 70% of trade is now with the EU, primarily Germany and Italy. Leading companies include Lek, Krka, and Gorenje.

After 1992, following the Law on Ownership Transformation, many of the previously 'socially owned' enterprises were transferred to private hands. Some went to investment funds or were sold for cash. But many others were transferred 'internally' to workers and management in exchange for ownership certificates, or for cash at heavily discounted rates.

So far, Slovenia has made little effort to attract foreign investment, and has discouraged companies who wanted to borrow abroad.

Agriculture has become steadily less important, employing no more than 5% of the workforce, mostly on small farms engaged in dairy farming and livestock rearing. They will now benefit from EU agricultural subsidies. Around half the country is covered in forests, mostly still state-owned.

Nowadays, most people work in services. Many are employed in trade and transportation: the high-tech port of Koper is a major outlet for goods from Austria. It also services trade between Eastern Europe and Asia: the port's largest customer is the South Korean car maker Daewoo. The capital, Ljubljana, is also at the crossroads of two of the EU's proposed new highways, from Venice to Kiev, and from Munich to Istanbul.

Koper serves as Austria's main port

Another important service industry is tourism, though so far most of the two million tourists are Austrians making day trips for cheap fuel, and Italians attracted by the casinos.

In May 2004, Slovenia joined the EU as the richest new entrant with a per capita GDP around 70% of the EU average.

By Balkan standards, Slovenia's independence struggle was relatively painless. It had no ambitions to take land from elsewhere, nor did it have ethnic problems. Following the declaration of independence in June 1991, the Yugoslav army made a half-hearted attempt to reassert control, but the Slovenes fended them off in a largely bloodless ten-day war.

Since independence, politics has generally been consensual, based on broad coalitions. In the first post-independence parliamentary election in 1992 Slovenes opted for familiar figures. The largest single party was the centre-left Liberal Democracy of Slovenia (LDS), led by Janez Drnovsek, a member of the previous communist establishment who had briefly served as president of Yugoslavia. Drnovsek formed a coalition that included the Slovene Christian Democrats (SKD).

At the end of 1996 voters once again made the LDS the largest party, though with fewer seats. The government that emerged early in 1997, again headed by Drnovsek, also included the populist, centre-right Slovene People's Party and the smaller Democratic Party of Slovene Pensioners.

In April 2000, the government collapsed when the coalition lost one of its members. Normal service was resumed, however, after the October general election in 2000 when Drnovsek and LDS returned at the head of a three-party coalition.

Drnovsek soon switched jobs. In 2002 he was elected president and was replaced as head of the LDS and prime minister by Anton Rop.

In the 2004 elections, however, the LDS was defeated and replaced by a coalition led by Janez Jansa of the centre-right Slovenian Democratic Party.

Solomon Islands

Australian intervention stops an ethnic-based civil war

Land area: 29,000 sq. km.
Population: 0.5 million—urban 17%
Capital city: Honiara, 35,000
People: Melanesian 95%, Polynesian 3%, Micronesian 1%, other 1%
Language: Solomon Islands Pidgin and many other languages
Religion: Christian
Government: Constitutional monarchy
Life expectancy: 62 years
GDP per capita: $PPP 1,753
Currency: Solomon Islands dollar
Major exports: Timber, fish, copra, palm oil

The Solomon Islands consists of several hundred islands stretched over approximately 1,500 kilometres. Many have steep mountain ranges and more than 60% of the country is forest or woodland. The main islands are Guadalcanal, Malaita, New Georgia, Makira, Santa Isabel, and Choiseul.

The population is almost entirely Melanesian, mostly living in scattered rural communities. They have more than 80 languages, but most also speak Solomon Islands Pidgin. The official language is English.

Most islanders rely on subsistence agriculture. Their principal cash crops are coconuts, palm oil, and cocoa. But the main source of export income has been timber—which has provided more than 40% of government income. Logging accelerated during the 1990s and timber was soon being felled at more than twice the sustainable rate. In recent years there has been some respite since the market for timber has shrunk.

Fishing, primarily for tuna, is also an important source of income and food: the islanders have one of the world's highest per capita consumptions of fish. The largest private-sector employer is the Solomon Taiyo tuna-canning plant. The main export market is Japan.

The country's head of state is the British monarch, represented by a governor-general who must always be a Solomon Islander. In 1999, this was Sir Moses Pitakaka. Politics tends to be organized around people and issues rather than ideas.

Following the 1997 elections, Bartholomew Ulufa'alu emerged as prime minister. In 1999, however, the country was shaken by ethnic violence in Guadalcanal in a struggle over land rights and jobs. Matters came to a head in June 2000 when the Malaita Eagle Force seized the capital forcing Ulufa'alu to resign and replacing him with Mannasseh Sogavare of the Social Credit Party. He did not last. A general election in December 2002 was won by the People's Alliance Party led by Sir Alan Kemakeza.

By July 2003, however, further ethnic violence led to chaos and an Australian-led military intervention— the 'Regional Assistance Mission to the Solomon Islands'.

The elections in 2006 provoked further violence after suggestions that the Chinese community had influenced the MPs' appointment as prime minister of Kemakeza's successor Snyder Rini. He was forced to resign and, after a number of MPs crossed the floor, he was replaced by Mannasseh Sogavare.

Somalia

A 'failed state' that somehow seems to function, and now in the hands of 'Islamic Courts'

Land area: 638,000 sq. km.
Population: 8 million—urban 35%
Capital city: Mogadishu, 0.8 million
People: Somali 85%, Bantu, Arab, and other 15%
Language: Somali, Arabic, Italian, English
Religion: Sunni Muslim
Government: None
Life expectancy: 47 years
GDP per capita: $PPP 600
Currency: Shilling
Major exports: Livestock, bananas (1990)

Somalia is largely semi-desert. It has rough grasslands suitable for pasture but little arable land. The highest part of the country is in the north, where rugged mountain ranges face the Gulf of Aden. To the south of these lies the Hawd plateau, beyond which there are flat sandy plains with extensive dunes along the Indian Ocean coastline.

Somalis have always been desperately poor and the incessant wars of recent decades have made them even poorer. Health standards are low and deteriorating. One-quarter of children die before their fifth birthday. Only one-third of the population have access to safe water. In most of the country the education system has virtually collapsed, and less than 10% of children enrol in primary school. There is also the ever-present threat of famine, whether from poor rains or from a collapse in the distribution system caused by the fighting. Dismal conditions in and around the refugee camps have also resulted in frequent outbreaks of cholera, which is now endemic.

By mid-2001, 350,000 people had been internally displaced by the conflict and a further 450,000 were living in refugee camps in Kenya and elsewhere.

Through the civil war people still struggle to make a living. The most important source of income is rearing livestock—goats, sheep, and cattle—which accounts for around 40% of GDP. Two-thirds of the population are nomadic or semi-nomadic herders. Most of the others are farmers, growing sorghum and maize, chiefly in the river valleys. The most lucrative crop, and an important export, has been bananas. Indeed, some of the clan battles have been fought over control of the 120 banana plantations. Although Somalia, with its long coastline, could be a significant producer of fish, most of the catch is taken by foreign vessels.

There are parts of the country that function fairly well commercially and individual companies can work if they employ enough security guards. In 2004, Coca-Cola, for example, opened a heavily fortified bottling plant.

Since independence in 1961,

Somalia has been riven by ethnic and clan divisions. Following the assassination of a newly elected president in 1969, the army took over, led by General Mohammad Siad Barre. But his rule became increasingly authoritarian.

Throughout the 1980s, several clans made an informal alliance against Siad Barre. There were three main groups: the United Somali Congress (USC—Hawiye clan); the Somali Patriotic Movement (SPM—Darod clan) in the centre and south of the country; and the Somali National Movement (SNM—Issaq clan) in the north. By January 1991, they had successfully driven Siad Barre from Somalia.

The USC claimed to have established an interim government but this was immediately rejected by the SPM, which started fighting USC forces in the south, and by the SNM, which in the north proclaimed independence for the 'Somaliland Republic'. Then the USC itself split between factions led by Ali Mahdi Mohammad and General Mohamed Farah Aideed, who fought for control of Mogadishu. Meanwhile, the country was stricken by a famine that killed 300,000 people.

An uncertain peace was established in 1992 with the arrival of a US-led UN peacekeeping force. However, this soon became bogged down and in 1995 the UN's mandate ended and its forces withdrew. By then Somalia had ceased to exist as a state.

Famine in the midst of war

Nevertheless efficient local administrations have emerged. Of these the most coherent is Somaliland in the north, which declared its independence in 1991 and whose president, Dhair Riyale Kahin, was elected in 2003. Its capital, Hargeisa, is relatively prosperous with wage rates higher than in some other African countries and now has a new university. The population is around two million and most government revenue comes from customs duties. Services such as electricity and telecommunications are run by the private sector.

Another region, in the north-east, which has been autonomous since the end of 1998, though it has not declared itself as a state, is Puntland.

There have been numerous attempts at peace agreements. One in 2000 set up a Transitional Administration but this failed to exert any control. The most recent in 2004 brought together all the main clans and minorities in Kenya and, with cash from the EU, set up a 275-strong Federal Transitional Parliament (FTP) whose members were chosen by warlords. In January 2006, this established itself in an abandoned warehouse in Baidoa.

In May 2006, however, Islamist militias, controlled by a 'Union of Islamic Courts', started an assault on the warlords and soon seized Mogadishu. They were welcomed by many as offering some form of stability—though they apply Shariah law fiercely, they appear less brutal than the warlords they have displaced. They have also denied links with al-Qaeda. While they could take over the whole country and work with the FTP, their success might also provoke an attack from Ethiopia which worries about its own Islamic militants.

South Africa

Post-Mandela South Africa is trying to heal the old wounds of apartheid

Land area: 1,221,000 sq. km.
Population: 47 million—urban 57%
Capital city: Pretoria, 1.0 million
People: Black 79%, white 10%, coloured 9%, Indian 2%
Language: Afrikaans, English, Ndebele, Pedi, Sotho, Swazi, Tsonga, Tswana, Venda, Xhosa, Zulu
Religion: Christian 76%, Islam 2%, other 2%, unspecified 2%, none 18%
Government: Republic
Life expectancy: 48 years
GDP per capita: $PPP 10,346
Currency: Rand
Major exports: Gold, diamonds and other minerals, metals

South Africa has narrow coastal plains in the east, west and south. These rise to a chain of mountains that curves around the coast. The mountains are at their highest in the east where the Great Escarpment culminates in the Drakensbergs in the north. Within these mountains lies the vast plateau that makes up around two-thirds of the country.

Three-quarters of the population are black. The largest groups are the Zulu and the Xhosa, though official status has been extended to nine African languages. Of the white population, just over half speak Afrikaans, while the rest largely speak English. The 'coloured' population is of mixed African, Asian, and European descent and most of these speak Afrikaans. The majority of Asians originate from India and are English-speaking. In addition, there are thought to be up to eight million unauthorized immigrants from neighbouring countries.

The government has been struggling to redress the injustices of apartheid. More than two million houses have been built, and many more homes in black residential areas now have piped water and electricity. But there is still a long way to go. Health standards for blacks are still similar to those in other African countries, while for whites they are comparable to those in Europe.

The government has also promoted black economic empowerment. But this remains a very unequal society: the richest 20% of the population get 62% of national income and the poorest 20% get only 4%. Unemployment is high at 28%.

Exacerbating these problem is the devastation being caused by HIV/AIDS. Around 20% of the population are infected and some estimates suggest that, as a result, by 2050 the population will fall by one-quarter. South Africa was slow to address the epidemic because of President Mbeki's dissident view that AIDS was caused not by the HIV virus but by poor health and poverty. Now the policy is to give anti-retroviral treatment to all who need it, though this too is happening only slowly.

Another major social problem is crime: some 20,000 people are murdered each year and rape and robbery are also at high levels.

South Africa's economic development in the past depended on its mineral wealth. It has the

De Beers dominates the world diamond market
world's largest reserves of gold, along with many other minerals such as platinum and manganese, as well as diamonds, coal, and iron. Diamond mining is controlled by the De Beers company which has also dominated the world diamond market.

South Africa's manufacturing companies, which are responsible for around 20% of GDP, process these minerals, as well as agricultural produce, and make a wide range of consumer goods. Some have suffered as a result of exposure to international trade, though others including the car industry have expanded.

Agriculture still employs one-quarter of the workforce. The main crop is maize, most of which is still produced on white-owned farms. Land reform has been limited. A 1994 law allows people to claim ancestral property seized during the apartheid years. But by the end of 2004 only 3% of agricultural land had been redistributed.

Like more advanced industrial economies, South Africa has a sophisticated service sector. This includes financial services which contribute 20% of GDP and a major tourist industry that each year welcomes more than six million

visitors.

South Africa's post-apartheid era started when Nelson Mandela strode from prison in 1990. In the first multiracial election in 1994 his once-banned African National Congress won 63% of the votes in the National Assembly, which duly elected him president. The main post-mortem on the apartheid years was through a Truth and Reconciliation Commission, headed by Bishop Desmond Tutu, which granted amnesties to many perpetrators of violence.

When Mandela retired, his successor, Thabo Mbeki, led the ANC to even more decisive victories in the 1999 and 2004 elections, with more than two-thirds of the vote. The ANC presided over steady, if not spectacular economic growth and has made some progress in delivering social services. But it has remained popular primarily because it is the party of liberation and has kept the country together without major racial violence or a mass exodus of minority groups.

ANC dominance could be unhealthy. But as yet there is little sign that it is abusing this power, and South Africa, unlike many other African countries, has many other democratic checks and balances through the courts, the media, businesses, and pressure groups.

The main opposition party is the mainly white Democratic Alliance, led by Tony Leon, formed in 2000 from parties that represented both Afrikaaners and white liberals, but in 2004 it attracted only 15% of the votes. The Inkatha Freedom party, led by Mangosuthu Buthelezi, rivals the ANC for black votes but only in Kwazulu-Natal.

Spain

Spain is now accommodating its dissident regions and making rapid economic strides

Land area: 505,000 sq. km.	
Population: 42 million—urban 77%	
Capital city: Madrid, 3.0 million	
People: Spanish, though with strong regional identities	
Language: Castilian Spanish 74%, Catalan 17%, Galician 7%, Basque 2%	
Religion: Roman Catholic	
Government: Constitutional monarchy	
Life expectancy: 80 years	
GDP per capita: $PPP 22,391	
Currency: Euro	
Major exports: Cars, chemicals, fruit, vegetables	

Spain, Western Europe's second largest country, is dominated by the Meseta, the vast and often barren central plain that covers more than two-fifths of the territory. Most of the people who live on the Meseta are in its centre, in Madrid, which was established there for political reasons. Otherwise, the most densely populated areas are those that encircle the plain.

This dispersal of population contributes to strong regional identities. Though Spain is racially fairly homogenous, it has major groups, such as the two million Basques in the north and the six million Catalans in the north-east, who are determined to defend their distinctive language and culture.

In future, however, there are likely to be fewer Spaniards of any kind. The average fertility rate has fallen to only 1.1 children per woman of childbearing age, and after around 2010 the population will start to fall.

Spain used to be a country of emigration. But now the flow is in the other direction. By 2004, Spain had attracted 3.7 million official immigrants (8.4% of the population) along with thousands more unauthorized immigrants. Most come from other EU countries, though the largest single group is from Morocco from which many people risk a dangerous clandestine journey across the Straits of Gibraltar.

Only 5% of the workforce are engaged in farming, but the country is a major exporter of wine, fruit, vegetables, and also of olive oil of which Spain is the world's largest producer. The rural population is shrinking fast, in some regions by 5% per year, and much of the land is turning to desert—left untended and subject to droughts and forest fires.

Rural communities on the coast also have the option of fishing—Spain's 17,000 or more boats are the largest fleet in Europe—but fishing too seems to be in decline. Of these boats, the 2,000 deep sea vessels that fish in foreign waters take the bulk of the catch.

Rather more healthy is Spanish

industry. The Basque country has long been a centre of heavy industry and machine tools. And Catalonia has thousands of small companies thriving in sectors such as textiles and shoes. But the most striking success in recent years has been the car industry, the third largest in Europe, turning out around three million cars per year of which 80% are exported. All the companies are owned by foreign transnationals.

Spain exports 80% of its car production

Agriculture and industry are now eclipsed by services, which now account for more than 60% of GDP. Of these, tourism is one of the most important. More than 50 million people arrive each year, making this the world's second most popular destination, after France. Most visitors head for the Mediterranean beaches of the Costa Brava and the Costa del Sol. However, the industry may now be in decline, faced with competition from Eastern Europe at the cheap end of the market and Asia at the top end.

Despite Spain's economic advance, major problems persist. The most serious is unemployment, which has been falling but was still 10% in 2004.

The most difficult political problem has been in the Basque country, where 38% of voters in 2000 chose parties that favour independence. The smallest of these is Herri Batisuna. Since 1969 its terrorist arm, ETA, has killed over 800 people. In 2006, however, ETA declared a ceasefire and the government gave the region greater autonomy.

Spain's recent renaissance dates from the end of the dictatorship of General Francisco Franco who died in 1975. The first free elections for 40 years were held in 1977, and in 1978 a new constitution established a constitutional monarchy that also involved devolution to 17 autonomous regions. Regional governments now spend more than the central government. Those to which most power is now devolved are the Basque country, Catalonia, and Galicia.

The successful transition to democracy owes a huge debt to King Juan Carlos—especially his role in resisting a 1981 coup attempt. But one of the most significant political figures, opening the country up both economically and socially, was Felipe González, whose centre-left Partido Socialista Obrero Español (PSOE) governed from 1982 to 1996.

González and the PSOE were replaced after the 1996 election by the centre-right Partido Popular (PP), led by José Maria Aznar who was re-elected in 2000. One of Aznar's most unpopular policies was support for the US-led war on Iraq, and this led to his downfall. On 11 March 2004, Islamic terrorists planted bombs in the Madrid railway system killed over 200 people. Aznar mistakenly blamed ETA.

In the elections three days later on 14 March the voters punished the PP by returning as the largest party the PSOE, now led by José Luis Rodríguez Zapatero. Zapatero confirmed his election promise that he would withdraw troops from Iraq.

His administration has also pressed ahead with liberal legislation, on gay marriage, for example. And he has been fairly liberal on the economic front, reforming labour laws. But, keeping close to the centre ground, he remains popular.

Sri Lanka

Sri Lanka's 20-year-old civil war continues, despite a series of ceasefires

Land area: *66,000 sq. km.*
Population: *20 million—urban 21%*
Capital city: *Colombo, 2.3 million*
People: *Sinhalese 74%, Moor 7%, Indian Tamil 5%, Sri Lankan Tamil 4%, other 10%*
Language: *Sinhala 74%, Tamil 18%, English 8%*
Religion: *Buddhist 70%, Muslim 8%, Hindu 7%, Christian 6%, other 9%*
Government: *Republic*
Life expectancy: *74 years*
GDP per capita: *$PPP 3,778*
Currency: *Sri Lankan rupee*
Major exports: *Textiles, garments, tea, gems, chemical and rubber products*

One-sixth of Sri Lanka consists of the Central Highlands in the south-centre. From these highlands a series of plains spread out, though rather than being flat they are traversed by a variety of ridges and valleys. Despite its relatively small size, Sri Lanka has distinct variations in rainfall, so there is a dry zone and a wet zone. The dry zone in the north and east relies for rainfall mostly on the annual monsoons. The wet zone is in the south-west. This is the heart of the country, with year-round rainfall and most of the cultivable land, as well as the bulk of industry and two-thirds of the people.

Many live close to the coast. In December 2004, however, a tsunami devastated much of Sri Lanka's coastline and killed 31,000 people.

Three-quarters of the population are Sinhalese and Buddhist. But there are also significant minorities.

The largest of these are the Tamils, which in turn are divided into two groups: half are the original 'Sri Lankan Tamils', who are generally better educated and live in the north of the island. The other half are 'Indian Tamils', descendants of people who came later during the colonial period to work on the tea plantations. Another, smaller Tamil-speaking minority are the 'Moors', who are descendants of Arab traders. Sri Lanka's ethnic battles have primarily been between the Sinhalese and the Sri Lankan Tamils.

Despite the conflict Sri Lanka has been a human development success story. Education and health services are mostly free. Over 90% of the population are literate and 93% have access to basic health facilities. Notably, Sri Lanka's successes have been shared equally between men and women. Nevertheless, unemployment remains high and around one-quarter of the population live below the poverty line.

As a result many have migrated in search of work. Around 900,000 Sri Lankans are employed overseas, the majority of them women domestic workers in the Middle East, who remit more that $1.7 billion per year.

In 1977, the government of Sri Lanka became the first in South Asia to liberalize and diversify its economy. Since then, it has seen a rapid expansion of manufacturing. This is dominated by the garments industry which now accounts for one-third of employment and produces more than half of exports. Competition is increasing from other low-cost producers but the industry should survive. Other important industries include footwear, food, chemicals, and rubber and petroleum products. Much of this was based on foreign investment, though this has slowed because of the conflict.

Agriculture remains important, employing around one-third of the labour force. Two-thirds of this is **The world's leading tea exporter** directed towards the domestic market, particularly the cultivation of rice, though one-quarter of the country's rice needs have to be imported. Sri Lanka's main agricultural exports— tea, rubber, and coconuts—are grown in the wet zone. Tea production is mostly on large plantations: Sri Lanka is the world's leading tea exporter. But rubber and coconuts, which are small-holder crops, have both stagnated.

A more buoyant source of export income has been the mining of gems, including sapphires and rubies, as well as semi-precious stones.

The roots of Sri Lanka's civil war lie in the 1950s when the government changed the official language from English to Sinhala. Tamils argued that this and other measures discriminated against them. By the 1980s discontent had erupted into inter-communal violence as Sri Lankan Tamils embarked on a struggle for an independent Tamil homeland—Eelam, which is around one-third of the island in the north and east. Their armed group is the Liberation Tigers of Tamil Eelam (LTTE) with 4,000 fighters, many of them children. In response, militant Sinhalese formed their own violent organization, and the security forces fought against both. Since then the civil war has killed at least 50,000 people. Despite a ceasefire agreement in 2002, in 2006 the civil war started up again killing at least 1,000 more and displacing 135,000 others.

To its credit, Sri Lanka has remained a democracy. Through the 1980s and early 1990s the government was in the hands of the United National Party (UNP). But the 1994 elections were won by a coalition, the People's Alliance (PA). The principal partner was the Sri Lanka Freedom Party (SLFP) whose leader Chandrika Kumaratunga was also elected president. She and the PA were re-elected at the end of 1999.

In 2001, however, her coalition fell apart and after a fresh election the UNP, led by Ranil Wickremesinghe, took over. He negotiated a ceasefire with the LTTE, but Kumaratunga accused him of conceding too much and in February 2004 she dissolved parliament. The SLFP won the most seats in the subsequent election and formed a new coalition, the United People's Freedom Alliance (UPFA).

In 2005, the Supreme Court ruled that Kumaratunga could not stand again. The UPFA candidate, Mahinda Rajapakse, defeated Wickremesinghe in a presidential election.

Sudan

As one war in Sudan ended, another humanitarian disaster erupted in Darfur

Land area: 2,506,000 sq. km.
Population: 35 million—urban 39%
Capital city: Khartoum, 925,000
People: Black 52%, Arab 39%, Beja 6%, other 3%
Language: Arabic, Nubian, Ta Bedawie, many other languages
Religion: Sunni Muslim 70%, indigenous beliefs 25%, Christian 5%
Government: Republic
Life expectancy: 56 years
GDP per capita: $PPP 1,910
Currency: Sudanese pound
Major exports: Sesame, cotton, livestock, groundnuts

Sudan has three main geographical regions. The north, covering around one-third of the country, consists largely of an arid, rocky plain. The centre has low mountains and sandy desert. The more tropical south has extensive swamps and rainforests. One of the country's major features is the River Nile. The Blue Nile and the White Nile flow from the south, joining at Khartoum to form the Nile itself, which flows north into Egypt.

Sudan is ethnically very diverse, with more than a dozen major groups and hundreds of subgroups. The largest, and politically the most dominant, are the Sunni Muslim Arabs who live in the north and centre. Most of the other groups are black Africans who predominate in the south and west and are either Christians or animists. The southern region has around one-quarter of the population.

Sudan has a low level of human development. Around 40% of the population are illiterate and only half the country's children enrol in primary school. Health standards too are very poor. Hospitals and clinics are primarily in the urban areas, while in the rural areas millions suffer from infectious diseases, particularly malaria and guinea worm. Poverty has been compounded by civil war which has led to several famines, driven hundreds of thousands of refugees into neighbouring countries, and displaced four million within Sudan. Since 1983, an estimated two million people have died in warfare.

Sudan is heavily dependent on agriculture, which accounts for one-third of GDP and employs two-thirds of the workforce. Sudan could be a major agricultural producer. Most people are still working at the subsistence level, growing sorghum and millet, particularly in the south and also in parts of the centre and west. But Sudan also has a mechanized farming sector—the result of extensive government investment. Three-quarters of this is rain-fed, particularly in the area around the Blue Nile. The remainder of the mechanized sector relies on irrigation for food and cash crops. This includes the vast Al-Gezira irrigation scheme south of Khartoum, between the White and Blue Niles, which waters the land of more than 100,000 tenant farmers. Originally this was used for cotton, but now more for sesame.

Sudanese industry is limited, consisting largely of processing agricultural crops—refining sugar, for example, and producing cotton textiles. But industry generally has been restricted by low investment and by a shortage of skilled workers, since many Sudanese have emigrated.

One of the most promising future areas is oil. Large deposits were

Prospects of oil income

discovered in the south in the early 1980s but production and the laying of pipelines have been disrupted by rebel attacks. Output is now increasing and could provide income of up to $2 billion per year.

Sudan's north-south civil war had its roots in the rebellion of a southern army corps in 1955 who were protesting at northern domination. A military coup in 1969, headed by Colonel Jafar al-Nimeiri, appeared to offer some respite and in 1972 the war stopped. But in 1983 it flared up again. Colonel John Garang led the rebellion of the southern Sudanese People's Liberation Army (SPLA).

Nimeiri introduced Islamic shariah law, but following food riots in 1985 he was ousted in a bloodless coup.

A subsequent civilian government was replaced in 1989 by another Islamic-dominated military coup led by Brigadier Omar Hassan al-Bashir. He banned all political parties— except effectively for the National

Islamic Front, now called the National Congress (NC), which is linked with the Muslim Brotherhood. Al-Bashir ruthlessly suppressed opposition and stepped up the war against the south.

Al-Bashir was elected president in 1996 in an election boycotted by the opposition. In 1998, a new constitution legalized other parties. In 1999 Al-Bashir suspended parliament and imposed a state of emergency. Then in 2000 he was overwhelmingly re-elected for a further five-year term.

Peace was achieved starting in 2002 with the 'Machakos Protocol' which included an undertaking to share power and hold a referendum on secession in the south within six years. The deal was sealed in 2005 and Garang became deputy president, only to die a few weeks later in a helicopter crash and be replaced by his deputy Salva Kiir.

Meanwhile another war had broken out in 2003 in Darfur in the west. An uprising by the black 'Sudan Liberation Army' protesting against the marginalization of the region met with a brutal response from a government-backed Arab militia, the Janjaweed. This led to horrific violence and accusations of genocide.

A ceasefire was supposedly agreed in 2004 backed by peacekeeping troops from the African Union. Another was signed in May 2006. But both have been undermined in part by splits among the Darfur rebels. And Al-Bashir, having given concessions to the south, is unwilling to extend them to the west and has refused to allow in UN peacekeepers.

By mid-2006 up to 300,000 people had died and two million more had been displaced.

Suriname

Fractious politics, corruption, and drug smuggling

Land area: 163,000 sq. km.
Population: 0.4 million—urban 76%
Capital city: Paramaribo, 243,000
People: East Indian 37%, Creole 31%, Javanese 15%, Bosneger 10%, other 7%
Language: Dutch, English, Sranang Tongo, Hindustani, Javanese
Religion: Hindu 27%, Protestant 25%, Roman Catholic 23%, Muslim 20%, indigenous beliefs 5%
Government: Republic
Life expectancy: 69 years
GDP per capita: $PPP 4,100
Currency: Suriname dollar
Major exports: Alumina, aluminium, rice

Suriname has a narrow, marshy coastal plain, but most of the country consists of a vast plateau along with ranges of mountains covered with dense tropical rainforests.

Suriname's complex ethnic mix reflects its colonial history. Today's creole population are descendants of African slaves whom the Dutch brought to work on the sugar and coffee plantations. They were subsequently replaced by indentured workers from India and from Java—whose descendants now make up the majority of the population. Others include the Bosnegers, descendants of escaped slaves, and Amerindians.

Suriname is well endowed with natural resources. The most important of these is bauxite, which is mined by US- and South African-owned companies. All of this is processed within Suriname into alumina, and some into aluminium, taking advantage of cheap hydroelectric power at Afobaka, where the dam has created a huge artificial lake.

Bauxite provides around three-quarters of export income. In addition, there are reserves of gold, nickel, and silver that have attracted the attention of Canadian companies.

Only 15% of the labour force work in agriculture, producing rice, fruit, and vegetables. But Suriname's vast tropical hardwood forests have attracted Asian logging companies, to the alarm of environmentalists.

A less official export is cocaine: Suriname is one of the main drug transit routes to the Netherlands.

Politics in Suriname has frequently been marked by violence and overshadowed by the military. In the 1996 National Assembly elections, the largest party was the pro-military National Democratische Partij, led by former military dictator Desi Bouterse. Later the National Assembly elected Jules Wijdenbosch as president.

The general election of May 2000 was won by the National Front (NF), a coalition of three ethnic parties, which used its majority to elect as president Ronald Venetiaan of the Nationale Partij Suriname.

Venetiaan had an unnerving start when it was discovered that the country's gold reserves had disappeared. He managed to stabilize the economy but had little impact on reducing poverty or corruption.

In the 2005 general election the NF lost some seats but eventually it allied itself with other parties and just managed to have Venetiaan re-elected president.

Swaziland

Africa's longest surviving absolute monarchy now stricken by HIV/AIDS

Land area: 17,000 sq. km.
Population: 1 million—urban 24%
Capital city: Mbabane, 60,000
People: African 97%, European 3%
Language: Siswati, English
Religion: Zionist 40%, Catholic 20%, Muslim 10%, other 30%
Government: Monarchy
Life expectancy: 33 years
GDP per capita: $PPP 4,726
Currency: Lilangeni
Major exports: Soft-drink concentrates, sugar, wood pulp, cotton yarn

Swaziland can be divided into four regions from west to east. In the west is the Highveld, which rises to 1,400 metres and covers one-third of the country. This descends first to the Middleveld, and then the Lowveld, before reaching the Lubombo Mountains, which form the eastern border.

Almost all of the people are Swazi. Heavy investment in education has resulted in an 80% literacy level. Health standards are lower, with a relatively high infant mortality rate, and they are set to fall further still as a result of HIV/AIDS: in 2005, around one-third of adults were carrying the virus. By 2010 life expectancy will fall to 27 years.

Most economic activity is based directly or indirectly on agriculture. Landholdings are of two forms. Around 60% is 'Swazi National Land', held in trust by the king and used communally for rain-fed subsistence crops and livestock as well as for cotton. The rest is 'Title Deed Land', most of which is owned by corporations that irrigate it to grow sugar cane, citrus fruits, and pineapples. These companies, like many businesses, are in the hands of trusts controlled by the royal family.

Industrial activity is centred on processing agricultural goods. In 1986, Coca-Cola moved its concentrate plant from South Africa to Swaziland, taking advantage of cheap sugar. Other similar companies have followed, and Swaziland's leading export is now soft-fruit concentrate.

Swaziland achieved independence in 1968 as a constitutional monarchy. This did not last. In 1973 King Sobhuza II suspended the constitution in what has been called the 'king's coup' and banned political activity. In 1986 his successor, King Mswati III, kept most of these restrictions in place and now rules as an absolute monarch with a lavish high-spending lifestyle.

There is also some quasi-democratic representation through the 65-member House of Assembly. The elections of 2003 were typically constrained, allowing neither parties nor political gatherings, but did deliver a more diverse group of members.

Opposition comes mostly from trade unions, churches, students, and groups that form the Swaziland Democratic Alliance (SDA). In recent years there has also been some anti-government violence.

A new constitution was approved in 2006, but offered no serious changes; indeed in some respects it gave the king even more power.

Sweden

Swedish social democracy has lost some of its gloss, but still has striking achievements

Land area: 450,000 sq. km.	
Population: 9 million—urban 83%	
Capital city: Stockholm, 765,000	
People: Swedish, Lapp (Sami)	
Language: Swedish	
Religion: Lutheran 87%, other 13%	
Government: Constitutional monarchy	
Life expectancy: 80 years	
GDP per capita: $PPP 26,570	
Currency: Krona	
Major exports: Manufactured goods, electrical machinery	

Sweden can be divided into three regions. The largest, in the north and centre, is Norrland whose mountains and forests constitute around 60% of the country. In the far south is Götaland, which includes the Småland Highlands and, on the southern tip, the plains of Skåne. Between these, in the south-centre, is Svealand, a lowland area with numerous lakes.

Sweden's people until recently were ethnically quite homogenous—the largest minorities being people of Finnish origin and small numbers of the reindeer-herding Sami in the north. From the 1970s, however, Sweden's liberal immigration and asylum policies welcomed people from further afield. In 2004 around 12% of the population were foreign born. Previously immigrants tended to come from other Nordic countries, but in the 1990s many people arrived as refugees.

Sweden's people enjoy one of the world's highest standards of living. Between the 1930s and the 1970s, Sweden created an extensive welfare state. Around 85% of healthcare expenditure, for example, comes from public sources. This has ensured that the benefits of economic growth have been equitably distributed.

Some of the credit for this must go to a strong civil society, including well-organized trade unions and a large number of voluntary organizations. Sweden has also created other institutions like the 'ombudsman', which many countries have since copied. Public-sector spending has been reduced in recent years but in 2004 still accounted for over 56% of GDP, and the government employed 30% of the workforce. Most of this expenditure is based on local taxes spent by local governments.

Sweden's early industrialization had been based on raw materials like wood and iron ore, though it has neither coal nor oil and is heavily dependent on hydroelectric power. Minerals are still important: Sweden produces most of the EU's iron ore. But Sweden's investment in education also ensured that the country could move rapidly to higher levels of technology. In the internet age Sweden has the advantage that many of its people speak English well, and Sweden has one of the world's highest internet usage rates: 71% of the population are online.

This small country has also produced some of the world's leading manufacturing companies including SKF, the world's largest producer of ball-bearings, and the Swiss–Swedish engineering group ABB. Some leading companies like Saab and Volvo have had their car divisions bought up by US companies, but Volvo and Scania are among the world's largest producers of trucks. And one of the most striking successes has been the telecommunications company Ericsson. As a result, Sweden has become a major trading country. In 2004, exports were 36% of GDP. Remarkably, around 40% of the shares on the Stockholm stock exchange are controlled by one family, the Wallenbergs.

Sweden is also one of the world's most generous aid donors,

Sweden is a generous aid donor giving more than 0.8% of GDP in official development assistance—one of the highest proportions in the world.

Sweden's successful socio-economic model started to come under question in the 1970s and 1980s. Economic growth had slowed and could no longer support public expenditures. By the early 1990s the economy was deep in recession and unemployment had shot up to 8%. This forced the government to cut back on welfare spending. By the standards of other countries, these changes have been modest. The economy has since revived and

unemployment in 2005 was 6%.

Sweden is a constitutional monarchy with King Carl Gustaf XVI as the ceremonial head of state. Since the Second World War, the country has usually been governed by the Social Democratic Party (SAP), though often in co-operation with other parties. From 1969, the SAP was headed by Olaf Palme who in 1986 fell victim to one of the world's most unexpected assassinations.

He was replaced as prime minister by Ingvar Carlsson. By the 1991 election, however, Sweden was deep in recession and the SAP lost heavily. Carlsson resigned and Carl Bildt took over as prime minister, heading a coalition led by his own Moderate Party along with the Centre, Liberal, and Christian Democrat Parties.

After the 1994 elections the SAP returned, forming a minority government, but Carlsson resigned the following year to be replaced by Goran Persson. In the 1998 elections, the SAP lost ground and had to enter into an informal alliance with the Left Party (the former communists) and the Greens.

This pattern continued after the 2002 elections with Persson returning as prime minister at the head of a minority administration that depends on the Left and Green parties.

Though Sweden joined the EU in 1995, support is at best lukewarm. In September 2003, despite the government's recommendation, Swedes rejected joining the euro zone.

In the elections in 2006 the voters were clearly ready for a change and gave a narrow victory to the centre-right Alliance coalition, led by Fredrik Reinfeldt of the Moderate Party.

Switzerland

Switzerland has now joined the UN, but still hesitates about the European Union

Land area: 41,000 sq. km.	
Population: 7 million—urban 68%	
Capital city: Bern, 123,000	
People: German 65%, French 18%, Italian 10%, Romansch 1%, other 6%	
Language: German 65%, French 18%, Italian 10%, Romansch 1%, other 6%	
Religion: Roman Catholic 42%, Protestant 35%, Muslim 4%, other 19%	
Government: Republic	
Life expectancy: 81 years	
GDP per capita: $PPP 30,552	
Currency: Swiss franc	
Major exports: Machinery, precision instruments	

Switzerland is one of Europe's most mountainous countries. More than two-thirds of its territory, to the south and east, is covered by the towering peaks of the Alps. Another eighth, running along the north-west border with France, is covered by the less dramatic Jura range. Lying between these, and occupying most of the rest of the country, is a plateau around 400 metres above sea level interspersed with hills. It is within this belt of land, stretching from Lake Geneva in the south-west to the Bodensee on the north-east border with Germany, that most industrial and agricultural activity takes place, and where most people live.

Switzerland's population is formed by a conjunction of three European cultures: German, French, and Italian. A highly decentralized form of government, based on semi-autonomous states or 'cantons', has enabled these cultures to coexist and thrive. Switzerland thus has three official languages. The German-speaking majority live mostly in the north and east; the French in the south-west; and the Italian in the south-east. There is also a fourth, though not official, set of dialects—Romansch—which are also spoken in the cantons bordering on Italy.

Cultural diversity is further intensified by a large immigrant community, which makes up around 20% of the population. Around one-fifth are Italian, and another 14%, more recent arrivals, are from former Yugoslavia. These workers are very diverse: some are highly paid international business managers and bureaucrats; others are seasonal agricultural workers or hotel staff.

Poorly endowed with productive natural resources, Switzerland has relied for survival on the skills and ingenuity of its people. As with most richer countries, around two-thirds of the workforce are employed in service industries. In Switzerland, however, one of the most significant of these is banking: Swiss banks manage about a third of all private financial assets invested across borders, and financial services are about 9% of GDP and employ 3% of the workforce. Switzerland's banking secrecy laws have contributed to this success, though they have come under criticism. Following accusations of collaboration with the Nazis in the Second World War, and of keeping the deposits of their victims, two banks agreed to pay $1.3 billion to settle Holocaust-related claims.

Industry in Switzerland has been weighted towards higher levels of technology in such areas as chemicals, pharmaceuticals, and the manufacture of precision instruments and watches. Engineering and food processing are also important.

The perennial strength of the Swiss franc hampers exports, but heavy investment in research has kept Switzerland at the technological leading edge. A notable achievement was that of the Swatch watch company in fending off the digital challenge from Japan. For a small country, Switzerland is also home to many multinationals, including CIBA, ABB, and Nestlé.

Swatch repels the Japanese challenge

Agriculture, whether growing crops on the plain or rearing livestock in the mountains, employs less than 4% of the workforce but is very productive and meets around 60% of local needs. It is also highly protected, and the expensive food produced encourages many Swiss to stock up in neighbouring countries.

The beautiful Swiss countryside also draws in many tourists—over 11 million per year—though many are increasingly discouraged by the expense.

The most remarkable aspect of Switzerland, however, is its form of direct democracy. The central government is weak. First, because much of the power resides in the 23 cantons, each of which is responsible for education, hospitals and taxation, and has its own judiciary. Second, because most vital decisions are subject to referendums, of which there are three or four each year.

This narrows the territory for national party politics. Most parties are stronger at the cantonal level, but they do elect members to the bicameral federal assembly which since 1959 has been controlled by the same broad four-party coalition—though the populist anti-immigrant Swiss People's Party has recently become the most prominent member. The assembly in turn elects a seven-member federal council to serve as the executive. Of these, one person is chosen to serve one year as president—a term of office so ephemeral that many Swiss are hard-pressed to name their head of state.

This decentralized system has produced stable if conservative government that has maintained Swiss neutrality and economic independence. Switzerland finally joined the UN in 2000.

In a referendum in 2001 Switzerland again rejected EU membership, though views are divided: broadly, urban dwellers and French-speakers are in favour, while rural dwellers and German-speakers are likely to be against. The country does, however, co-operate closely with the EU, for example allowing workers from all EU countries.

Syria

Syria is coming under increasing pressure as a result of the war on terrorism

Land area: 185,000 sq. km.			

Land area: 185,000 sq. km.
Population: 18 million—urban 50%
Capital city: Damascus, 3.9 million
People: Arab 90%, Kurd, Armenian, and other 10%
Language: Arabic
Religion: Sunni Muslim 74%, Alawi, Druze, and other Muslim sects 16%, Christian 10%
Government: Republic
Life expectancy: 73 years
GDP per capita: $PPP 3,576
Currency: Syrian pound
Major exports: Oil, phosphates

Syria has three main regions. One is a narrow fertile coastal strip that has year-round supplies of water. To the east of this is a mountainous zone with two ranges running parallel to the coast. The rest of the country is the Syrian Desert. Most people live in the section of land between Damascus in the south and the second city, Aleppo, near the Turkish border, though the most densely populated land is along the coast.

The population is mostly Arab and Muslim—split between the majority Sunni community and others including the Alawi, a branch of the Shia sect. Syria has oil but is not wealthy. The population is growing rapidly and the healthcare system is struggling to keep pace. Education standards, however, are relatively high.

Agriculture is still an important part of the economy, employing around one-third of the labour force.

Two-thirds of the cultivable land is in private hands and it is evenly distributed among small farmers, thanks to extensive land reform in the 1960s. Nevertheless, the government exerts a strong influence since it controls the prices—often supporting farmers by paying more than the world price for crops.

Farmers devote around two-thirds of the land to wheat and barley for local consumption, but more significant is cotton, which generates around half of agricultural GDP and provides around one-tenth of export earnings. Some 70% of Syria's agriculture relies on the often erratic rainfall; the remaining irrigated land is mostly in the coastal strip.

Industry is also a major employer. The government has invested in heavy industries, but most activity nowadays is in lighter manufacturing such as cotton textiles, which employ one-third of the industrial workforce.

The government has tried to encourage greater private, and even foreign, investment, and since 1991 the proportion of the economy in private hands has risen from 35% to 70%. But liberalization has slowed in recent years. One major obstacle remains the state-controlled banking system. Even the government prefers to use banks elsewhere, particularly in neighbouring Lebanon.

Syria still depends heavily on oil, which provides around 70% of export income. Oil has been extracted in quantity only since the 1980s; nevertheless the reserves are much smaller than those in other countries,

Syria's oil will soon run out
especially the Gulf, and could be exhausted within 15 years, not merely reducing export income but also turning the country into a net importer. Production is controlled by Al-Furat Petroleum, which is a joint venture between the state-owned Syrian Petroleum Company and three foreign corporations.

Syria also has gas reserves which have been developed by foreign investors, Conoco and Elf Aquitaine, to supply local energy markets, hopefully releasing more oil for export. There are also plans for a major pipeline for gas exports.

The other major mineral export is phosphates. A further source of foreign exchange has been expatriate Syrian workers in the Gulf countries whose remittances are equivalent to 20% of export earnings.

Since 1963, the government of Syria has been the exclusive preserve of the socialist Ba'ath (Resurrection) Party, though the current regime dates from a coup in 1970 when a faction representing the minority Alawi sect seized control. In 1971 the coup's leader, Hafez al-Assad, was elected president, a post he held until his death in June 2000.

Although nominally a republic with a parliament and a council of ministers, Syria is in practice a dictatorship with the president exerting control through the army and the intelligence services. Military spending is heavy—11% of GDP.

The Assad family and other Alawis control the government but most of the business community is Sunni—a division that is a source of tension. The government is nervous about future liberalization since this would put too much power into the hands of Sunni merchants.

Syria's control has also effectively extended over Lebanon. Syria sent in troops in 1976 to help quell the civil war and subsequently retained a presence in what became a client state —though under international pressure withdrew in 2005.

Syria considers itself to be a centre of Arab nationalism and Damascus has been host to a number of dissident Palestinian groups. Syria has also been one of the most implacable opponents of Israel, with whom it has fought wars in 1967, 1973, and 1982.

Hafez Assad was succeeded by his son Bashar who was nominated by the People's Assembly and confirmed as president in a referendum in July 2000. Bashar's arrival could have marked a new era and he embarked on some economic reforms. In 2004 he even considered political reforms that might lead to a multiparty democracy.

Since then the window of opportunity seems to have closed. Assad tried to give the impression that his reform efforts were being held back by the 'old guard' in the government and the army. But this now rings hollow as he has replaced them with equally hardline characters. Syria seems merely to have acquired a new but less competent dictator.

Taiwan

Taiwan is the most financially robust of the Asian tigers, but its relationship with China remains precarious

Land area: 35,980 sq. km.
Population: 23 million
Capital city: Taipei, 2.6 million
People: Taiwanese 84%, mainland Chinese 14%, aborigine 2%
Language: Mandarin Chinese, Taiwanese
Religion: Buddhist, Confucian, and Taoist 93%, Christian 5%, other 2%
Government: Republic
Life expectancy: 77 years
GDP per capita: $PPP 27,600
Currency: New Taiwan dollar
Major exports: Machinery, electrical and electronic products

The eastern part of the island of Taiwan, around two-thirds of the territory, consists mostly of a series of mountain ranges running north to south that descend steeply to the eastern coast. To the west, they descend more gently to a broad coastal plain that is home to most of the population. The island is geologically active: in 1999 an earthquake killed 2,448 people.

Taiwan's original inhabitants were a diverse collection of Malayo-Polynesian groups. Subsequent waves of emigration from China, however, displaced most of these groups to the mountains and they now account for only around 2% of the population. Most people are descendants of the Chinese who had been arriving since the 17th century. A later addition, and the most powerful, however, are the descendants of the two million people

who arrived following the 1949 communist revolution. These were largely the élite of the former regime: the industrialists and the professionals. More recent arrivals, but very much at the bottom of the social ladder, are the 300,000 or so immigrant workers. In recent years, however, there has also been an increase in labour emigration to China: at least 2% of Taiwanese now live and work there.

The basis for Taiwan's fairly equitable pattern of development was an extensive land reform over the period 1949–53. Most of the land on the fertile plains is still worked by small and very productive family farms that ensure the country remains self-sufficient in rice. Nowadays, however, agriculture employs only 6% of the workforce.

Taiwan's industrial development after 1949 benefited from considerable US aid. The initial emphasis was on import substitution, but from the 1960s the USA obliged the Taiwanese to open their economy and produce goods for export. This initiated a steady upward progression: in the 1960s, light assembly work and garments; in the 1970s, heavy industry; in the 1980s, TVs; and in the

1990s, computers. Manufacturing still accounts for around one-fifth of GDP.

Taiwan is now the world's third largest producer of information technology goods, mostly subcontracting for foreign brands—90% of the circuit boards for the world's personal computers, as well as 80% of the notebook computers, are made by Taiwanese companies. However most of these are not made in Taiwan. Taiwanese companies have shifted many operations to other countries, particularly China, not just for simpler products like TVs and VCRs, but also for computers; around two-thirds of IT hardware is now made in China.

Taiwan's industrial development is largely based on small producers—which account for around 80% of employment, and roughly half the economy. There is one company for every 18 people—the highest ratio in the world. Their effectiveness is based on extensive networking. Elsewhere, this might lead to corruption but this seems to have been less of a problem here. Taiwan did not suffer greatly in the Asian financial crisis, partly because its financial system is based on family loans and other unofficial sources, and because Taiwan has tight capital controls. It could not in any case turn to the IMF, since it is not a member.

Taiwan has a company for every 18 people

Taiwan would have a rosy economic future were it not for the looming threat from China. Since 1949, political life has been dominated by the question of the island's status. The Nationalists (the Kuomintang, or KMT), who fled the mainland, for decades asserted that the long-term aim was reunification—assuming that they would rule the whole of China.

The USA initially backed this position, but by 1979 had opened diplomatic relations with China and started to loosen ties with Taiwan. Taiwan is not a member of the UN and has diplomatic relations with only around 30 countries.

The USA's engagement with Taiwan was also weakened by decades of an authoritarian and corrupt KMT regime. Since then, however, Taiwan has become an open democracy, with direct elections for both parliament and the presidency.

In 1996 the presidential vote saw the re-election of the KMT's Lee Teng Hui who espoused reunification, but the presidential elections in 2000 produced a dramatic change. The KMT was split between an official and unofficial candidate, opening the way for Chen Shui-ban of the Democratic Progressive Party (DPP), which espouses independence.

The 2004 presidential elections also proved a dramatic affair. Though the KMT had dropped its reunification ambitions Chen had raised the stakes with simultaneous referendums on the relationship with China. The tension was further heightened when he was shot and lightly wounded the day before the election.

Chen, facing a single KMT candidate, won but only by a razor-thin margin, leading to unprecedented KMT street demonstrations demanding a recount. He survived these but in 2006 faced parliamentary proceeding to fire him because of allegations of corruption against members of his family.

309

Tajikistan

Tajikistan has suffered from civil war, corruption, and increasingly autocratic rule

Land area: 143,000 sq. km.
Population: 6 million—urban 25%
Capital city: Dushanbe, 562,000
People: Tajik 80%, Uzbek 15%, Russian 1%, other 4%
Language: Tajik, Russian
Religion: Muslim
Government: Republic
Life expectancy: 64 years
GDP per capita: $PPP 1,106
Currency: Tajikistan rouble
Major exports: Aluminium, electricity, cotton

roof of one house is the yard of the house higher up—heightening vulnerability to the earthquakes to which the region is prone.

This was always one of the poorest of the Soviet republics and a civil war from 1992 further undermined human development. Infant mortality is high and there is widespread poverty. Around one million Tajiks migrate at least seasonally to Russia each year, earning four or five times their $50 a month Tajik wage. Their remittances are though to be around one-fifth of GDP.

Ironically, for a country with so many rivers, one of the scarcest commodities is clean drinking water. The country's water treatment plants are in a state of disrepair and water supplies are frequently contaminated.

Agriculture occupies around two-thirds of the workforce, though the principal crop is not wheat but cotton, which is one of the economic mainstays, accounting for 20% of exports. Cotton demands extensive irrigation from the rivers and canals, but run-off from heavily-used pesticides and fertilizers for cotton production has contaminated groundwater and rivers. Farmers also rear livestock and grow cereals and a

Tajikistan is very mountainous. The country is largely a collection of valleys with rapidly flowing rivers fed by melting snow and glaciers. More than half this landlocked country is 3,000 metres or more above sea level.

Tajikistan was a forced creation of the Soviet Union in 1929 that took little account of ethnic divisions. Tajiks make up four-fifths of the population but there is also a substantial Uzbek minority, in addition to Russians and numerous other smaller groups. Most Uzbeks live in the more industrialized northern region, Khujand, which is connected only tenuously to the rest of the country. The Tajiks, who are mostly Sunni Muslims, largely live in the rural areas in the foothills of the mountains and in small communities stretched out alongside the rivers and irrigation canals.

In the narrower valleys, the flat

variety of other crops, but the country still remains heavily dependent on food imports.

Tajikistan's dramatic terrain also contributes directly to its two other main exports. Many of its fast-flowing rivers have been harnessed for hydroelectric power. The largest dam, at Nurek, delivers 11 billion kilowatt hours per year, and a second is now under construction. Much of this electricity is sold to neighbouring countries and accounts for one-quarter of export earnings.

Abundant hydroelectricity is also ideal for aluminium smelters, which absorb 40% of output—though all the aluminium oxide has to be imported. The Tursunzae smelter 65 kilometres west of the capital, Dushanbe, is one of the world's largest. Ageing equipment has contributed to a fall in output, but low world prices have discouraged investment.

Though it has no bauxite, Tajikistan does have rich deposits of other minerals, which are being extracted in co-operation with foreign mining companies. These include gold and silver, along with strontium, salt, lead, zinc, and fluorspar. The largest foreign investor is the British company Nelson Gold.

However, economic development is seriously hampered by corruption which the European Bank for Reconstruction and Development reckons to be the worst in Central Asia. This includes the trafficking of opium, en route from Afghanistan to Europe.

Tajikistan's fractured geography has contributed to its complex and fractious politics which aligns along not just ideological but also religious and regional fault lines. Since independence in 1991, there has been a struggle for power between communists and Muslims and also secular democrats. Former communists remained in power following independence, but Islamic groups embarked on a widespread armed struggle. Many fled the country and Islamic guerrillas established a base in Afghanistan. Around 25,000 Russian troops fought on the government's side while Iran backed the Islamic force. In addition there have been attacks from the Uzbeks.

Fractured geography; fractious politics

In 1994, a disputed presidential election resulted in a victory for the neo-communist Imomali Rahmonov at the head of his People's Democratic Party (PDP). In 1997 Russia and Iran supported a UN-brokered peace deal in which Rahmonov agreed to share power with the other parties, which had come together as the United Tajik Opposition (UTO), and to hold presidential and parliamentary elections. The UTO, for its part, agreed to demobilize.

Rahmonov duly rigged the November 1998 election and was re-elected with 96% of the vote in an impressive 98% turnout. The 2000 legislative election was also flawed, giving the PDP two-thirds of the seats.

Rahmonov's regime has been notable for press censorship, economic decline, and pervasive corruption. In 2003 a plebiscite approved constitutional changes to allow the president two more terms. In the 2005 parliamentary elections the PDP claimed an unlikely 75% of the vote. Political change seems remote.

Tanzania

Tanzania is politically stable, but development is slow and tarnished by corruption

Land area: 945,000 sq. km.
Population: 37 million—urban 35%
Capital city: Dodoma, 1.7 million
People: More than 100 African groups; Arab on Zanzibar
Language: Kiswahili, English, Arabic, and many local languages
Religion: Mainland: Christian 30%, Muslim 35%, indigenous beliefs 35% Zanzibar: Muslim
Government: Republic
Life expectancy: 46 years
GDP per capita: $PPP 621
Currency: Tanzanian shilling
Major exports: Coffee, manufactured goods, cashew nuts

Mainland Tanzania has a narrow flat coastal plain which rises to the vast plateau that makes up most of the country. But there is also some spectacular scenery, including in the north-east Africa's highest mountain, Kilimanjaro, at 5,895 metres. Much of the land is assigned to national parks and game reserves. The republic also includes islands in the Indian Ocean, two of the largest of which make up Zanzibar.

Tanzanians can belong to any one of 120 or more ethnic groups. None of these holds a dominant position, but effective efforts at 'nation-building' after independence, which included the promotion of Kiswahili as the national language, helped Tanzania to avoid serious ethnic conflict. The same language is also spoken in Zanzibar, though the population here—around one million—has a strong Arab component and is almost entirely Muslim.

Most Tanzanians are poor: around one third are thought to be living below the poverty line. One-third of children are malnourished and infant mortality is high. More than one-third of a million people die each year from malaria. HIV/AIDS is also now a major problem with around 7% of adults infected. The government has said it will make anti-retroviral drugs available to all who need them. Tanzanians do however have reasonable standards of education, a legacy of earlier socialist investment.

Most people live in the rural areas from subsistence agriculture though only 8% of the land is under cultivation; there are large fertile zones as yet untouched. The main food crops are maize, cassava, rice, sorghum, and beans. Most smallholders also grow one or more cash crops, including coffee, cotton, cashew nuts, and tobacco. Tea and sisal are the other main cash crops, though these are grown mostly on estates.

Many farmers also raise livestock, notably Zebu cattle along with sheep, goats, and poultry. There is also considerable potential for developing fishing offshore as well as sustainable exploitation of freshwater Nile perch in Lake Victoria.

In Zanzibar the main export crop is cloves, but agricultural productivity is low and output is not increasing fast enough to have an impact on poverty. Most manufacturing is based on processing agricultural commodities. Previously this was dominated by state-owned companies, though now many of these have been privatized.

Tanzania has a number of promising mineral deposits, which include diamonds as well as iron ore, nickel, and phosphates. However the main source of income is gold. There are three main gold mines, owned by Ghanaian, South African, and Australian companies, which have seen a steady rise in production.

Another area for development is tourism to the nature reserves and Africa's two largest game parks, as well as to the 'spice island' of Zanzibar. But annual tourist arrivals, around half of which come from the EU, are only half a million—far fewer than in neighbouring Kenya.

For the first quarter of a century after independence Tanzania was governed, first as prime minister

Julius Nyerere's influence

and later as president, by Julius Nyerere. He was a hugely influential figure nationally and internationally. Nyerere's brand of African socialism had some social and political successes in Tanzania but also notable failures, including his attempts to concentrate people into villages and promote co-operative development of the land. When Nyerere retired in 1985 (he died

in 1999), he was succeeded by Ali Hassan Mwinyi.

As a consequence of Nyerere's economic failures Mwinyi from 1986 had to sign Tanzania up to an IMF-directed structural adjustment programme. Donors also pressed for an end to one-party rule by the Chama Cha Mapiduzi (CCM) party.

Multiparty presidential elections were duly held from 1992. But multiparty democracy has yet to take root. The main alternative is the Civic United Front (CUF) but this does not provide strong opposition to the CCM.

For the 1995 elections the ineffective Mwinyi was succeeded as president by Benjamin Mkapa, and he and the CCM duly won the presidential and legislative elections in 2000. In turn, he was succeeded in 2005 by Jakaya Kikwete, who vowed to take a stronger line against corruption.

Semi-autonomous Zanzibar has representation in the mainland parliament, but also has its own president and internal administration. There has long been tension with mainland Tanzania. The CUF on Zanzibar accuses the CCM of sacrificing Zanzibar to the mainland, while the CCM says the CUF has plans to dissolve the union.

The 2000 elections in Zanzibar were violent and chaotic with the CCM presidential candidate Amani Abeid Karume declared the winner. 2001 saw a bombing campaign and an exodus of refugees while the CUF boycotted parliament. Peace was restored in the 'muafaka' agreement later that year. As the CCM promised, the 2005 elections were fairer, though they still returned Karume.

Thailand

A period of political chaos has been ended by a bloodless military coup

Land area: 513,000 sq. km.	
Population: 63 million—urban 32%	
Capital city: Bangkok, 5.7 million	
People: Thai 75%, Chinese 14%, other 11%	
Language: Thai, English	
Religion: Buddhism 95%, Muslim 4%, other 1%	
Government: Constitutional monarchy	
Life expectancy: 70 years	
GDP per capita: $PPP 5,840	
Currency: Baht	
Major exports: Food, machinery, manufactured goods	

Thailand's distinctive geographical shape extends over four main regions. In the far north are mountain ranges covered by forests of tropical hardwood, while the north-eastern area is much dryer—a barren plateau that makes up one-third of the country. The southern region, which stretches down the Malay peninsula, is mountainous with narrow coastal plains. But the core of Thailand, and the focus of most activity, is the central plain—a rich agricultural area criss-crossed with rivers and canals.

Thailand's people are relatively homogenous—held together by language, religion, and a deep respect for the monarchy. The majority of people speak Thai and almost all are Buddhist—most communities have ornate temples. Nevertheless, Thailand also has distinct ethnic groups. In the northern region these include the hill tribes and in the far south Malays,

who are Muslims and have affinities with Malaysia. But the largest group of non-Thai origin, even if now mostly assimilated, are the Chinese who are the driving force in many commercial activities.

Thailand has made rapid progress in human development in recent years. Literacy is high, and secondary school enrolment, previously low, is now rising. There is also virtually free basic health care—for the equivalent of $0.75 per hospital visit.

Although incomes have increased, millions still live in poverty, particularly in the north-east, and there are one million child labourers.

Thailand is a migration hub. Although many Thai workers have headed overseas, chiefly to Saudi Arabia, Brunei, and Singapore, around one million immigrants have also arrived from neighbouring countries, mainly Burma and Cambodia, to work on construction sites and sugar mills or in the 'entertainment' industry.

Thailand's population is still predominantly rural and 39% of the labour force still work in agriculture, mostly on small farms. Thailand's fertile land has enabled it to become the world's largest rice exporter.

Rapid economic growth has,

however, taken its toll on the environment—particularly on the forests. Forest cover fell from over 50% in the mid-1960s to less than 20% by 2004.

In recent years the most dramatic development has been the hectic process of modernization and urbanization. In the 1960s the government opened up the economy and invested in infrastructure. This helped to expand exports, first of agricultural products and then of manufactured goods and later moving into higher technology items.

Most of the country's industrial activity is focused on the sprawling capital Bangkok which, with its environs, now accounts for more than half the country's GDP—a rapid expansion that made the city a byword for pollution and traffic congestion. In recent years there have been considerable improvements.

Above and below Bangkok's traffic

Bangkok now boasts an overhead railway, the Skytrain, and a subway system, and pollution has fallen. More worrying is that the city, built over a swamp, is sinking.

Thailand's leading source of foreign exchange is tourism. More than 11 million people visit each year, attracted to beach resorts such as Phuket and Samui, and to the often raucous nightlife, as well as the more peaceful hills around Chiang Mai.

Thailand's boom came to a sudden halt in 1997 and triggered the Asian financial crisis. In 1996, exports had faltered and the currency came under attack. Since then, the currency and economy have stabilized and unemployment is officially only 2%.

Thailand was never colonized. It became a constitutional monarchy in 1932, and since 1946 has been ruled by the highly respected King Bhumibol. It has, however, had many autocratic or military governments and a number of coups. In 1992 the middle classes took to the streets—protesting against military-backed parties, and this, along with an intervention by the king, opened the way to civilian rule.

In 1997 the financial crisis brought down the government which was replaced by a coalition led by Chuan Leekpai of the Democratic Party.

But the elections in 2000 saw a victory for a new party, Thai Rak Thai (Thais Love Thais), led by millionaire businessman Thaksin Shinawatra. Thaksin had successful populist programmes, including a cheap healthcare system. But he showed little concern for human rights. A campaign against drug dealers resulted in over 2,000 extra-judicial killings. His popularity waned following Islamic terrorist attacks in the South but revived after his effective handling of the relief following the Indian Ocean tsunami that killed 5,374 people. As a result, in February 2005 he won an outright electoral victory.

But opposition was mounting, and in 2006, following a murky financial deal by his family and mass demonstrations, Thaksin called a snap election. This was boycotted by the opposition and subsequently annulled.

In September 2006, citing the danger of social chaos if Thaksin stayed, the army, led by General Sonthi Boonyaratglin, and with the tacit approval of the king, carried out a bloodless coup. The army says democracy will soon be restored.

Timor-Leste

A new nation thrown into chaos by an army rebellion

Land area: 15,850 sq. km.
Population: 0.8 million—urban 8%
Capital city: Dili, 60,000
People: Maubere, Chinese, Indonesian
Language: Tetun, Portuguese, Indonesian
Religion: Roman Catholic 90%, Musllim 4%, Protestant 3%, other 3%
Government: Republic
Life expectancy: 56 years
GDP per capita: $PPP 400
Currency: US dollar
Major exports: Coffee, oil, gas

Timor-Leste, formerly known as East Timor, comprises the eastern half of the Pacific island of Timor and the north-western enclave of Oecussi Ambeno. The territory consists largely of forested mountains that descend to coastal plains and mangrove swamps. Most people are Maubere, a mixture of Melanesian and Malay. Although their traditional beliefs are animist, the Catholic Church has also been important. The newly adopted official language is Portuguese—a strange and expensive choice since few people here, or elsewhere in Asia, speak it.

Most Timorese have made their living from agriculture, growing food crops such as sweet potatoes or corn, along with cash crops—especially coffee, which has been the leading export. Farmers on the coastal plains also grow rice and plantation crops such as rubber, tobacco, and coconuts. In addition the forests yield many kinds of timber, including sandalwood.

The main source of national income in the future will be oil and gas since there are considerable offshore reserves, the income from which enabled the government to run a budget surplus in 2004/05.

The island of Timor had been divided between the Dutch colonists in the west and the Portuguese in the east. The Portuguese hung on longer and East Timor declared its independence only in 1975. Indonesia promptly invaded. Within three months, 60,000 people had died. But the struggle continued—led by Xanana Gusmão's Frente Revolucionária do Timor-Leste Independente (Fretilin). Following the collapse of the Suharto regime in Indonesia, the East Timorese were offered a referendum. In 1999, despite massive Indonesian-backed violence, 80% of people voted for independence.

This led to an interim UN administration. Elections for a constituent assembly in 2001 were won by Fretilin. Presidential elections in April 2002 were won by Xanana Gusmão. Independence was declared.

For a few years there was an uneasy peace, though with tension between Gusmão and Fretilin and particularly prime minister Mari Alkatiri. Matters came to a head in March 2006 when Alkatiri sacked half the army for going on strike and brutally repressed their subsequent rebellion. Soon Dili was taken over by rioters and peace was only restored by the arrival of Australian forces. In July, Alkatiri resigned and was replaced by Nobel Peace Prize winner Jose Ramos Horta.

Togo

Togo has yet again failed to install a democratic government

Land area: 57,000 sq. km.
Population: 6 million—urban 35%
Capital city: Lomé, 729,000
People: Ewe, Mina, Kabré
Language: French, Ewe, Mina, Dagomba
Religion: Indigenous beliefs 51%,
Christian 29%, Muslim 20%
Government: Republic
Life expectancy: 54 years
GDP per capita: $PPP 1,696
Currency: CFA franc
Major exports: Coffee, cotton, phosphates

Within its elongated shape, Togo has a variety of geographical regions. Inland from the thin coastal belt with its series of lagoons there is a plateau region. This extends to the Togo Mountains, a chain that crosses the country from south-west to north-east. Beyond the mountains to the north-west there is another plateau drained by the River Oti.

Togo has more than 30 ethnic groups. The largest, who live mostly in the south, are the Ewe, a group also to be found in Benin and Ghana. The population density is greatest in the south and continues to grow fairly rapidly—by 2.6% per year.

This has put pressure on public services, particularly education. Togo has maintained higher educational standards than neighbouring Francophone countries but the system is deteriorating. Health services too are under strain and around one-

quarter of children are malnourished. Togo is also afflicted by HIV/AIDS which by 2005 had infected 3% of adults. Services are particularly poor in the rural areas.

Two-thirds of Togolese make their living from agriculture. Most are smallholders, growing such crops as cassava, millet, yams, and maize. In a normal year, Togo is self-sufficient in these crops, though it still needs to import rice. In the plateau regions and particularly in the north many people raise cattle and sheep.

Farmers can also supplement their incomes with a range of cash crops. In the mountains along the border with Ghana they grow cocoa and coffee, while to the east they grow cotton. Cotton has become particularly important and production, which takes place mostly on small farms, has more than doubled over the past decade and accounts for around one-quarter of export earnings. However, the intensity of cotton production is causing environmental damage. Efforts to further liberalize the cotton industry by selling Sotoco, the state marketing board, have been frustrated because no-one wants to buy it.

Industrial activity is dominated by phosphates, of which Togo is one of the world's largest producers. Togo's deposits are extensive and compact, and conveniently located just 40 kilometres from the capital, Lomé, which is also the main port. However output has fallen over the past few years, partly as a result of ageing equipment and inefficient management but also because Togo's deposits have a cadmium content above acceptable levels for the EU.

Togo also has a small manufacturing sector. Much of this involves processing agricultural goods such as cotton, palm oil, and coffee for export, but there are also other enterprises making cement, beer, and textiles for local consumption. Some of these companies are to be found in an export-processing zone which by 2000 had attracted 45 companies, from Europe, South Korea, and India, employing around 7,000 people, mostly in light manufacturing.

For almost 40 years Togo was ruled by Gnassingbe Eyadéma, who was appointed president following a military coup in 1967. Eyadéma, who was from the minority, northern Kabré group, headed a military government but later set up his own political party, the Rassemblement du peuple togolais (RPT). Since in 1979 the RPT was the only legal political party Eyadéma was duly elected.

By 1991, with a wave of democracy sweeping across Africa, and following considerable donor pressure, Eyadéma was forced to include in the government members of opposition groups, most of whom are Ewe.

In 1992, this government came under attack from the army, provoking a nine-month general strike. In the ensuing violence more than 200,000 people were forced to flee the country. In 1993 Eyadéma was re-elected in a fraudulent election that resulted in the withdrawal of most development aid.

The 1998 election was worse. The opposition were denied access to the media and the electoral lists were tampered with. When it became clear that despite this the chief opposition candidate, Gilchrist Olympio, was winning, the paramilitary police seized the ballot boxes, the interior minister declared Eyadéma the winner, and Olympio fled into exile.

Following the election, repression was stepped up and hundreds of

Corpses washed up on the beaches people were killed by the security forces. Corpses were washed up on the beaches of Togo and neighbouring Benin for days afterwards.

International mediation started in July 1999. Eyadéma, under economic pressure from the EU, established an electoral commission but then amended the code it produced so the opposition refused to participate in the legislative elections of 2002. This new RPT-dominated parliament then removed the two-term restriction for presidents, allowing Eyadéma to stand again in 2003, and changed the rules so as to exclude Olympio. In another rigged election Eyadéma won yet again.

Eyadéma died in February 2005 and the military initially tried to install his son Faure Gnassingbé as president. Following international pressure they eventually retracted and a presidential election was held in April. As usual, this was rigged, giving Gnassingbé 60% of the vote. Now he claims to want a dialogue with the opposition.

Tonga

Tonga is still dominated by its king but change is in the air

Land area: 750 sq. km.
Population: 0.1 million—urban 34%
Capital city: Nuku'alofa, 21,000
People: Polynesian
Language: Tongan, English
Religion: Christian
Government: Constitutional monarchy
Life expectancy: 72 years
GDP per capita: $PPP 6,992
Currency: Pa'anga (Tongan dollar)
Major exports: Copra, squash, vanilla

Tonga comprises an archipelago of more than 150 Pacific islands; some coral, some volcanic.

Around two-thirds of the population live on the main island, Tongatapu. The islands are vulnerable to cyclones—the most recent of which was Cyclone Ami in 2003. Tonga is also threatened by rising sea levels and a steady increase in salination.

Tongans are ethnically fairly homogenous and although they live in a lower-income country they enjoy good access to sanitation, clean water, and health care. Now they are starting to suffer from 'lifestyle diseases', including obesity and diabetes. They also have a high rate of literacy—close to 100%. Fertility remains high, but population growth has been slowed by high rates of emigration, particularly to New Zealand.

Most people still rely on agriculture, which accounts for over one-quarter of GDP. They grow food and a narrow range of cash crops—coconuts, squash pumpkins, vanilla, and melons. Tonga's foreign exchange income has, however, become over-reliant on one commodity—squash pumpkins—which go primarily to Japan.

In the past, Tonga has benefited from foreign aid, but this is likely to dry up because Tonga, despite its protests of poverty, no longer qualifies as a 'least-developed country'.

The other main source of foreign exchange, equivalent to around half of GDP, is migrant remittances from overseas workers. This too may shrink because fewer people are leaving. So further progress will be difficult unless Tonga diversifies its economy.

Tourism, with around 40,000 visitors per year, is also developing, offering the opportunity to enjoy the scenery and see hump-back whales.

Tonga is effectively an absolute monarchy. From 1965 until his death in 2006 the the country was dominated by King Taufa'ahau Tupou IV. He was succeeded by his son Ulukalala Lavaka Ata. The king appoints the cabinet and the prime minister. There is also a legislative assembly, the Fale Alea, but only nine of its 30 members are directly elected; the others include the king and his cabinet, as well as nine members elected by the traditional nobles.

The elections in March 2005 produced little change in the make-up of the parliament. But there was an innovation in that two of the nine people's representatives have since been appointed to the king's cabinet. This and other changes are in response to national and international pressures for democratic reform—which, following the change of monarch, are likely to intensify.

Trinidad and Tobago

The oil will soon run out. Gas should fill some of the gap but more jobs are needed

| | Miles | 20 |
| 0 | Km | 36 |

Land area: 5,000 sq. km.
Population: 1 million—urban 75%
Capital city: Port of Spain, 345,000
People: Indian 40%, Black 38%, mixed 21%, other 1%
Language: English, Hindi
Religion: Roman Catholic 26%, Hindu 23%, Anglican 8%, Baptist 7%, Muslim 6%, other 30%
Government: Parliamentary democracy
Life expectancy: 70 years
GDP per capita: $PPP 10,766
Currency: Trinidad and Tobago dollar
Major exports: Oil

These two islands are the southernmost of the Caribbean chain. But they can also be considered as an extension of South America: Trinidad, which has 94% of the land area, is only 12 kilometres from the coast of Venezuela. Trinidad's northern range of mountains essentially continues the coastal range of the Andes, which surfaces again as the island of Tobago. The southern quarter of Trinidad is oil-bearing land and includes the pitch lake at La Brea, which is the world's largest natural source of asphalt.

Tobago is less industrially developed. Its chief geographical feature is its main ridge which slopes down to extensive coral reefs that are popular with divers.

The country is racially very diverse—a product of its sugar-producing past. The main groups are the blacks who are descendants of the original African slave labour force, and the East Indians who are descendants of the indentured Indian workers who largely replaced blacks on the plantations after the abolition of slavery. In addition, there are smaller European and Chinese communities. The people of Tobago have long felt dominated and neglected by Trinidad.

Standards of human development are reasonably good though health and education services are uneven and young people have limited employment opportunities.

Agriculture still offers some employment but crops are now mostly for local consumption. Sugar for many years was a major cash crop and export earner and is still important for some farmers, but output has declined sharply and will fall further. There is also a small fishing industry.

Since the middle of the 20th century economic life has instead largely been restructured around oil and gas. Trinidad is a major offshore producer and also a refiner of crude oil imported from other countries in the region. In the 1970s, the sharp rise in oil prices transformed Trinidad and Tobago into one of the richest countries in the region. Current high oil prices have also boosted revenues, some of which go into a Heritage and Stabilisation Fund which in 2006 stood at $1.3 billion.

Oil production is now declining as the fields become exhausted but fortunately there are ample reserves of natural gas and production has increased steadily. This is used for local consumption and power generation as well as being liquefied for export.

Just as important, the gas now serves as feedstock for the manufacture of petrochemicals or fertilizers. The country has become the world's largest exporter of ammonia and methanol. Foreign investment in these plants and the growth of ancillary enterprises has helped reduce unemployment which by 2005 had fallen to 7%.

Two-thirds of the workforce are now employed in services. Some work in the tourist industry, though a smaller proportion than in most Caribbean countries. Except when attending the annual carnival, relatively few visitors spend much time in Trinidad which now suffers from rising levels of crime, much of which is associated with drugs. The chief long-stay attraction is Tobago which has many unspoilt beaches as well as a variety of plants and wildlife in the rainforests.

Tourists head for Tobago

With oil reserves likely to last no more than ten years, the government is anxious to diversify into other sectors, including agribusiness and marine, financial, and information technology services.

Politics in Trinidad has usually been conducted along ethnic lines.

There have also been disputes between the two islands which led in 1980 to the formation of the Tobago House of Assembly. For many years the dominant party was the People's National Movement (PNM), which has generally had the support of blacks as well as the Chinese, those of mixed race, and non-Hindu Indians. The PNM saw the country through independence and held power for decades, effectively excluding the Indian community.

This pattern was broken in 1986 when the National Alliance for Reconstruction (NAR), a multiracial coalition led by Arthur Robinson, won most seats in the house of representatives. But the NAR's austerity measures proved unpopular and the coalition soon fell apart. In 1989, one of its factions formed a new party with largely Indian support: the United National Congress (UNC).

In the 1991 elections, the PNM, led by Patrick Manning, regained power, only to lose it narrowly in the 1995 elections, following which Basdeo Panday became the first person of Indian extraction to become prime minister, as the UNC formed a coalition with the NAR.

In 1997 the country had its first presidential election. This was won by former prime minister Arthur Robinson. But the outcome of the subsequent legislative elections was less clear. It took a series of closely fought ballots in 2000, 2001, and 2002 before the PNM, still led by Patrick Manning, finally gained a majority.

In 2006 Panday was jailed for two years for corruption and replaced as leader of the UNC by Kamla Persad-Bissessar.

Tunisia

Tunisians have traded democratic freedoms for economic and political stability

Land area: 164,000 sq. km.	
Population: 10 million—urban 64%	
Capital city: Tunis, 728,000	
People: Arab 98%, other 2%	
Language: Arabic	
Religion: Muslim 98%, other 2%	
Government: Republic	
Life expectancy: 73 years	
GDP per capita: $PPP 7,161	
Currency: Tunisian dinar	
Major exports: Textiles, food, petroleum products	

Tunisia can be divided into four main geographical regions. The northern one-third of the country consists of spurs of the Atlas Mountains but also encompasses fertile valleys and plains. These mountains descend further south to a broad plateau and further south still to a series of shallow salt lakes, known as shatts. Beyond these lies the Sahara Desert, which occupies around two-fifths of the territory.

Tunisians are almost entirely Arab, since the original Berber inhabitants have been assimilated. By African standards, they enjoy a relatively high standard of living and a low rate of population growth.

There is still some poverty, and unemployment is around 14%, but mindful of the potential for militant Islam among disaffected youth, the government has maintained food subsidies and largely free health services. Primary education is free and compulsory though literacy is still only 74%.

Tunisia's economy is broadly based. The country has successfully moved on from agriculture and from dependence on oil and phosphates and now has a number of manufacturing industries that employ one-fifth of the workforce. Some of the fastest growth has been in textiles, which are now the leading export, though this industry is facing increasing competition from Asia.

Tunisia has been an oil producer since 1966. There have been promising new finds but proven reserves will only last ten more years. Phosphates too now make up a much smaller proportion of exports since they are used locally as raw material for fertilizers and chemicals.

One-fifth of the workforce is employed in agriculture. Most farms are fairly small and largely worked by hand. The main crop in the fertile northern plains is wheat, along with barley and vegetables, but harvests are vulnerable to erratic rainfall.

The drier parts of the country are used to grow dates and particularly olives, of which Tunisia is one of the largest producers. The main restriction is a lack of water since the country is already using around 80% of its water for irrigation. One-quarter of the land is forest, or used for grazing animals, but livestock-raising too is vulnerable to drought.

Of the service industries, one of the most important is tourism. Tunisia's coastal resorts are favourite package-holiday destinations for Europeans. Tourism, with around six million visitors a year, accounts for 6% of GDP but revenues are fairly static. The government takes care to prevent tourists finding out about human rights abuses, and regularly bans the foreign press. The country's hotels and resorts will need considerable investment if they are to attract higher-spending visitors.

Tunisia's political system is less diverse than its economy. Since independence in 1956, it has been ruled by the same party and has had only two presidents, neither of whom have been enthusiasts for political pluralism.

Only two presidents since 1956

Tunisia's first president was Habib Bourgiba, at the head of what was subsequently called the Parti socialiste destourien. Initially, Bourgiba was relatively progressive, promoting the rights of women, for example. But he grew increasingly autocratic, jailing opposition leaders and clamping down on the media. In 1975, he had himself elected president-for-life. The strongest opposition came from the underground Mouvement de la tendence Islamique (MTI). To combat the MTI, Bourgiba in 1986 appointed as interior minister General Zine el-Abidine Ben Ali. He was so successful at crushing the MTI that in 1987 Bourgiba gratefully appointed him prime minister.

This was a mistake. Ben Ali had Bourgiba declared senile, removed him from office, and in December 1987 assumed the presidency himself. At first Ben Ali seemed fairly liberal. He released hundreds of detainees and legalized opposition parties. He also renamed the party as the Rassemblement constitutionnel démocratique (RCD). In the 1989 elections, the RCD won most of the seats and Ben Ali was elected unopposed as president.

Ben Ali's enthusiasm for consensus was short lived. By the early 1990s, he was again persecuting the MTI, by then renamed Hizb al-Nahda (party of the awakening), jailing hundreds of its members and driving others into exile. He also took control of other opposition parties and restricted the press and trade unions.

In the 1994 elections, the RCD won almost every seat and in the presidential election Ben Ali was again elected unopposed, with 99% of the vote. To appear more democratic he subsequently decreed that minority parties would automatically get 20% of seats regardless of their share of the vote. Elections in 1999 and 2004 produced similar results.

State repression has spread beyond Islamic activists to include many secular groups, particularly human rights activists. Amnesty International reports hundreds of incidents of police intimidation, arbitrary arrest, and torture.

Democracy seems a remote prospect. There are a couple of minor opposition parties, and some rumblings of discontent. But most people have effectively traded democracy for political stability.

Turkey

Turkey is making peace with Greece and is now a candidate for the EU

Land area: 779,000 sq. km.
Population: 70 million—urban 66%
Capital city: Ankara, 3.2 million
People: Turkish 80%, Kurdish 20%
Language: Turkish, Kurdish, Arabic
Religion: Muslim
Government: Republic
Life expectancy: 69 years
GDP per capita: $PPP 6,772
Currency: Turkish lira
Major exports: Garments, textiles, fruit, vegetables

Turkey is predominantly mountainous. The lowlands are mostly confined to coastal areas around the Black Sea, the Aegean, and the Mediterranean. Turkey straddles the border between Europe and Asia—though its European part is quite small. The Asian part, known as Anatolia, has at its heart the central Anatolian plateau, which is encircled by mountains. Western Anatolia consists of long mountainous ridges and deep valley floors, though the highest mountains are in the east.

The majority of the population are Turkish and almost all are Sunni Muslim. Over recent decades Turkey has been industrializing rapidly and more than two-thirds of the population live in cities. But by the standards of other industrial countries, levels of human development are low. In 2003, adult literacy, for example, was only 88%, and significantly lower for

women. And outside the cities, health services are poor. Until recently, many Turks were emigrating to the EU, particularly to Germany. The exodus has more or less ceased, but there are still some two million Turks in the EU, whose annual remittances although declining are still around $1 billion. However, there is also an inflow: around 2.5 million foreigners are living in Turkey, of whom one million are unauthorized.

Within Turkey, there is also a significant minority of Kurds—one-fifth of the population. The Kurds are one of the world's largest ethnic groups without a state of their own, though eastern Anatolia, where most Turkish Kurds live, together with neighbouring regions of Iran and Iraq, is referred to as Kurdistan. The government of Turkey has long tried to repress Kurdish nationalism and from 1984 to 1999 fought the guerrillas of the Kurdish Workers' Party in a war that cost more than 30,000 lives. In 1999 the PKK declared a ceasefire but resumed its campaign on a smaller scale in 2004.

With more than half the workforce employed in service industries, and one-fifth in manufacturing, Turkey has the characteristics of an advanced industrial economy. Many of the largest enterprises, including iron, steel, and chemicals, are state run. But the private sector has now become more significant, particularly in food processing, and in garments and textiles which make up one-quarter of exports. Turkey is also the world's second largest exporter of pasta.

Tourism is also a major source of employment and income, with 21 million visitors bringing in around $18 billion per year. Most of the tourists on packaged holidays come from Germany or the UK.

Turkey's diverse landscape has sustained steady growth in agriculture which still employs one-third of the workforce. Although levels of technology are lower than in Western Europe, output has kept pace with population, so Turkey is largely self-sufficient in food, and has crops such as wheat, sugar beet, cotton, and tobacco, some of which are exported.

Turkey has always had an ambivalent relationship with Europe. On the one hand, Turkey is strategically important, and has been a key member of NATO, for which

Ambivalent relationship with Europe

it was a missile launching pad during the cold war. On the other hand, it has a poor record on human rights, particularly with respect to the Kurds. Added to this has been the dispute with Greece over Cyprus.

These difficulties had made the EU nervous of inviting Turkey as a candidate for membership. Negotiations finally started in 2005, but with some EU countries vowing to put Turkish accession to a referendum —and with continuing human rights abuses—enthusiasm is waning.

Turkey's recent political history has followed an unsteady course. A series of weak coalition governments have been subject to military intervention: since 1945 there have been three military coups. The military intervened yet again after the 1995 election when the party with the largest proportion of votes—21%— was the pro-Islamic Welfare Party (RP). Eventually in 1997 they forced its leader Necmetting Erbakan to resign in a 'soft coup'.

A subsequent three-party coalition proved short lived so in 1999 Turkey had a further general election. This resulted in another coalition which had the Democratic Left Party (DSP) as the leading party, along with the Motherland Party (Anap) and the right-wing Nationalist Action Party (MHP). The new government was led by Bulent Ecevit.

Ecevit had some successes, including a new loan from the IMF and the declaration from the EU that Turkey could be a candidate for accession. But disagreements with members of his coalition and his own party forced another election in 2002. This produced a surprising result: a landslide victory for a centre-right party with some Islamic origins—the Justice and Development Party. Recep Tayyip Erdogan is the party leader and subsequently became prime minister.

Erdogan says he wants to get Turkey into the EU so has made some gestures towards the Kurds and has made efforts to resolve the Cyprus question. But following repressive 'anti-terror' legislation in 2006 and more persecution of the Kurds, Turkey's commitment to reform is looking much less convincing.

Turkmenistan

An autocratic leader is spending scarce money on grandiose monuments

Land area: 488,000 sq. km.
Population: 5 million—urban 45%
Capital city: Ashkhabad, 412,000
People: Turkmen 85%, Uzbek 5%, Russian 4%, other 6%
Language: Turkmen 72%, Russian 12%, Uzbek 9%, other 7%
Religion: Muslim 89%, Eastern Orthodox 9%, unknown 2%
Government: Republic
Life expectancy: 62 years
GDP per capita: $PPP 5,938
Currency: Manat
Major exports: Gas

Turkmenistan is largely a vast, sandy desert, with some low mountains to the south. The country borders on the Caspian Sea, but the main rivers, which are along the northern and southern borders, flow into neighbouring countries to the east.

Turkmenistan does, however, have many canals to provide water for drinking and irrigation, including the Garagum canal, one of the world's longest, which takes water from the Amu Darya River on the eastern border to the capital, Ashkhabad, 1,400 kilometres to the south. The scale of the diversion has, however, created major ecological problems, including the drying up of the Aral Sea into which the Amu Darya flows.

The Turkmen, who are Sunni Muslims, were formerly a nomadic people. During the Soviet era much of their culture and lifestyle was repressed, and now they are mostly settled along the rivers and canals. Following the collapse of the Soviet Union, there were considerable population movements: between 1989 and 1995, many Russians left, while over 300,000 Turkmen arrived from other republics. Turkmenistan also has a substantial Uzbek minority living along the northern border.

The population grew swiftly over this period, though more recently the birth rate has been falling. The population was fairly well educated but the quality of schooling has been slipping. Health services too have deteriorated as the budget has fallen sharply. Officially unemployment is zero, since the state guarantees work for all; in fact some observers say it could be as high as 70%.

Agriculture, based on irrigation, employs more than half the workforce and makes up one-quarter of GDP. The largest crop is cotton, followed by wheat. But production has been falling. The land is of poor quality and farmers have little incentive to increase output. There has not been much progress in land reform and the state, which is the monopoly buyer, pays far less than the world prices for wheat and cotton.

Production is in any case fairly inefficient: much of the irrigation water in the canals evaporates before it can be used and excessive application of fertilizers for cotton has built up toxic residues in the soil.

In the past, weaknesses in agriculture have been compensated for by production of gas. The Garagum Desert in the east holds a huge basin of natural gas—1.3% of world reserves. This is responsible for more than half of GDP and three-quarters of export earnings.

The largest gas producer is the state-owned Turkmenneftegaz which has seen sales increase following new contracts with Russia and Ukraine. These exports rely on a now ageing pipeline system which needs more investment. Sales to other countries would need new pipelines, such as the 1,500-kilometre one proposed to cross Afghanistan to Pakistan, and possibly extending to India. Originally proposed in the mid-1990s, this seems plausible now that Afghanistan is somewhat more stable.

Turkmenistan avoided economic liberalization and instead has been promoting a policy of

Using gas funds to invest in industry

import-substituting industrialization—using gas revenues for funds. The government has been investing in food processing and textile manufacture, as well as in a cellulose plant and steel mills. But private investors are slower to spend, deterred by massive corruption and excessive bureaucracy. The private sector accounts for only 25% of GDP.

Gas revenue has also gone into construction. Much of this is for the gas industry itself, and also for the exploitation of oil. But there have also been a number of fairly grandiose projects, including a new international airport, a marble presidential palace, and an enormous mosque.

All this activity has been directed by dictatorial president, and prime minister, Saparmurad Niyazov. He became leader of the Turkmen Communist Party in 1986 and tried to resist economic and political liberalization. Following independence in 1991, the party re-emerged as the Democratic Party, which is still the only legal political organization. The party itself does not have a heavy workload; it meets once or twice a year to rubber-stamp presidential decisions.

Niyazov has developed probably the world's most shameless personality cult. He has had himself proclaimed as 'Turkenbashi'—leader of the Turkmen, and one of his most striking monuments is a column topped by a revolving golden statue of himself. His heavyweight secret police, the KNB, have helped him to suppress all forms of dissent.

Having held a rigged referendum in 1994, Niyazov was already due to stay in power until 2004. Then in early 2000 his dutiful parliament requested that he declare himself president-for-life. Niyazov was happy to oblige—though has promised to hold a presidential election in 2010.

For the 2004 parliamentary elections, all the candidates were members of the Democratic Party, though somewhat surprisingly the reported turnout was only 77%—as opposed to the 100% that is usually announced.

Uganda

Uganda now has a multiparty democracy but its president is reluctant to step aside

Land area: 236,000 sq. km.
Population: 27 million—urban 12%
Capital city: Kampala, 1.2 million
People: Baganda 17%, Ankole 8%, Basogo 8%, Teso 8%, Langi 6%, other 49%
Language: English, Luganda, Swahili, Bantu languages, Nilotic languages
Religion: Roman Catholic 33%, Protestant 33%, Muslim 16%, indigenous beliefs 18%
Government: Republic
Life expectancy: 47 years
GDP per capita: $PPP 1,457
Currency: Ugandan shilling
Major exports: Coffee, gold, fish, cotton

Uganda lies on the equator, but because of the altitude the climate is relatively mild. The country forms part of the East African Plateau, most of which lies between 1,000 and 2,000 metres. To the west the boundary is formed both by mountain ranges and by the Western Rift Valley which incorporates lakes Edward and Albert. The eastern and north-eastern borders are also marked by mountains.

Most of the southern border runs across the world's second largest freshwater lake, Lake Victoria, which is shared with Kenya and Tanzania. Uganda's landscape shows great variety, from swampy river banks, to forests, to snowy peaks.

Uganda's population is also very diverse. It comprises dozens of major groups, but these can be divided into two main groups—the Bantu and the Nilotic. The Bantu, who live mostly in the south, make up around 70% of the

population. The largest of these are the Ganda, also called the Baganda. The Nilotic group, mostly found in the north, include the Teso. While the Bantu groups have provided most of the political élite, the Nilotic people have in the past dominated the armed forces.

More than half the population live below the poverty line. And they have one of the world's lowest life expectancies, a result partly of poor health services, but mainly of the AIDS epidemic. Uganda has, however, been a pioneer in the fight against the disease, with a pervasive and frank information campaign. This has helped reduce HIV prevalence detected in ante-natal clinics from 29% in 1992 to 7% in 2005.

Nevertheless, Uganda has already made some progress in reducing poverty and also has a donor-funded Poverty Eradication and Action Plan to boost the incomes of the poor.

More than 80% of the workforce depend on agriculture, generally working on small family farms growing basic food crops such as cassava, plantains, sweet potatoes, millet, and sorghum, mostly for their own consumption. They also grow a number of cash crops, notably coffee, which is the leading source of export income, and cotton, as well as raising cattle, sheep, and goats. In addition, a number of estates also produce tea.

With ample access to inland water Uganda is also able to catch large quantities of freshwater fish, notably Nile perch, which provide the second largest source of export income.

Manufacturing is mostly for domestic use and consists largely of processing agricultural output.

One of the most serious challenges to security comes from a rebel group in the north, the Lord's Resistance Army (LRA), led by Joseph Kony, who claims communication with the spirits, which has been killing thousands of people, kidnapping children to fight as soldiers, and generally wreaking havoc. In 2005 Kony and his commanders were indicted for war crimes by the International Criminal Court. Probably in response, in 2006 they started to engage in peace talks. A smaller rebel group, the Allied Democratic Forces (ADF), also makes sporadic attacks in the west.

A bizarre and violent cult

Following independence in 1962, Uganda suffered decades of dictatorship and violence, much of it linked to rivalries between regions and ethnic groups. This included the reign of terror of General Idi Amin, who from 1971 killed around 300,000 people and expelled 70,000 people of Asian origin—followed by the despotic rule of Milton Obote.

State-sponsored violence only stopped with the victory of a rebel group the National Resistance Army (NRA) led by Yoweri Museveni. In 1986 he was sworn in as president, representing the NRA's political arm, the National Resistance Movement (NRM). In an effort to subdue sectarianism, Museveni banned multiparty politics. Instead he introduced a 'movement' or 'no-party' system. Members of the constituent assembly were instead to be elected as independents, though in practice almost all have been NRM supporters. In 1996, Museveni was elected president with a large majority.

Museveni remained popular for a long time largely because he ushered in political stability and rebuilt Uganda's wrecked economy—removing many controls and welcoming private investment. Aid and stability produced steady economic growth.

But growth subsequently slackened and donors started to fret about defence expenditure which had risen as a result of Uganda's unsuccessful intervention in the civil war in the neighbouring Democratic Republic of the Congo. They also criticised rising levels of corruption which were costing the country around $100 million a year.

A referendum in July 2000 voted to retain the no-party system. But under donor pressure, the parliament in 2005 approved constitutional amendments to return the country to multiparty politics. They also however removed the two-term limit on the presidency, allowing Museveni to stand again.

For the 2006 elections the NRM will become a party, the National Resistance Movement Organization. A group of three opposition parties have united as the Forum for Democratic Change.

Ukraine

People power arrives in an 'orange revolution' that forces an election rerun

Land area: 604,000 sq. km.
Population: 48 million—urban 67%
Capital city: Kiev, 2.6 million
People: Ukrainian 78%, Russian 17%, other 5%
Language: Ukrainian, Russian
Religion: Ukrainian Orthodox and others
Government: Republic
Life expectancy: 66 years
GDP per capita: $PPP 5,491
Currency: Hryvna
Major exports: Metals, food, chemicals, machinery

Ukraine consists almost entirely of gently rolling plains with the highest part in the west. The broad Dnieper River flows through the centre of the country, draining into the Black Sea in the south, north of the Crimean peninsula.

The majority of the population are ethnic Ukrainians, speaking a language closely related to Russian. Ukraine never existed as an independent country until the break-up of the Soviet Union, and its population distribution reflects its union with Russia. The population in the rural west is almost entirely Ukrainians while that in the more industrial east adjacent to Russia has a higher proportion of ethnic Russians. Since there are strong cultural and language links between the two communities, there is relatively little tension. Only in the Crimea, where around two-thirds of the population are ethnic Russians, are there separatist inclinations.

Since independence, Ukraine's population has shrunk by four million. Death rates rose as a result of declining standards of nutrition and health. Infectious diseases such as diphtheria, cholera, and TB surged, and health problems were compounded by alcoholism. In the early 1990s, falling numbers were offset to some extent by immigration from other former Soviet republics, particularly into the Crimea. Since then, however, the country has suffered net emigration.

Among the largest permanent departures have been of the Jews to Israel, Germany, or the USA. At the same time, there is steady labour migration to neighbouring countries, especially to the Czech Republic, where Ukrainians are prepared to work for desperately low wages.

This is a response to widespread poverty in Ukraine, where the GDP fell by more than 60% between 1990 and 2000. Poverty tends to be worst in the old smokestack, industrial, and coal mining areas of the east; people living in the rural west do at least have the option of farming their family plots.

Industry has gone into steep decline but still employs more than one-fifth of the labour force. Between 1990 and 1997, industrial output fell by more than two-thirds, though in recent years there has been some growth. Certain areas of heavy industry, such as steel or chemicals, managed to sustain output, but others collapsed, particularly those related to defence and those producing low-quality consumer goods that could not compete with imports.

Ukraine has a significant mining sector and is the world's fifth largest producer of iron ore. But its coal industry is rapidly disappearing and though it has deposits of oil and gas these remain under-developed.

The majority of enterprises are now in private hands. But even in some of these the government has retained a controlling share and privatization generally has been very slow. Foreign investors have mostly stayed away, discouraged by an unpredictable legal system and corruption. Some compensation is that the black economy is thought to be equivalent to half the official GDP.

Agriculture, which also employs around one-fifth of the workforce, is in a similarly difficult situation. Ukraine is blessed with rich black soil—'chernozem'—and

Blessed with rich black soil

was considered the breadbasket of the Soviet Union. But output has fallen by half since independence—a result both of the loss of its Soviet market and a general shortage of investment and inputs. Ukraine was also slow to break up its collective farms: the necessary legislation was only passed in 1999,

though by 2000 most had disappeared.

Ukraine was one of the final Soviet republics to declare its independence. Its first president was Leonid Kravchuk. A former communist, he had little appetite for economic reform. In 1992, he appointed as prime minister the more reform-minded Leonid Kuchma. Kravchuk fell out with Kuchma and sacked him the following year, but in 1994 he conceded a presidential election—which he lost to Kuchma.

Kuchma struggled with a parliament still dominated by communists and his regime was uninspiring, notable chiefly for enriching his family and friends, and for extensive political violence. Kuchma was re-elected in 1999 only after a ruthless campaign against the communist candidate.

The presidential elections of November 2004, however, were to prove dramatic. Kuchma could not stand again, so nominated his prime minister, Viktor Yanukovych, to run against the popular and more reform-minded opposition candidate, former prime minister Viktor Yushchenko.

By most accounts the election went to Yushchenko, so officials supporting Yukanovich tried to rig the ballot. It did not work. Hundreds of thousands of orange-clad demonstrators took to the streets, eventually forcing the Supreme Court to order a repeat election in December which Yushchenko won comfortably. But the 'orange' group of parties proved fractious. And following parliamentary elections in 2006 in which most seats went to Yanukovich's Party of the Regions he was eventually appointed prime minister.

United Arab Emirates

One of the world's wealthiest countries that is a regional magnet for shoppers

Land area: *84,000 sq. km.*
Population: *4 million—urban 85%*
Capital city: *Abu Dhabi , 1.7 million*
People: *Emiri 19%, other Arab and Iranian 23%, South Asian 50%, other 8%*
Language: *Arabic*
Religion: *Muslim*
Government: *Federation of monarchies*
Life expectancy: *78 years*
GDP per capita: *$PPP 22,420*
Currency: *Emirian dirham*
Major exports: *Oil, natural gas, re-exports, fish, dates*

Apart from the Al-Hajar mountains in the Musandam peninsula along the eastern border with Oman, the territory of the United Arab Emirates, which stretches south from the Persian Gulf, is flat, barren, and low-lying—a mixture of sandy desert, gravel, and salt flats.

The UAE is a federation of seven sheikhdoms. By far the largest is Abu Dhabi, which covers three-quarters of the territory and includes the capital city. The six others lie east of Abu Dhabi. They are Dubai, Ajman, Sharjah, Umm al-Qaywayn, Ra's al-Khaymah on the Persian Gulf, and Al-Fujayrah, which faces the Gulf of Oman side.

Most people live on or near the coast. Only 20% are UAE citizens, who are mostly Sunni Muslims, though there are Shia minorities in Sharjah and Dubai. The rest of the population are immigrant workers,

unskilled and professional, chiefly from the Indian sub-continent who make up 98% of the private-sector workforce. Almost all national workers are in government employment. In an effort to reduce the imbalance, the government periodically stops issuing new work permits. However the efforts seem to be fairly half-hearted. Unemployment is low and occasional crackdowns that have expelled unauthorized workers have generated labour shortages.

In the past, the people of this area made their living from subsistence agriculture and fishing, though with little good land or supplies of fresh water the agricultural prospects are limited. A combination of land reclamation and irrigation from underground aquifers allows farmers to grow dates for export and enables the UAE to be self-sufficient in fruit and vegetables. As a result of such efforts, total production increased six-fold during the 1990s. Nevertheless, most grains have to be imported.

The fortunes of the UAE changed forever in 1958 with the discovery of oil and gas in Abu Dhabi, and later in Dubai, where production started in 1969. Sharjah is the only other emirate with significant deposits. Abu Dhabi's reserves are still substantial and could last one hundred years, though those in Dubai and elsewhere are running out more rapidly. Abu Dhabi also has substantial gas reserves which could last for another 150 years.

But the UAE's wealth stems not just from current oil income but also from income from investments in foreign bonds and equity markets, particularly those of the Abu Dhabi Investment Authority. In total the foreign assets of the UAE are probably around $250 billion.

Both Abu Dhabi and Dubai have steadily been diversifying away from oil into such areas as refining and petrochemicals. Dubai has done so more extensively. It has, for example, the world's largest single-site aluminium smelter and has a major duty-free zone at Jebel. This zone is used largely for small-scale local assembly, and as a distribution centre for other parts of the region, and now hosts more than 2,500 companies.

To promote less tangible commerce Dubai in 2000 opened an Internet City as a centre for e-commerce, which has drawn in companies like Microsoft and Cisco. There is also a Media City designed to attract publishers and broadcasters—though

Dubai Internet City

it is unclear how they will cope with local censorship. Abu Dhabi has made similar efforts to create duty-free zones but has been less successful.

Tourism is another useful source of income, and around five million people visit the UAE each year. Dubai's port, airport, and glittering array of hotels, restaurants, shops, and brothels draw people from all over the region. It has also built dramatic artificial islands to house luxury apartments. Even more startling is the indoor ski resort.

The UAE is a federation. But the federation is a weak one: the federal government deals with defence, foreign policy, and immigration but only 30% of total UAE expenditures goes through the federal budget, and most of this is contributed by Abu Dhabi. The rest is spent by individual emirates who control their own natural resources and regulate local business.

Abu Dhabi's economic clout gives it the greatest political power. Dubai essentially accepts this supremacy in exchange for economic autonomy.

The rulers of the emirates meet as the Supreme Council and elect the president. They always choose the Emir of Abu Dhabi, who is currently Sheikh Khalifa bin Zayed al-Nahyanal-Nahayan. He was appointed in 2004, following the death of his father. The Supreme Council also appoints the Council of Ministers in which the larger emirates take the most important portfolios.

Sheikh Khalifa seems to be quite popular and does not face major opposition. He can also rely on his brother, the younger and more energetic Sheikh Mohammed bin Zayed al-Nahyan. In 2006, Sheikh Mohammed bin Rashid al-Maktoum became the new ruler of Dubai and was appointed federal prime minister.

The UAE has no political parties but there does not seem to be much agitation for greater democracy, nor is there much sign of Islamic militancy. Legislative activity is handled by the 40-person Federal National Council, which is appointed and has only advisory powers.

United Kingdom

Though a leading member of the EU, the UK often also aligns itself closely with the USA

0	Miles 300
0	Km 480

Land area: 245,000 sq. km.
Population: 59 million—urban 89%
Capital city: London, 7.4 million
People: English, Scottish, Irish, Welsh, and ethnic minorities
Language: English, Welsh, Gaelic, Urdu, Hindi
Religion: Protestant, Roman Catholic, Muslim, Sikh, Hindu, Jewish
Government: Constitutional monarchy
Life expectancy: 78 years
GDP per capita: $PPP 27,147
Currency: Pound
Major exports: Manufactured goods, fuels, food

Scotland, England, and Wales, together with Northern Ireland make up the nation-state that is the United Kingdom (UK). The first three are also referred to as Great Britain. Of the four, Scotland in the north is the most mountainous, though to the west Wales too has an often rugged landscape, and there are lower mountain ranges in England, mostly in the north. The lowlands and plains that make up around half the UK's territory have rich soil and a temperate climate that allow productive agriculture.

Over 80% of the population are English. The Scottish, Welsh, and Irish, however, tend to have stronger national identities, including their own languages—though these are spoken by only a small minority. The UK has also received immigrants from former colonies—in the 1950s and 1960s from the Caribbean, and later from Africa and South Asia. Ethnic minorities make up around 8% of the population; they are exposed to racism and are poorer than average.

The British still benefit from the egalitarian welfare state created after the Second World War, though health and education services have now come under financial pressure. Meanwhile inequality is rising: the richest 10% now have more than half the wealth. There are also regional contrasts: wages in the South East are 20% higher than elsewhere.

The UK was the cradle of the industrial revolution, but nowadays manufacturing industry accounts for only 15% of GDP and some industries, such as coal, have virtually disappeared. British industry was boosted in the 1960s and 1970s by the discovery of oil and gas in the North Sea though by 2005, as oil production fell, the country once again had become a net oil importer. The UK has also been an attractive base for multinational companies and in 2005 was the world's largest recipient of foreign investment.

Agriculture accounts for less than 2% of GDP and of the workforce, but it is highly productive and grows around two-thirds of national needs. Outbreaks of 'mad-cow' and foot and mouth diseases and the use of genetically modified crops have, however, undermined confidence in intensive farming. The fishing industry supplies two-thirds of UK needs.

The UK has a very diverse service sector. This includes a leading global financial centre in the City of London and the world's third largest stock market. The UK is also the world's fifth most popular tourist destination. The tourist industry employs two million people, generating 5% of GDP annually and bringing in 30 million foreign visitors.

The UK is a constitutional monarchy ruled since 1952 by Queen Elizabeth II. For the past 60 years, British politics has been the preserve of two main parties: the pro-business Conservative Party and the trade union-backed Labour Party, with the party now called the Liberal Democrats some way behind.

The most radical government was the 1979–90 Conservative administration of Margaret Thatcher, which privatized many state-owned enterprises and broke the power of the trade unions. Thatcher's popularity peaked with victory over Argentina

Thatcher's legacy in the Falklands War in 1982 but she eventually alienated too many people and in 1990 was replaced by John Major, who surprisingly won the 1994 general election. In 1997, however, with his party split on the EU, Major suffered a crushing electoral defeat.

The victor was Tony Blair, who had purged his party of the far left and created a New Labour Party that would be neither socialist nor capitalist. One of his first challenges was Northern Ireland which for decades had been a battleground between Catholic and Protestant terrorists. In 1998, diplomacy prevailed with the Good Friday Agreement that allowed for a new directly elected assembly. In 1999, both Wales and Scotland also acquired similar assemblies.

With the Conservatives still in disarray, Blair won a further landslide election victory in 2001. For this he could also thank his chancellor (finance minister), Gordon Brown, who had fostered economic growth and reduced poverty. During the second term Brown invested in health and education.

Some of Labour's authoritarian policies have dismayed its traditional supporters, notably the bowing to pressure from the tabloid press to clamp down on asylum seekers.

Blair had taken successful foreign policy initiatives, intervening in Bosnia and Sierra Leone, but his wholehearted support in 2003 for the US-led invasion of Iraq was far less popular—dividing the country, and pitting the UK against France and Germany. The UK also found itself in the terrorist firing line with an attack on London transport in July 2005.

Blair and Labour won a third term in 2005, though with a reduced majority. In the same year the Conservative opposition acquired yet another new leader in David Cameron.

Blair says he will step down before the next election to allow time to bed in a new Labour leader, presumed to be Gordon Brown.

United States

The lone superpower has one of the world's most dynamic economies, but is becoming a more unequal society

Land area: 9,364,000 sq. km.
Population: 292 million—urban 80%
Capital city: Washington DC , 5.1 million
People: White 82%, black 12%, Asian 4%, Amerindian 1%, other 1% (Hispanics are 13% but may be of any ethnic group)
Language: English, Spanish
Religion: Protestant 52%, Roman Catholic 24%, other 14%, none 10%
Government: Republic
Life expectancy: 77 years
GDP per capita: $PPP 37,562
Currency: Dollar
Major exports: Capital goods, automobiles, industrial supplies and materials, consumer goods, food

Within its vast area, the USA has most types of landscape and climate—from tropical swamps in Florida, to the deserts of New Mexico, to the snow-capped peaks of the Rocky Mountains. The east of the country consists of the Atlantic coastal plain, which extends down from New England then broadens further south to include Florida and the Gulf coast. West of this plain are the Appalachian mountains, and then the vast areas of the prairies to the north and the Mississippi basin to the south.

West of the prairies are the Rocky Mountains and then a complex system of mountains and plateaux that stretches down to the Pacific coast. From north to south, the western landscape includes the forests of Washington state, through the beaches of California, to the semi-desert of Arizona in the south. The main land

mass has 48 states. The remaining two are Alaska, which lies beyond Canada, and Hawaii in the Pacific Ocean.

The diversity of the US landscape is complemented by the ethnic diversity of its population. Most people are white—descendants of European immigration from the 17th century onwards. Added to these were millions of blacks who came mainly to the southern states as slaves. But the USA has always been a magnet for immigrants and today 12% of the population are foreign born. Since the 1960s, following changes in legislation, immigrants come from a wider range of nationalities. In 2005, of the 1.1 million legal arrivals, the highest proportion came from Mexico (14%), India (7%), China (6%), Russia (5%), and the Philippines (5%). Millions more arrive illegally—there are 12 million undocumented immigrants, half from Mexico, whom enterprises are only too happy to use as cheap and exploitable labour.

The US population is generally mobile and dynamic, but it does suffer from rising inequality: between 1980 and 2004 the proportion of national income going to the richest 20% of the population increased from 44% to 50%. In 2004, 13% of the population lived below the poverty line.

Black and Hispanic families earn 40% less than white families. The mortality rate for black infants is twice the rate of whites, and black adults are twice as likely to be unemployed and far more likely to be in jail.

Most people are educated in public schools, and many are college graduates: in 2004, 49% of Asians, 28% of whites, 18% of blacks, and 13% of Hispanics. Health care relies more on private provision, so the 15% of the working population who are uninsured can find themselves in difficulties—one reason why life expectancy in the USA is lower than in other rich countries.

Americans are notable for churchgoing: 40% claim to attend a service each week. But they are also more likely than people in other countries to find themselves in jail: the prison population is more than two million.

The USA has the world's largest and most diverse economy, built on a rich endowment of natural resources that include deposits of most important minerals, including copper,

Almost two cars per household

gold, aluminium, and coal. It is also a major oil producer, though the reserves are depleted so the USA has to import most of its oil. Much of US life is constructed around the automobile—averaging 1.8 cars per household.

Another vital US resource has been agricultural land. The USA produces two-fifths of the world's maize and soybeans—mostly through the

agribusiness corporations that have displaced small farmers. One-quarter of agricultural output is exported—and around half in the case of cotton, wheat, and rice. This enormous output has been sustained despite shrinkages in the farm labour force and in agricultural land area. But it does rely on vast subsidies—averaging $18 billion annually between 2002 and 2005. US cotton farmers, for example, receive half their income through subsidies.

Abundant natural resources, combined with an energetic immigrant population and a large national market, have also made US industry dynamic and innovative. Major industries include steel, automobiles, chemicals, and aerospace. And although other countries have caught up in basic production, the USA has set the pace in more advanced industries such as electronics, telecommunications, and computers.

However, the US economy, like that of all industrialized countries, is now dominated by services. In 2005 less than 2% of the labour force worked in agriculture, and 14% in manufacturing, with 7% in construction, while 79% were in trade or other kinds of service.

Since it has a huge internal market, the USA does not need to depend heavily on exports of goods and services which are only around 10% of GDP. Nevertheless its sheer size still makes it the world's largest exporter. The most important recent trade development was the signing in 1994 of the North American Free Trade Agreement with Canada and Mexico. Even so, there is often strong resistance to free trade from industries such as steel that lobby for protection.

Politics in the USA is based on competition between two main parties: the Democrats and the Republicans. While there is considerable overlap in terms of support and policy, historically the Democrats have represented working people and ethnic minorities, and taken more liberal attitudes on civil rights and welfare. The Republicans have been more likely to represent big business and to adopt socially conservative positions on gay rights or abortion, and will want to minimize the size of government.

Hovering around the government are thousands of interest groups with

Professional lobbyists

highly professional lobbying machinery. One of the most

significant recent influences has been the Christian Right.

The US government is based on a separation of powers between the executive (the President), the judiciary (the Supreme Court), and the legislature (the Congress). Since this is a federal republic, individual states also wield considerable power through their governors and state legislatures.

The 1980s were dominated by the Republican presidency of former movie star Ronald Reagan. He turned the country sharply to the right, but a combination of vast defence expenditure and tax cuts created huge budget deficits.

Reagan was succeeded in 1988 by his vice-president George Bush. Having presided over the 1990 Iraq war, Bush lasted only one term, being defeated in 1992 by Democrat Bill Clinton.

Clinton had a number of successes,

both in foreign and economic affairs, and was re-elected in 1996, but came close to being forced out of office as a result of a sexual relationship with a White House intern.

The presidential election of 2000 resulted in a close and disputed Republican victory for one of Bush's sons, George W. Bush. The younger Bush started out with a right-wing agenda, making deep tax cuts for the rich and disdaining international treaties, notably the Kyoto protocol on global warming.

Although prior to his election Bush appeared to have relatively little interest in foreign affairs he was suddenly thrust into the international arena on September 11, 2001 by the al-Qaida terrorist attacks on the World Trade Center and the Pentagon which killed nearly 3,000 people.

Bush responded with an international 'war on terror'. This first involved overthrowing the Taliban regime in Afghanistan, which had harboured al-Qaeda. More controversially, in 2003 with UK support he attacked Iraq on the mistaken assumption that its ruler Saddam Hussein was concealing weapons of mass destruction.

In the November 2004 presidential elections Bush faced a strong challenge from Democrat John Kerry. Despite widespread opposition to the Iraq war, Bush won convincingly with 51% of the vote and the Republicans also made significant gains in both houses of Congress.

But with the situation in Iraq deteriorating, and a weak response in 2005 to Hurricane Katrina which devastated New Orleans, his popularity has fallen steeply.

Uruguay

Uruguay had a reputation for prosperity and freedom that it is now trying to restore

| 0 | Miles | 200 |
| 0 | Km | 320 |

ARGENTINA

Salto

Paysandú

Rivera • BRAZIL

Las Piedras

■ **Montevideo** • ATLANTIC OCEAN

Land area: 177,000 sq. km.
Population: 3 million—urban 93%
Capital city: Montevideo, 1.3 million
People: White 88%, mestizo 8%, black 4%
Language: Spanish
Religion: Roman Catholic 66%,
Protestant 2%, Jewish 2%, other 30%
Government: Republic
Life expectancy: 75 years
GDP per capita: $PPP 8,280
Currency: Peso
Major exports: Wool, textiles, beef, rice

Lodged between two much larger neighbours, Brazil and Argentina, Uruguay consists mostly of plateaux and rolling grasslands—the pampas—that are ideal for cattle raising. The highest parts of the country are in the hills of the Cuchilla Grande in the south-east. Another of Uruguay's natural advantages is its fresh temperate climate which, combined with long sandy beaches, makes this a popular holiday destination.

Uruguayans are largely of European origin, predominantly Spanish and Italian. The small black population has immigrated from Brazil. Uruguay has long been one of the most urbanized countries of Latin America.

From the 1930s, Uruguay built up a comprehensive system of social welfare and by the late 1980s around one-quarter of the labour force was in government service. The welfare system has been eroded in recent years, and the social security system has been partly privatized. But Uruguayans still have high levels of health and education—even university education is free. By Latin American standards, this is also a relatively egalitarian country and one of the few where inequality has not increased over the past decade. The country has a large urban middle class, while most of the poor people live in the rural areas where many still lack safe water and sanitation.

Uruguay's wealth was built on the raising of cattle and sheep on the pampas. And although agriculture now accounts for only 6% of GDP, and employs only one-tenth of the workforce, it still plays a central role in the economy since meat, wool, and hides make up two-fifths of the country's exports and also supply vital raw materials for industry. Uruguayans themselves are also serious meat eaters.

Most of the land is occupied by large ranches. Until 2000 the herd was largely disease-free and Uruguay has been very successful at selling meat to the USA and the EU. In recent years, it has also discovered a new market for wool in China. Most cereal output is for local consumption, but rice, which is chiefly grown in the department of Rocha on the southern coastal plain, is becoming an increasingly important export to Brazil.

Manufacturing and commerce are concentrated in Montevideo, which generates more than half the country's GDP. A high proportion of this is still in the public sector, which seems to have proved more efficient than in other countries. The state company ANAP refines petroleum, as well as producing alcohol and cement. Most manufacturing is for local consumption, but recently-planted forests of eucalyptus and pine have been supplying mills making pulp and paper for export.

The country's other major foreign-currency earner is tourism, mostly to the beaches around the exclusive up-market resort of Punta del Este to the east of Montevideo. More than two million visitors arrive each year, almost all of these from Argentina.

Argentines heading for the beach

Uruguay's trade links with its heavyweight neighbours have been further strengthened by the creation of Mercosur, a common market for the countries of the 'Southern Cone', whose secretariat is in Montevideo. There are also plans to build an $800-million giant bridge across the River Plate to Buenos Aires which would promote even greater flows of trade and tourism.

Political power in Uruguay has traditionally alternated between the Partido Colorado and the Partido Nacional (also known as the 'Blancos'). This division does not correspond to any ideological divide;

more a matter of personalities and history. People tend to vote for one party or the other based on family tradition. Such ideological divisions as do exist tend to be between factions within each of the parties.

The pattern of control by either the Colorados or the Blancos was broken in the 1970s. Left-wing groups, notably the Tupemaros urban guerrillas, established a strong presence and eventually provoked a coup by the military who ruled from 1973 to 1984 in a remarkably vicious fashion.

When democracy was restored, the Colorados, led by Julio María Sanguinetti, won the first presidential election in 1985. He was replaced in 1989 by Luis Albert Lacalle of the Blancos.

Throughout this period, however, a left-wing coalition, the Encuentro Progresista–Frente Amplio (EP–FA) had steadily been building support. So although Sanguinetti duly won the 1994 presidential election, the 1995 congressional elections produced a three-way split that required a Blanco and Colorado coalition.

The EP-FA candidate, Tabaré Vázquez, narrowly lost the 1999 presidential elections to the Colorado candidate Jorge Batlle and again the Colorados and the Blancos formed a coalition in congress—though this came apart in 2002, leaving a minority Colorado administration.

In the 2004 election, however, EP–FA finally succeeded, gaining an historic congressional majority and having Vázquez elected president. He immediately restored relations with Cuba and has ambitous plans for social and economic reform.

Uzbekistan

Very dependent on cotton and gold, and making little progress towards democracy

Land area:	447,000 sq. km.

Land area: 447,000 sq. km.
Population: 26 million—urban 37%
Capital city: Tashkent, 2.4 million
People: Uzbek 80%, Russian 6%, Tajik 5%, Kazakh 3%, other 6%
Language: Uzbek 74%, Russian 14%, Tajik 4%, other 8%
Religion: Muslim 88%, Eastern Orthodox 9%, other 3%
Government: Republic
Life expectancy: 67 years
GDP per capita: $PPP 1,744
Currency: Som
Major exports: Cotton, gold

Uzbekistan has some mountainous terrain in the east, and a few fertile oases, but around 80% of the territory is a sandy plain that merges in the south with the Kyzylkum Desert. The country is doubly landlocked—at least two frontiers from the sea.

Most of the population are ethnic Uzbeks and live in the rural areas. Uzbeks are Sunni Muslims and tend to be more devout than those in neighbouring countries. Like other former Soviet Republics, Uzbekistan also has a substantial Russian minority. The Russians live mostly in Tashkent and other cities, though many have also left: over the period 1990–99, 845,000 people emigrated. The next largest group, the Tajiks, are to be found in older cities like Samarkand.

Standards of health are poor, and recent years of economic crisis have caused health services to deteriorate

further. Expenditure on education has also been falling, as has school enrolment. Wages too have been falling, and inequality has been rising steeply.

Housing has been privatized and more than one million people own formerly state-owned apartments.

Around 40% of the population work in private agriculture, though all farmers have to lease their land from the government. People are allocated small plots of land for lifetime use. These have helped maintain food production and account for around 80% of meat and milk production. Farmers have the advantage of fertile land and a fairly mild climate, but since rainfall is light they rely heavily on irrigation.

The most important cash crop is cotton. Uzbekistan is the world's second largest exporter of cotton, referred to locally as 'white gold'. Though most cotton producers are private or belong to co-operatives, they can sell only to the state—which pays well below the world price so they have little incentive to increase efficiency or output. With so much of the land devoted to cotton, the country does not produce sufficient grain and is a substantial importer.

Heavy and inefficient use of water for cotton production has also drawn so much water from the country's two main rivers that the Aral Sea into which they drain has shrunk dramatically—by one-third between 1974 and 1995—with severe health and environmental damage in the surrounding areas.

Over-use of irrigation is drying up the Aral Sea

Industry too has been strongly connected with agriculture, notably through the Tashkent tractor factory. Rather than opening up to the outside world, Uzbekistan has pursued a strategy of import-substitution—borrowing from abroad to invest in heavy industry such as steel and in sugar production. Foreign investment is limited. There is a Daewoo car plant but this is largely for assembly operation.

The most lucrative industrial activity is gold mining which in recent years has helped to underpin the economy. Most of this takes place at the huge state-owned open-cast Muruntau mine. Uzbekistan is the world's ninth largest producer of gold.

The government has declared in principle that it is moving more towards a market economy, but progress has been slow. Unemployment is officially quite low but this is probably because enterprises are reluctant to restructure and to fire excess labour, and the real rate is probably closer to 20%.

Politics in Uzbekistan is dominated by the president, Islam Karimov, who is effectively the dictator of a police state. Karimov was the communist leader before the break-up of the Soviet Union. He opposed independence, but when it came he successfully had himself elected president in a 1991 vote generally assumed to have been rigged.

Uzbekistan does have a legislative body, the Supreme Assembly, but in practice this does little more than carry out presidential decisions. The two leading opposition parties are banned and their leaders are exiled.

The only effective opposition centres around Islamic groups. Mosques that were closed during Stalin's era have been allowed to reopen, but only under strict control. The clergy are regularly harassed by the police and by the national security service. Two groups that have refused to co-operate are the non-violent Hizbut-Tahrir and an armed guerilla group, the Islamic Movement of Uzbekistan (IMU), though this is now much weaker.

For the presidential election in 2000 Karimov arranged a contest by setting up a token opponent, though he still received 92% of the vote. In 2002 a dubious referendum extended the presidential term to seven years.

The US has aligned itself with Uzbekistan in the war on terror and has an air base here, but has had little success in promoting democracy. The European Bank for Reconstruction and Development has cut its lending because of human rights abuses.

Despite promises of greater openness, in the parliamentary elections in 2005 no opposition groups were allowed to register as parties. In May 2005, troops opened fire on demonstrators occupying a square in the city of Andizhan, killing at least 200. Many others have been jailed.

Vanuatu

Vanuatu is still divided along colonial lines

Vanuatu comprises around 80 Pacific islands of either volcanic or coral origin, many of which are still covered by tropical rainforests. The country is vulnerable to natural disasters, including volcanic eruptions, earthquakes, and tidal waves.

Most of the population, who are called 'Ni Vanuatu', are of Melanesian extraction and have a pidgin-based lingua franca, Bislama, along with around 105 other languages.

Vanuatu's society remains divided by its strange colonial heritage. As the New Hebrides until 1980, the country was ruled jointly by the British and French—who had established their own schools and churches. As a result, the élite population tend to consist of either French-speaking Catholics or English-speaking Protestants, with the latter in the majority.

The population as a whole is also fairly sharply divided between the 80% who live in scattered rural communities from subsistence or small-scale agriculture and the urban dwellers. The farmers grow food crops, such as taro and yams, as well as cash crops such as coconuts, cocoa, and squash for export. The cattle they raise also provide export income through sales of high-quality beef.

Tourism, mostly from Australia and New Zealand, is an important source of income. Vanuatu has also

Land area: 12,000 sq. km.
Population: 0.2 million—urban 23%
Capital city: Port Vila, 34,000
People: Melanesian 98%, other 2%
Language: English, French, Bislama
Religion: Christian
Government: Republic
Life expectancy: 69 years
GDP per capita: $PPP 2,944
Currency: Vatu
Major exports: Copra, beef, cocoa

been establishing itself as an offshore tax haven, and financial service companies contribute around one-tenth of GDP. In addition there is a small manufacturing sector, mostly involved in processing agricultural and forestry products.

After independence, the political parties organized themselves along language lines—though subsequently they have splintered. The main anglophone party is the Vanua'aku Party (VP), while the main francophone party is the Union des partis modérés (UMP).

The VP won the first election but subsequently split several times to create first the Melanesian Progressive Party (MPP) and later the National United Party (NUP). In the 1980s and 1990s Vanuatu was governed by a series of unstable coalitions.

In 1999 Barak Sope of the MPP took over as prime minister at the head of a coalition including the MPP and the NUP. But he was imprisoned for forgery and replaced by Edward Natapei of the VP, who retained the post after the 2002 election. Following a snap election in July 2004, Serge Vohor of the UMP emerged as prime minister, but he was soon replaced by Ham Lini of the UMP who has since fended off numerous no-confidence votes.

Venezuela

Venezuela has been polarized by its populist president while poverty increases

Land area: 912,000 sq. km.
Population: 26 million—urban 88%
Capital city: Caracas, 1.8 million
People: Spanish, Italian, Portuguese, Arab, German, African, indigenous people
Language: Spanish
Religion: Roman Catholic
Government: Federal Republic
Life expectancy: 73 years
GDP per capita: $PPP 4,919
Currency: Bolívar
Major exports: Oil, bauxite, aluminium, steel

The central one-third of Venezuela consists of open grasslands, the 'llanos'. To the south-east, and along the borders with Guyana and Brazil, are the Guiana Highlands, an isolated area of heavily forested plateaux and low mountains.

Despite rapid population growth—since 1958 from five million to 25 million—Venezuela remains sparsely populated. The majority of people live in the north and north-west around the northern tip of the Andes, and within this area most are now concentrated in the cities that are home to 88% of the population.

In the past Venezuela has been one of the richer South American countries, with an extensive system of social security. But poverty remains high, especially in the 'ranchos', the slums that ring Caracas. At the end of 2005, 38% of the population lived in poverty and 15% in extreme poverty. Poverty was heightened by the country's worst natural disaster in December 1999, when massive mudslides killed at least 30,000 people and made 250,000 homeless.

At the same time, social services have deteriorated. Literacy is high, but schools are in poor shape as a result of underinvestment. And as poverty has increased, so enrolment and attendance have been dropping. The health system too is in crisis and many hospitals have closed.

Despite rising poverty levels Venezuela continues to attract immigrants from Colombia to work as agricultural labourers or as illegal gold miners. There are thought to be around two million unauthorized immigrants—75% from Colombia.

Venezuela used to be an agricultural country—though only a small proportion of the land is suitable for arable farming. Nowadays, however, only around 10% of the labour force work in agriculture, many of them raising cattle, and around 70% of food has to be imported. Landholding is very concentrated: 70% is owned by 3% of proprietors—though there are now plans for reform.

Agriculture was pushed into the background after 1914 by the discovery of oil near Maracaibo in the north-west. Venezuela rapidly became the world's first major oil exporter. Even today, its proven reserves are among the world's largest. Oil usually accounts for three-quarters of export income and half of central government revenue. In addition, Venezuela has significant quantities of natural gas— the world's eighth largest reserves—as well as substantial reserves of bauxite, coal, iron ore, and gold.

Oil revenues accelerated Venezuela's economic and industrial development, much of which was in the government's hands. The government nationalized the oil industry in 1975, creating PdVSA which is now the largest employer. Then it invested heavily in other sectors such as steel, cement, and petro-chemicals, but abundant oil money reduced the incentive to establish a broadly based economy.

Public expenditure and investment have usually also fluctuated along with international oil prices—rising when prices are high, then crashing again. The worst crash was in the early 1980s when the country was hit

Government spending crashes with the oil price
by falling oil prices at a time of rising interest rates and the Latin American debt crisis. But even when oil prices recovered after 1998 the economy failed to recover and growth since then has been slow.

In recent decades Venezuela has usually been governed by one of two main political parties: the centre-left Democratic Action (AD) and the centre-right Independent Political Organization Committee (COPEI). But the political process has been

deeply corrupt and presidential candidates have now deemed it wise to break away from their own parties and stand under fresh colours.

Hugo Chavez, a former military officer who organized a failed coup in the early 1990s, won the 1998 presidential elections on a populist platform at the head of his Movimiento Quinta República (MVR). He then organized a referendum to establish a constituent assembly to write a new constitution that would enable him to usurp the powers of the parliament.

Chavez is a return to the old-style Latin American 'caudillo', or strongman, politics, using public expenditure to win votes. He has boosted spending on programmes for the poor, who mostly support him, while the middle classes are against.

With a new constitution that offered him the prospect of two six-year terms, and a higher oil price that allowed him to increase public spending a little, Chavez called an election in May 2000. He was duly re-elected with 60% of the vote, but since then the country has become increasingly polarized around efforts to unseat him. In 2002 there was a two-month national strike demanding his resignation followed by an unsuccessful military coup.

In 2004, the opposition parties and other groups, united as Coordinadora Democrática, organized a petition for a recall referendum. This was held in August 2004 and resulted in a victory for Chavez with 59% of the vote, though the opposition claimed fraud.

For the 2006 elections he will be opposed by Manuel Rosales who has his own party, A New Time.

Vietnam

Vietnam has partially liberalized its economy, but the Communist Party remains firmly in charge

Land area: 332,000 sq. km.	
Population: 82 million—urban 26%	
Capital city: Hanoi, 2.8 million	
People: Vietnamese 90%, other 10%	
Language: Vietnamese, Chinese, English, French, Khmer, tribal languages	
Religion: Buddhist, Taoist, Christian, indigenous beliefs	
Government: Communist	
Life expectancy: 71 years	
GDP per capita: $PPP 2,490	
Currency: Dong	
Major exports: Textiles, oil, rice, marine products, coffee, footwear	

Around two-thirds of Vietnam consists of the Annamite mountain chain, which snakes down the length of the country from north to south. The two main lowland areas are the Red River delta in the north, which includes Hanoi, and the much larger Mekong delta in the south, which includes Ho Chi Minh City. These two are linked by a narrow coastal plain.

Most of the population are ethnic Vietnamese who live in one of the two deltas. Also to be found there are many of the Khmer minority, and more than one million people of Chinese origin, chiefly in the urban areas. In the mountain areas, the ethnic composition is more complex, with more than 50 different groups. Known collectively by the French term 'Montagnards', they have struggled to preserve their traditional lifestyles. Because of

population pressure in the deltas the government has been resettling people in the uplands. There are also many Vietnamese overseas: 750,000 people left after the end of the war.

The Vietnamese have had good standards of health though public services are now more limited and people have to pay for most care. Education, however, has benefited from increased spending and many more children now go to school.

Poverty was reduced dramatically between 1990 and 2003, from 50% to less than 10%. However, there was also a rise in inequality, and a widening gap between the cities and the countryside. Around one-third of children are malnourished.

A major problem nowadays is corruption. Liberalization has also seen a growth in the trafficking of drugs from Laos and Myanmar.

Two-thirds of people work in agriculture and most people now work on their own account as households. Their principle crop, particularly in the deltas, is rice. Yields are high and Vietnam is now the world's second largest rice exporter. The main constraint is the lack of land. Much of the additional land brought under cultivation, particularly in the highlands, has been used for cash crops such as tea, cotton, coffee, rubber, and sugar. Coffee has been a notable success: yields are high and Vietnam is now the world's second largest exporter. Rubber too has done well. Vietnam also has ample fishing resources and exports seafood.

Rural communities rely on wood for 80% of their fuel but the countryside has suffered from deforestation. The government has had to close many logging companies and has banned log exports.

Since 1986, Vietnam has undertaken a process of economic renovation—'doi moi'—with greater emphasis on market forces and price incentives. This led to a rapid growth in industry, particularly in light manufacturing areas such as garments, toys, and footwear, as well as electronics. There is also a growing tourist industry, with around three million visitors per year.

'Doi moi' boosts output

In recent years, the economy has been growing at around 8% annually and much of this expansion was driven by transnational companies. Investment accelerated after 1994 when the USA lifted its trade embargo, and by 2002 foreign direct investment accounted for one-third of industrial output. Intel, for example, is to build a new microchip plant in Ho Chi Minh City.

On the other hand, privatization has proceeded relatively slowly. Central state-owned enterprises are responsible for 40% of industrial output. There are also many locally-run state enterprises.

Vietnam has offshore oil deposits and is an oil exporter, but production is now falling. It does, however, have plentiful supplies of coal and much hydroelectric potential.

Over the period 1959–75, Vietnam was embroiled in one of the 20th century's major wars—referred to in Vietnam as the 'American War'. The communist forces of the north emerged victorious and since then the Communist Party has retained control.

Although the country has an elected National Assembly, which can hold government officials accountable, the country is managed by three people: the general secretary of the Communist Party, Nong Duc Manh; the prime minister, Phan Van Khai; and the president, Tran Duc Luong.

There has however been some liberalization. The 'doi moi' policy represented some slackening of socialist economic ideology. And the constitution, as amended in 1992, no longer charges the government with 'building socialism' but establishes it more as a manager of the economy. In addition it strengthens the powers of the National Assembly.

In terms of economic policy the government is divided between the conservatives and the economic liberals, but both wings would probably put the brakes on economic reform if they felt this would loosen the party's political grip.

There is little by way of opposition though there are occasional protests, as in 2001 when ethnic minorities demonstrated for freedom of religion and against the loss of their land to Vietnamese settlers. The National Assembly too is becoming somewhat more vocal, though even today 90% of representatives are party members.

Virgin Islands (US)

One of the most popular US tourist destinations

Land area: 340 sq. km.
Population: 0.1 million—urban 49%
Capital city: Charlotte Amalie
People: Black 76%, white 13%, other 11%
Language: English, Spanish, creole
Religion: Baptist 42%, Roman Catholic 34%, Episcopalian 17%, other 7%
Government: Dependency of the USA
Life expectancy: 79 years
GDP per capita: $PPP 14,500
Currency: US dollar
Major exports: Refined petroleum products

The US Virgin Islands in the Caribbean comprise three main islands: St Croix, St John, and St Thomas, and more than 50 other islets or cays. The most southerly island, St Croix, is the largest. But the most developed, with the capital, is St John.

The islands' varied scenery includes rugged mountains, dense subtropical forests, and mangrove swamps, along with white, sandy beaches and coral reefs. The climate is mild—though the islands are regularly hit by hurricanes.

The Virgin Islanders have a mixed African and European heritage. The Danes were in charge for a couple of centuries and they settled the country with African slaves to work on the sugar plantations. But with an eye to their strategic value, the US bought the islands for $25 million in 1917 and the dominant culture is now American. The islands have one of the highest standards of living in the Caribbean and have also attracted immigrants from other Caribbean islands, particularly Puerto Rico.

The main source of income is tourism, which employs around two-thirds of the workforce. More than two million visitors arrive each year, most of whom come from the USA, many on cruise ships, heading primarily for the beaches and the duty-free shopping.

On the agricultural front, sugar cane has given way to a wide range of fruit and vegetables grown for local consumption and to meet the demand from tourists.

The principal industrial activity is oil refining, but there are also factories making pharmaceuticals and electronic goods which take advantage of duty-free entry into the USA.

The islands are an unincorporated territory of the USA. They send one member to the US House of Representatives, though he or she lacks full voting rights. The islanders also elect a governor, as well as representatives to their 15-seat legislature.

Both the Democratic and Republican parties are represented locally. A Democrat, Charles Turnbull, was elected governor in 1999 and re-elected in 2002. The member of congress, Donna Christian-Christenson, also a Democrat, was re-elected in 2002. Democrats also have a majority in the legislature—other seats being filled by Republicans and a local party, the Independent Citizens' Movement.

There have at times been debates about the future status of the islands, but in 1993 the islanders voted in a referendum to retain the status quo.

Western Sahara

Still waiting for a referendum on independence

Western Sahara's territory on the north-west coast of Africa is largely flat and almost entirely desert. The western two-thirds of the country is occupied by Moroccan troops, while the eastern third is controlled by the independence movement, Polisario.

The Saharawi are a mixture of Arab, Berber, and black African descent. Three-quarters of the population, around 170,000 people, are now living in refugee camps in western Algeria. The camps are fairly well run; even so, there appear to be serious health and nutritional problems. In recent years, the country has also been flooded with around 200,000 Moroccan settlers and more than 100,000 Moroccan soldiers.

In the past, the Saharawi have survived largely as nomadic herders, with a little agriculture along the coast. But this is an area of huge potential. In 1963, large phosphate deposits were discovered at Boucraa in the north. Total reserves have been estimated at 10 billion tons. In addition, there are rich coastal fishing grounds which are currently exploited by Spanish trawlers.

Western Sahara was a Spanish colony. An armed liberation movement, Polisario, was formed in 1973 but when Spain relinquished the territory in 1976 it handed the northern two-thirds to Morocco and

Land area: *266,000 sq. km.*
Population: *0.3 million*
Capital city: *L'ayoune*
People: *Saharawi, Moroccan*
Language: *Arabic*
Religion: *Muslim*
Government: *Government in exile*
Life expectancy: *Not available*
GDP per capita: *Not available*
Currency: *Moroccan dirham*
Major exports: *Phosphates*

the southern third to Mauritania—though the International Court of Justice said that the Saharawi were entitled to independence. Years of brutal warfare ensued with Polisario working from bases in Algeria. Mauritania withdrew in 1976 and Morocco seized their portion too.

Polisario's guerrilla attacks did not shift Morocco, but their diplomatic offensive proved more successful. By the mid-1980s, more than 60 countries had recognized them as the legitimate government. The real breakthrough, however, came in 1990, when the UN managed to achieve a peace plan that would lead to a UN-supervised referendum on independence based on a 1974 census. A ceasefire was declared in 1991.

Since then there has been little progress on the referendum. Polisario, and its main supporter, Algeria, wants the electorate to be limited to anyone on the 1974 census, while Morocco, relying on its settlers, wants it to include anyone born there. The UN has arrived at a list of 86,000 voters but also has claims from another 130,000 people in Morocco. Since the referendum process seems to have ground to a halt, the UN has also suggested partitioning the territory into a Moroccan north and independent south.

Yemen

Despite recent discoveries of oil, Yemen remains poor—and prone to violence

Land area: 528,000 sq. km.
Population: 20 million—urban 26%
Capital city: San'a, 1.7 million
People: Arab
Language: Arabic
Religion: Muslim
Government: Republic
Life expectancy: 61 years
GDP per capita: $PPP 889
Currency: Yemeni riyal
Major exports: Oil

Yemen has three main regions. It has narrow desert coastal plains along both the Red Sea and the Gulf of Aden. These plains rise steeply to highlands that reach 1,500 metres. Then to the north the highlands descend to a desert area that covers half the country and merges with the 'empty quarter' of Saudi Arabia.

Almost everyone is Arab but the mountainous terrain has dispersed the population so there can be differences in dialects of Arabic, heightened by historical divisions between north and south. In the 1830s, the territory had been carved up between the Ottoman empire in the north and the British empire in the south. After independence, the government in the north remained more conservative while the government in the south pursued Soviet-style socialism that diluted some aspects of Islam— particularly the restrictions it imposed on women. The two countries united in 1990 but the social and political differences remain.

As a whole, Yemen is one of the world's poorest countries. Literacy is low and around half of children are malnourished. Population growth is high at over 3% annually.

Around half of people rely on agriculture. Most of the country's rainfall is in the mountain areas, so farmers have had to build elaborate systems of terracing, growing basic subsistence crops such as sorghum and potatoes, as well as some cash crops, including fruit and high-quality 'mocha' coffee for export. But around one-third of agricultural production is of 'qat'—a plant whose chewable leaves contain a mild amphetamine. In addition, the mountains support large numbers of sheep and goats.

With little land to spare and few other opportunities for employment, more than one million Yemenis have left to work in richer neighbouring countries—though many were expelled from Saudi Arabia and Kuwait because of Yemen's tacit support for Iraq in the 1990 Gulf War.

Some respite from the country's poverty came from the discovery of oil in the 1980s. By the mid-1990s, around 70% of government revenue was coming from oil. Yemen also has natural gas reserves, and there are plans to expand liquefaction plants to enable more gas to be exported.

The government is also attempting to capitalize on the country's strategic location by establishing a new Aden Free Zone. This will include a container port and an industrial estate, as well as a larger airport.

Another important source of foreign exchange has been tourism for the hardier travellers—around 80,000 of whom visit each year to explore the country's rich historical heritage. The

Adventure holidays in Yemen

industry was badly hit by a spate of kidnapping of tourists in 1999 by tribesmen wanting to draw attention to a lack of amenities. Although tourists have rapidly been released unharmed, Yemen now has a reputation as an overly adventurous destination.

Yemen's unification in 1990 was the culmination of a period of increasing co-operation between the Yemen Arab Republic in the north and the People's Democratic Republic of Yemen in the south. President Ali Abdullah Saleh from the north—which had three-quarters of the total population—became president of a united Yemen, and his counterpart in the south, Ali Salim al-Bidh, became vice-president.

Unfortunately, unification coincided with the first Gulf War, which forced more than 800,000 Yemeni migrant workers to return home—leading to unemployment and violent political unrest. Parliamentary elections in 1993 delivered a majority for Saleh's General People's Congress (GPC)—a mixture of military, tribal, and other groups. Al-Bidh's Yemeni

Socialist Party came second—though with a majority in the south. An Islamic coalition, Islah ('gathering for Reform'), came third. But people in the south complained of harassment and 'internal colonization' and in 1994, with support from Saudi Arabia, they tried to secede. This caused a civil war, which Saleh's forces won after six months and 10,000 deaths.

Yemen looked for international aid and entered into agreements with the IMF that required it to liberalize the economy, but low oil prices and falling remittance income forced the government to cut many services.

Even so, there was scarcely any opposition for the 1999 presidential election, largely because the Socialist Party boycotted it, and Islah nominated Saleh. Unsurprisingly, he won, with 96% of the vote. In 2001 a referendum extended the presidential term to seven years and introduced a new assembly. In the 2003 assembly elections GPC increased its already very large majority.

Meanwhile, much practical political power around the country remains dispersed among tribal leaders who have official control over certain areas and have ample and sophisticated arms. The government is struggling to subdue those involved in kidnapping.

Yemen has also been affected by the growth in the number of Islamic terrorists who have reportedly been using Yemen as a base and who in 2000 sank a US warship in Aden. Since then Saleh has co-operated closely with the US, but at a domestic political cost: over the past few years there has been a spate of violent attacks on the government.

Zambia

Zambians suffer from low copper prices, poverty, and now HIV/AIDS

Land area: 753,000 sq. km.
Population: 11 million—urban 36%
Capital city: Lusaka, 1.4 million
People: Bemba, Tonga, and many others
Language: English, Bemba, Tonga, and about 70 other languages
Religion: Christian, Muslim, Hindu
Government: Republic
Life expectancy: 38 years
GDP per capita: $PPP 877
Currency: Kwacha
Major exports: Copper, cobalt, tobacco, lead

Most of Zambia consists of a plateau at around 1,200 metres above sea level. The highest point is in the Muchinga Mountains in the north-east. The territory is mostly open grasslands with occasional trees. The main river is the Zambezi, whose energies are tapped by the huge Kariba dam.

Zambians can belong to any of 70 or more ethnic groups, most of whom speak Bantu languages, though the official language is English. There are also small numbers of Europeans.

Three-quarters of Zambians live below the national poverty line and half are classified as 'extremely poor'. In the past one of the more successful areas was education: two-thirds of children are enrolled in primary schools and adult literacy in 2003 was 68%. In recent years, however, with budget cuts and low and often late pay for teachers, and rising levels of poverty, educational standards have been falling. Zambia had also made progress in health, but here too services have deteriorated.

But Zambia could start to reverse the trends. In 2005 donors wrote off $3.9 billion of the country's foreign debt and most of the savings are going into health and education. In 2006 charges in rural clinics were scrapped.

This still leaves the spectre of HIV/AIDS: in 2005, one-fifth of the population were HIV-positive. As a result, life expectancy has dropped by around 15 years to 38 and around half a million children have been orphaned. One-quarter of households are now thought to be caring for one or more orphans. Still, it looks as though infection might have peaked and the government is providing free anti-retroviral therapy.

Most people rely to some extent on agriculture. Zambia could be a major food producer: only around one-tenth of the land is arable though only about half of this is currently being worked. One of the main constraints for local food availability is the pattern of landholding. Much of the best land is held by large commercial farmers, usually white. While they do also grow the main food crop, maize, low prices have encouraged large farms to devote more land to tobacco, cotton, sugar, and flowers for the export market.

Meanwhile the subsistence farmers work on lower-quality communal land growing maize, sorghum, millet, and other basic crops, and in some cases cotton. Efforts to transfer communal land to private ownership have been blocked by traditional leadership.

One of the most important sources of income and employment is copper, which provides two-thirds of export income. The 'copperbelt' is in the north-west of the country, bordering on the Democratic Republic of the Congo. This is the most industrialized part of Zambia, and also produces cobalt, coal, lead, and zinc. Mineral income dropped during the 1990s. This was partly a reflection of falling international prices but also of poor management and low investment.

Privatization of the mines from 2000 was a tortuous business but

Tortuous privatization

one of the largest owners now is an Indian company, Sterlite, and companies from Switzerland, China, and elsewhere are also involved. Higher international prices have encouraged fresh investment and output has increased.

After independence Zambia also invested heavily in manufacturing industry in areas such as food processing, chemicals, textiles, and tobacco. Initially most of this activity was in state hands, but many enterprises have been sold off.

Tourism is also growing rapidly, mainly to game parks and Victoria Falls. Over a million visitors arrive each year—many diverted from crisis-hit Zimbabwe.

Zambia's initial decades of independence after 1964 were dominated by its first president, Dr Kenneth Kaunda. In 1972, Kaunda outlawed political parties except for his own United National Independence Party (UNIP). Opposition had to come from the Zambian Congress of Trade Unions, led by Frederick Chiluba. The economic situation deteriorated in the 1980s, provoking strikes and riots. In 1990, after further unrest Kaunda allowed other parties to organize.

Frederick Chiluba had by then created a new party, the Movement for Multiparty Democracy (MMD), and won the ensuing presidential election. Chiluba and the MMD certainly liberalized the economy but had less appetite for democracy and in 1996, Chiluba introduced a new constitution that prevented Kaunda from standing for election. UNIP boycotted the elections that year, giving Chiluba and the MMD an easy victory.

Chiluba wanted a third term, but did not get his way, so chose Patrick Levy Mwanawasa as his successor. Mwanawasa narrowly won the 2001 presidential election just ahead of Anderson Mazoka of the recently formed United Party for National Development (UPND). The election result was disputed and Mwanawasa had to survive a legal challenge to it. He soon acquired greater popularity, however, with an anti-corruption drive that included charging Chiluba and others with the theft of state funds.

Mwanawasa will stand again in 2006 but, following the death of Mazoka, against a new UPND candidate. However, Mwanawasa's health too is poor.

Zimbabwe

A country facing economic collapse and increasing political repression

Land area: *391,000 sq. km.*
Population: *13 million—urban 35%*
Capital city: *Harare, 1.4 million*
People: *Shona 82%, Ndebele 14%, other African 2%, white 1%, other 1%*
Language: *English, Shona, Sindebele*
Religion: *Syncretic 50%, Christian 25%, indigenous 24%, other 1%*
Government: *Republic*
Life expectancy: *37 years*
GDP per capita: *$PPP 2,443*
Currency: *Zimbabwe dollar*
Major exports: *Tobacco, food, gold*

Zimbabwe's dominant topographical feature is the High Veld, a broad ridge around 1,500 metres above sea level that runs from south-west to north-east, and covers around one-quarter of the territory. The land drops on either side of this to the plateau of the Middle Veld, and then to the Low Veld, which reaches the frontiers at the Zambezi River in the north-west and the Limpopo in the south-east. Though it is in the tropics, Zimbabwe's elevation ensures a fairly mild climate.

The main population group are the Shona, who settled in the north and west of Zimbabwe more than 1,000 years ago and have primarily been farmers. The smaller Ndebele group arrived in the 19th century and lived initially as pastoralists in the south and east. With different forms of livelihood, the two coexisted fairly peaceably. More disruptive was the arrival from South Africa at the end of the 19th century of a small number of whites. Disappointed by not finding much gold to mine, instead they took the best land—the 'white highlands'—a seizure that rankles to this day.

When black majority rule was achieved in 1978, the government invested heavily in education and primary enrolment ratios soon shot up to 100%. There was similar progress in health with a steady rise in life expectancy. Since then, the country has been on a downward spiral. Economic collapse led to cuts in government expenditure, and fees were introduced for education and health services. Primary enrolment has fallen to 79% and clinics and hospitals are undermanned and badly equipped.

Health has also been badly affected by the AIDS epidemic: by 2005, one-fifth of adults were HIV-positive and life expectancy had fallen to 37 years.

By the standards of Sub-Saharan Africa, Zimbabwe has a fairly diverse economy. Sanctions imposed during the final years of white rule in the 1970s had forced the country to develop its own manufacturing industry, including steel production. Though output declined when price controls and protection were removed, even by 2004 manufacturing still made up 14% of employment. Zimbabwe also has an important mining industry, particularly for gold, which makes up 15% of exports.

Agriculture remains central to the economy, and accounts for 15% of employment. Before the current crisis, in years of good rainfall, Zimbabwe's farmers could even grow enough food to export, as well as producing cash crops such as sugar and cotton, and particularly tobacco, which is the major export earner.

Agriculture has always been overshadowed by questions of land distribution. After independence the government, nervous of provoking

Skewed patterns of landholding

'white flight', left many of the large landholdings intact. Whites made up less than 1% of the population but owned one-third of arable land—leaving black farmers to work the drier, lower-lying territory.

Both industry and agriculture are now in a poor state. The economy was undermined initially by drought and collapses in the prices of tobacco and gold, but mostly now due to general mismanagement and corruption: since 2000 the economy has been contracting sharply, in 2005 by 7%. Since 2001, however, agricultural output has dropped by one-third. By 2006, inflation was over 1,000%.

When Zimbabwe achieved black majority rule in 1980, Robert Mugabe became prime minister at the head of the Zimbabwe African National Union-Popular Front (ZANU-PF). In 1985, the party was re-elected and in 1987 the parliament replaced the office of prime minister with that of

an executive president. Since then, Mugabe and ZANU have won all assembly and presidential elections.

Mugabe amended the constitution 14 more times—generally to augment his own power as Zimbabwe degenerated into a one-party state and his corrupt and repressive regime became increasingly unpopular.

Early in 2000 a more effective opposition emerged as the Movement for Democratic Change (MDC) with the backing of trade unions, churches, and others. In 2000, in a shock result, Mugabe finally lost the latest in his series of constitutional referendums.

The prospect of a powerful opposition provoked an aggressive response. ZANU zealots started attacking and killing MDC members, and Mugabe, to gather popular support, condoned the violent seizure of white commercial farms by 'veterans' of the civil war.

Despite this and widespread intimidation and violence ZANU almost lost the June 2000 parliamentary election to the MDC. In 2002 in a doubtful election Mugabe was returned as president.

By 2003 the land seizures had led to famine for half the population. Mugabe stepped up the repression, frequently arresting, and sometimes torturing, opposition MPs.

The 2005 parliamentary elections were duly rigged and ZANU maintained its grip. In May 2005 the government launched 'Operation Drive Out Rubbish'—bulldozing thousands of poor people's homes.

Now with the MDC weak and increasingly divided, the only real opposition comes from churches and trade unions.

Smaller countries

American Samoa

Pacific nation relying on the USA, tuna, and tourism

People: *58,000, of whom 92% are Samoans, bilingual in Samoan and English.*
Life expectancy: 76 years
Government: *Administered by the USA.*
Capital: Pago Pago
Economy: *GDP per capita: $PPP 5,800.*
Main export: canned tuna

American Samoa is a US territory in the South Pacific. It consists of three main high volcanic islands and two atolls. Most people live on the main island of Tutuila, which also has a US naval base. More than 90% of the land is communally owned.

Many Samoans have emigrated to the USA. This has been offset by immigration from Asian countries such as Taiwan and South Korea, though immigration restrictions are becoming tighter.

The territory is economically highly dependent on the USA, which funds half the government budget. Around one-third of the workforce are employed by the government. Another one-third work in the two tuna canneries that produce the main export. Goods can enter the US duty-free providing the finished product has at least 30% locally originating material, so foreign investors can gain a foothold in the US market.

Although the US president is head of state, citizens elect their governor as well as members of the parliament, the Fono. In 2000 they re-elected a Democrat, Tauese Sunia, as governor. When he died in March 2003 he was replaced, as acting governor, by Togiola Tulafono.

Andorra

Distinctive co-principality between France and Spain

People: *71,000, speaking Catalan, French, Spanish and Portuguese.*
Life expectancy: 84 years
Government: *Constitutional co-principality.*
Capital: Andorra la Vella
Economy: *GDP per capita: $PPP 24,000.*
Main export: electricity

Andorra is a tiny, mountainous country lodged in the eastern Pyrenees between France and Spain. Less than one-third of the population are native Andorrans. Most are Catalan Spanish, with some French, and the majority live in and around the capital, Andorra la Vella.

The main industry is tourism, which makes up over 80% of GDP. Millions are drawn each year to the spectacular scenery and the ski resorts. Duty-free shopping is also an attraction for the Spanish and French. Andorra uses both the French franc and the Spanish peseta. It is also distinctive in that it has neither income nor value-added tax, though there are municipal charges.

Andorra has a unique constitution. By tradition, the heads of state are two 'princes', the French president and the Bishop of Urgel in Spain. But this is largely a ceremonial arrangement. In 1993, Andorra adopted a new constitution, reducing the powers of the two princes and giving full sovereignty to the Andorrans, who elect a 28-member general council. The 2005 elections returned the Partie liberal, with Albert Pintat Santolària as general council president.

Anguilla

Politically lively British Caribbean dependency

People: 14,000, most of whom are black or mulatto. Language: English.
Life expectancy: 77 years
Government: Dependency of the UK.
Capital: The Valley
Economy: GDP per capita: $PPP 5,800.
Main export: lobsters

Anguilla is a flat coral and limestone island in the eastern Caribbean. Most of its people are black or mulatto. The climate is dry and the soil is not very fertile so agriculture is limited. In the past, the main source of income has been fishing, but Anguillans nowadays rely more on remittances from migrants, luxury tourism, and more recently on financial services.

In 1969, Anguilla broke away from a self-governing federation with neighbouring St Kitts and Nevis—in a strange incident that required the dispatch of a squad of London policemen to restore order.

Subsequently, Anguilla again became a dependent territory with an 11-member house of assembly. The 1999 election returned Hubert Hughes as chief minister, but he has regularly quarrelled with the British government.

In early 2000, the island was plunged into a constitutional crisis when a combination of defections and boycotts meant that the House of Assembly could not summon a quorum. A new election in March 2000 resulted in a victory for the conservative United Front coalition led by Osbourne Fleming, who won again in February 2005.

Antigua and Barbuda

Tourist paradise tinged with crime and corruption

People: 69,000, mostly black and English-speaking. Life expectancy: 72 years
Government: Constitutional monarchy.
Capital: Saint John's
Economy: GDP per capita: $PPP 11,000.
Main exports: petroleum products and manufactures

Antigua is the larger and more populated island in this two-island state in the eastern Caribbean. Its most significant geographical features are its white, sandy beaches. It has a warm, dry climate.

The standard of living is high and most of the income comes from tourism: over 200,000 visitors arrive each year either on cruise ships or to stay in the islands' exclusive resorts. Agriculture and manufacturing are mainly for local consumption and to supply the tourist industry.

The country has also become an offshore financial centre and has more than 50 banks. Unfortunately, its lax regime has also made it an attractive money-laundering centre for the Russian mafia and Colombian drugs barons.

Antigua's head of state is the British monarch, but for more than 50 years the country was under the political sway of the Bird dynasty and the Antigua Labour Party (ALP), first Vere Bird, who died in 1999, and then his sons Lester and Vere Junior—a family seldom free of allegations of corruption. Dynastic rule finally ended in March 2004, with a victory for the United Progressive Party, led by Baldwin Spencer.

Aruba

A rocky Caribbean dependency of the Netherlands

> **People:** 71,000, ethnically diverse, speaking Dutch and Papiamento.
> Life expectancy: 79 years
> **Government:** Dependency of the Netherlands. Capital: Oranjestad
> **Economy:** GDP per capita: $PPP 28,000. Main export: oil products

Aruba is a flat rocky island in the Caribbean with few natural resources. Its people are a diverse ethnic mixture—with strains of Dutch, Amerindian, African, and Spanish. The majority are Roman Catholic. They have one of the highest standards of living in the Caribbean. Around one-quarter of the population are immigrants from the Netherlands, Venezuela, Colombia, and other Caribbean islands.

The chief industry is tourism which, with over 700,000 visitors per year, accounts for around 40% of GDP. But there is also a modest offshore financial services sector. In addition, the island refines some Venezuelan oil in a refinery owned by the US company El Paso.

Aruba was part of the Netherlands Antilles until 1986, when it was given 'status aparte', which involves having a Dutch governor but a locally elected assembly that deals with internal affairs. Earlier plans for political independence have now been shelved, though the Netherlands is keen to encourage financial independence. The 2001 and 2005 elections resulted in outright victories for the Movimentu Electoral di Pueblo, led by Nelson Oduber.

Bermuda

Atlantic island of wealth that could become independent

> **People:** 65,000. Black 55%, white 34%, other 11%. Language: English.
> Life expectancy: 78 years
> **Government:** Overseas territory of the UK. Capital: Hamilton
> **Economy:** GDP per capita: $PPP 69,000. Main export: pharmaceutical re-exports

Bermuda comprises one main coral-based island in the Atlantic. Bermudans enjoy a higher standard of living than their colonial masters in the UK. Two-thirds are black and the rest white; around one-quarter are foreign born. There have been violent racial tensions but conflict today is resolved through the political system.

Bermuda has little agriculture or manufacturing. One of the main sources of income, employing around one-third of the workforce, is the offshore financial sector, particularly insurance. Another is up-market tourism which directly or indirectly employs more than half the population. The island receives a quarter of a million visitors per year, though the industry has suffered several years of decline.

Bermuda is an internally self-governing British overseas territory. For 30 years the government was formed by the white-based United Bermuda Party. But the 1998 and 2003 elections were won by the black-based Progressive Labour Party. Independence was rejected in a 1995 referendum, but the issue could resurface since the PLP in 2003 replaced its leader by one more likely to pursue a break with the UK.

British Virgin Islands

A dependency of the UK that prefers the US dollar

People: 23,000. Most are black or mulatto with some whites and Asians.
Language: English. *Life expectancy:* 77 years
Government: Overseas territory of the UK.
Capital: Road Town
Economy: GDP per capita: $PPP 39,000.
Main exports: rum, fish

There are 36 British Virgin Islands in the eastern Caribbean, but only 16 are inhabited. The four main islands are Tortola, Anegada, Virgin Gorda, and Jost Van Dyke. The islands are hilly with many lagoons and coral reefs.

The majority of Virgin Islanders live on Tortola. Some work in agriculture and fishing, but their main source of income nowadays is tourism; some land-based, the others on yachts. Tourism accounts for around half of GDP.

Because of close economic links with the neighbouring US Virgin Islands the currency is the US dollar.

Since the mid-1980s the British Virgin Islands has built up a large offshore financial services industry. By 2004 it had 544,000 companies registered locally. These include accountancy and legal firms, and banks and insurance companies, which are staffed by locals and expatriates.

As an overseas territory of the UK, the British Virgin Islands have an appointed governor and a 15-member legislative council. Following a government financial scandal, in the 2003 elections the National Democratic Party took control, with Orlando Smith as chief minister.

Cayman Islands

A congenial location for footloose international banks

People: 45,000. Mixed ancestry 40%, black 20%, white 20%, other 20%. *Life expectancy:* 80 years. *Language:* English
Government: Overseas territory of the UK.
Capital: George Town
Economy: GDP per capita $PPP 32,000.
Main export: turtle products

These are three coral-based islands in the Caribbean, mostly flat and occasionally marshy, with white, sandy beaches and coral reefs. The islanders are on average far richer than their British colonial rulers—and pay no income tax. There are also many immigrant workers who make up around half the workforce.

Tourism is the largest employer. Around 300,000 people stay each year in this up-market destination, with many more visting on cruise liners, the majority from the USA.

The Cayman Islands is also a major offshore financial centre and tax haven—with over 300 registered banks and trust companies. The country is a favourite location for company registration—with 65,000 enterprises claiming this as home.

This is a British overseas territory with an appointed governor, who since 2002 has been Bruce Dinwiddy. He works with an Executive Council which includes members of the Legislative Assembly.

In the 2005 election to the 15-seat Assembly the People's Progressive Movement (PPM), headed by Kurt Tibbetts, gained a majority. There is no pressure for independence, though there are plans for greater autonomy.

Christmas Island

Phosphates, crabs, and tourism in the Indian ocean

People: 1,500. Chinese 70%, European 20%, Malay 10%. Language: English.
Life expectancy: data not available
Government: Territory of Australia. Capital: The Settlement
Economy: GDP per capita: data not available. Main export: phosphates

Christmas Island in the Indian Ocean was left largely uninhabited until the discovery of phosphates at the end of the 19th century. Most subsequent mining work was done by immigrant labour, primarily from the Cocos Islands, Malaya, and Singapore.

Since 1958 Christmas Island has been administered by Australia. In 1985, the island established its first local assembly and in 1994 the islanders voted overwhelmingly in a referendum against independence. Christmas Island now has a shire council as an electoral district within Australia's Northern Territory.

As the phosphate mines became exhausted, the Australian government transferred ownership to a private operator and a new lease was signed in 1998.

The government also started to invest in infrastructure in order to build up a tourist industry—including a casino, golf courses, and facilities for scuba diving. The island is well known for its 120 million red land crabs which each year between October and January migrate from the forest to the coast.

Since 2000 Australia has operated an offshore detention centre here for processing asylum seekers.

Cocos Islands

Australian dependency that aims for Australian wealth

People: 574. Cocos Malays (who are Muslim) and Europeans.
Languages are Malay and English
Government: Territory of Australia. Capital: West Island
Economy: GDP per capita: data not available. Main exports: copra, coconuts

The Cocos Islands, officially called the Cocos (Keeling) Islands, comprise two main coral atolls in the eastern Indian Ocean. The only inhabited islands are Home Island and West Island.

Most of the Cocos islander community, who are Muslims, are to be found on Home Island. They are descendants of people who were brought from Malaya, East Africa, and other countries to work on the coconut plantations. Those on West Island are largely Australian government employees and their families.

Though there is some fishing and horticulture, their main source of income is still coconuts. Most of their work is organized by the Cocos Islands Co-operative Society, which supervises the production of copra and other activities.

The islands came under Australian control in 1955. In 1978, the Australian government purchased most of the land owned by the Clunies-Ross family and distributed it to the islanders. In 1984, in a UN-supervised referendum they voted to become a part of Australia. There is a non-resident Australian administrator as well as a seven-seat Shire Council.

Cook Islands

Pacific islands steadily losing people to New Zealand

People: *21,000, of whom 88% are full Polynesian. Life expectancy: 71 years*
Government: *Self-governing, in free association with New Zealand.*
Capital: Avarua
Economy: *GDP per capita: $PPP 5,000. Main exports: copra, fruit*

The Cook Islands consist of 15 islands and atolls scattered across two million square kilometres of the South Pacific Ocean.

Most Cook Islanders are of mixed Polynesian descent and speak Cook Islands Maori as well as English. More than half are on the island of Rarotonga. Many have also emigrated: there are 60,000 Cook Islanders in New Zealand and 20,000 in Australia.

Around one-fifth of the islanders depend on agriculture. There is also some industry, including the manufacture of clothing and footwear. However, most of the economy is based on tourism, with 65,000 visitors per year, and financial services, with over 1,200 registered companies, along with income from remittances.

The Cook Islands have full self-government in 'free association' with New Zealand, which deals with foreign affairs. The head of state is the British monarch. She and the New Zealand government are represented by a high commissioner, but most decisions are taken by the prime minister and the 24-member parliament. Following the 2004 elections, the prime minister was Jim Marurai of the Cook Islands Party.

Dominica

Caribbean island heavily dependent on banana exports

People: *69,000, predominantly black and English-speaking. Life expectancy: 75 years*
Government: *Republic. The president is elected by the legislature for a five-year term.*
Capital: Roseau
Economy: *GDP per capita: $PPP 5,500. Main export: bananas*

Dominica is a volcanic island in the Caribbean with rich fertile soil. Most people are of African or mixed descent and around 40% work on small farms, growing subsistence crops. They also grow bananas for export to the EU, primarily the UK, but remain highly dependent on preferential access. Other crops such as mangoes, grapefruits, avocados, and oranges do not offer such good returns.

Without a long-runway airport or extensive beaches, Dominica's tourist potential is limited, though there are facilities for cruise-ship visitors. A better prospect is eco-tourism, which can capitalize on the rich fauna and wildlife. Dominica also an offshore financial sector, with 8,600 registered companies.

For most of the period following independence in 1978, the government was in the hands of Dame Eugenia Charles and the conservative Dominica Freedom Party (DFP). She retired in 1995 and the DFP lost an election to the centre-left United Workers' Party. Following the 2000 election, Roosevelt Skerrits of the Dominican Labour Party formed a coalition with the DFP but won an absolute majority in 2005.

Faroe Islands

Fishing nation moving towards independence

People: 68,000. Faroese, speaking Faroese and Danish.
Life expectancy: 75 years
Government: Dependency of Denmark.
Capital: Tórshavn
Economy: GDP per capita: $PPP 22,000.
Main export: fish

The Faroes are a group of 17 rugged and largely treeless volcanic-based islands in the North Atlantic. Most Faroese live on the islands of Streymoy and Eysturoy.

The people are descendants of Viking settlers. The country's name means 'sheep islands' in old Norse, and raising sheep is still the main land-based activity, but the primary source of income is fishing, mainly for cod. Recently there have been some offshore oil discoveries. The islands also benefit from substantial subsidies from Denmark.

Since 1948, the Faroes have been a self-governing community within the kingdom of Denmark—though in order to protect their fishing interests they are not part of the EU. They have the oldest parliament in the world, the Løgting, founded more than 1,000 years ago, but they also have two seats in the Danish parliament.

Independence is on the agenda. The 2004 elections, as usual, produced a coalition government, and Joannes Eidesgaard was chosen as prime minister.

Most representatives support independence, which the Danish parliament has said it will grant—if and when the Faroese ask for it.

Falkland Islands

British territory suddenly prosperous from seafood

People: 2,967, of British origin, speaking English.
Government: Dependency of the UK.
Capital: Stanley. Territory also claimed by Argentina as the Islas Malvinas
Economy: GDP per capita: $PPP 25,000.
Main exports: wool, hides, mutton

The Falkland Islands consist of around 200 islands in the South Atlantic. This British dependency is disputed with Argentina, which claims it as the Islas Malvinas. The main islands are East and West Falkland.

The population are of British extraction. They have a British-appointed governor but the islands are self-governing with their own executive and legislative councils. Many people have made their living from agriculture, predominantly sheep farming. But the economy is now booming from selling licences to foreign vessels, bringing in $40 million per year. In addition there are 40,000 tourists annually.

The UK and Argentina had been trying to resolve their dispute when the Argentine military government invaded in 1982, provoking a war which ended when British troops reoccupied the islands. The UK maintains a military presence.

The islanders, who are now financially independent, want to maintain the status quo—a view the UK supports. Relations with Argentina improved following agreement on a 'sovereignty umbrella' to co-operate on practical issues, but have deteriorated again recently.

Gibraltar

Spain and the UK still dispute the status of 'the rock'

People: 28,000, mostly of British and southern European descent.
Life expectancy: 80 years
Government: British dependency, though claimed by Spain. Capital: Gibraltar
Economy: GDP per capita: $PPP 27,900.
Main exports: mostly re-exports

Gibraltar is a massive limestone and shale rock joined by a narrow isthmus to the coast of southern Spain. It holds a commanding position at the entrance to the Mediterranean.

Gibraltar's population reflects its long history as a British naval base: around two-thirds of the people are native Gibraltarians, of mixed British and southern European descent, while the rest are foreign workers.

The British military presence now accounts for only around one-tenth of Gibraltar's economy. More important are tourism, with more than five million visitors per year, and offshore financial services—the latest development being internet gambling.

Spain wants sovereignty over Gibraltar while the UK insists on respecting the wishes of the Gibraltarians who have refused to unite with Spain. Gibraltarians have self-government in all areas but defence, and are British citizens.

In July 2002 Spain and the UK announced that their preferred solution was joint sovereignty, but an unofficial referendum organized in 2002 by Gibaraltar's chief minister, Peter Caruana of the Gibraltar Social Democrats, found that 99% of Gibraltarians opposed the idea.

Greenland

The world's largest island, but not very green

People: 56,000. More than 85% are Greenlanders; most of the rest are Danish.
Life expectancy: 70 years
Government: Dependency of Denmark.
Capital: Nuuk
Economy: GDP per capita: $PPP 20,000.
Main export: fish

Two-thirds of Greenland lies within the Arctic Circle and most of the territory is covered by an ice cap which at its centre is eight kilometres thick and holds 10% of the world's total fresh water.

Greenlanders are mostly a mixture of the native Inuit and settlers from Scandinavia. Few pure Inuit remain. Traditionally they have survived by hunting seals, but this probably occupies only around one-fifth of the population. Today, more people make a living through fishing, mainly for shrimp. There has also been some zinc mining but this has now ceased, and left Greenland even more dependent on subsidies from Denmark which provide half the government budget.

Greenlanders are Danish citizens who elect two representatives to the Danish parliament—though Greenland withdrew from the EU in 1985 to protect its fishing interests. For internal affairs they have a Home Rule Parliament.

The 2001 elections resulted in a coalition between Siumut, a social democratic party that favours greater autonomy, and the socialist Inuit Ataqataqiit party. Hans Enoksen of Siumut is prime minister.

Grenada

A spice island of strategic importance to the USA

People: *89,000, mostly black, speaking English or a French patois.*
Life expectancy: 65 years
Government: *Constitutional monarchy.*
Capital: St George's
Economy: *GDP per capita: $PPP 5,000.*
Main exports: nutmeg, mace

Grenada is a volcanic island with a rugged mountainous interior and an attractive coastline that has many secluded bays and beaches. Most Grenadans are black, while a minority are of mixed race.

The country's fertile soil and abundant rainfall allow around one-quarter of the population to make a living as small farmers. Apart from growing food crops, they also specialize in nutmeg.

But the main source of foreign exchange nowadays is tourism. There are more than 130,000 annual stopover visitors, and a similar number arrive on cruise ships.

In recent years the largest bilateral donor to Grenada has been Taiwan, in exchange for political support. In 2004 Hurricane Ivan caused over $800 million-worth of damage.

Grenada, which achieved independence from the UK in 1974, has a history of political conflict. Turmoil in 1983 resulted in the killing of prime minister Maurice Bishop, and provoked a US invasion.

Democracy was restored, though it has been unstable. In 2003 the New National Party, headed by Keith Mitchell, was re-elected but with only a one-seat majority.

Isle of Man

From buckets and spades to banking in the Irish Sea

People: *75,000, Manx and British.*
Life expectancy: 78 years
Government: *Dependency of the British Crown, rather than of the UK.*
Capital: Douglas
Economy: *GDP per capita: $PPP 28,500.*
Main exports: tweeds, kippers

The Isle of Man lies halfway between England and Ireland in the Irish Sea. The highest point is Snaefell, at 620 metres, but most of the territory is low-lying and used for agriculture.

Its people, the Manx, are of Celtic descent. Health and welfare services are at least equal to those in the UK and incomes are now similar.

Until recently, the island relied heavily on budget tourism. But tourism has been eclipsed by financial services. Low tax rates have attracted more than 60 banks, and finance and related services now account for more than half the GDP. As a result, the Isle of Man's economy has been very healthy, with annual growth of about 5% combined with low inflation and unemployment under 1.5%.

The Isle of Man is not part of the UK; it is a crown dependency with a lieutenant-governor appointed by the British monarch. But it has considerable autonomy: its parliament, the Tynwald, makes its own laws and sets the tax rates. It also applies immigration controls: even British citizens need a work permit.

Politics is mostly consensual. The parliament elects a chief minister who in turn chooses a cabinet, but there are no political parties.

Kiribati

Small islands scattered over a vast area of ocean

People: 105,000, most of whom are Micronesians and Christians.
Life expectancy: 61 years
Government: Republic, with a president elected by the legislature. Capital: Tarawa
Economy: GDP per capita: $PPP 800. Main exports: copra, fish

Kiribati consists of 33 islands or atolls spread over three million square kilometres of the Pacific. Since most of the territory is less than a metre or two above sea level, it would be an early victim of global warming.

Its people, called the I-Kiribati, are mostly Micronesians and around one-third live on one atoll, Tarawa. Many I-Kiribati also work outside the islands, either elsewhere in the Pacific, particularly in Nauru, or as sailors on foreign vessels.

There is little fertile soil, so agriculture is limited, though farmers do grow taro as well as breadfruit, bananas, and especially coconuts to export copra.

Most people earn their living from fishing. Kiribati's control over such a huge area of the Pacific gives it considerable fishing potential, particularly for tuna, both from its own boats and by selling licences to foreign fleets. Even so, the country is heavily dependent on aid.

Kiribati, formerly the Gilbert Islands, gained independence from the UK in 1979. Politics is personality based. There is has 39-member parliament. Anote Tong was elected president in 2003, narrowly beating his brother Harry.

Liechtenstein

Commuters head for one of Europe's richest nations

People: 33,000. Alemani, though many people commute in from neighbouring countries. Life expectancy: 80 years
Government: Constitutional monarchy. Capital: Vaduz
Economy: GDP per capita: $PPP 25,000. Main export: machinery

Liechtenstein is a tiny country between Austria and Switzerland. The western one-third lies along the valley of the River Rhine. The eastern two-thirds is mountainous, formed by the foothills of the Alps.

The people are descendants of the Alemani tribe and many still speak the Alemani dialect, which is a variant of Swiss German.

Liechtenstein has some agriculture, primarily raising livestock. But more important are precision engineering factories, tourism, and the financial services industry. Attracted by Liechtenstein's low rates of corporate taxation, and secretive banks, more than 60,000 companies are registered there. These activities also employ people from neighbouring countries—every day more than one-third of the workforce commute in.

Liechtenstein is a principality, with Prince Hans Adam II as head of state. In 2001, elections to the single-chamber parliament gave a narrow overall majority to the conservative 'Progressive Citizens' Party in Liechtenstein', with Otmar Hasler as prime minister. Following a 2003 referendum, however, the country moved closer to being an absolute monarchy.

Marshall Islands

Reductions in US aid will have serious repercussions

People: 60,000. Micronesian, speaking English or Marshallese dialects.
Life expectancy: 70 years
Government: Republic in free association with the USA. Capital: Majuro
Economy: GDP per capita: $PPP 2,300
Main exports: fish, coconut oil

The Marshall Islands consist of 30 low-lying islands, along with hundreds of islets, scattered across two million square kilometres of the Pacific. All are threatened by global warming. They include Bikini atoll, which was used as a US nuclear test site, and the largest, Kwajelein, which was a military testing range.

One-third of the population live in the slums of Majuro. Many Marshallese still suffer from diseases linked to radiation exposure, though they are entitled to compensation from a trust fund.

On the outer islands, most people survive through subsistence farming, fishing, and remittances from people working on other islands. Those on the larger islands have benefited from considerable US aid.

Since 1986, this has been an independent country in 'free association' with the USA, which looks after defence and other matters. In 2003 the country signed a new compact with the USA, guaranteeing funding of around $800m per year.

There is a 33-member legislative assembly. Since 2000, the president has been Kessai H. Note, the first time the job has not been filled by an island chief.

Monaco

A minuscule, and not very democratic, city-state

People: 32,000, of whom almost half are French; only 16% are Monegasque.
Life expectancy: 80 years
Government: Constitutional monarchy, ruled by the Grimaldi family.
Capital: Monaco
Economy: GDP per capita: $PPP 27,000

Embedded in France's Mediterranean coast, and less than two square kilometres in surface area, Monaco is the smallest country in the UN, though landfill and reclamation from the sea have increased the area by 20% over the past fifty years. Monaco is highly dependent on France: the currency is the French franc and almost half its residents are French citizens attracted by lower taxes.

The primary industry is tourism—four-star and upwards. The casino in the Monte Carlo district is one attraction, and the largest employer; the harbour is always packed with luxury yachts. Many banks and insurance companies find this a comfortable location. Monaco also has some industries, including pharmaceuticals and electronics. More than 25,000 people commute in each day from Italy and France.

Since 2005, Monaco's hereditary head of state has been Prince Albert II —the latest in the line of succession of the Grimaldi family. There is a legislative national council, but executive power is in the hands of an unelected, and always French, minister of state whom the Prince chooses from a list presented to him by the French president.

Montserrat

Still suffering the effects of a disastrous volcanic eruption

People: 9,245. Black, with a few expatriates.
Language: English.
Life expectancy: 79 years
Government: Overseas territory of the
UK. Capital: temporary capital is Brades
Economy: GDP per capita: $PPP 3,400.
Heavily dependent on aid

Montserrat is a volcanic, forest-covered island in the Eastern Caribbean. Its Soufrière Hills include a volcano that erupted in 1997 laying waste to the southern two-thirds of the island, including the capital Plymouth, the airport, and the port. The volcano is still active so almost everyone now lives in a safe zone in the north. The population includes indigenous black residents and some white expatriates.

Rehabilitation got off to a controversial start when the British government argued that Montserrat was asking for too much aid. Relations have continued to be fractious, but aid has financed new houses, roads, and services in the safe zone and there could be a new capital at Little Bay. Agriculture now offers limited employment opportunities. Construction offers an alternative, including that of a new airport, which could help revive the tourist industry.

Montserrat is a UK overseas territory, with its own executive council and a nine-member legislature. In the 2001 election the New People's Liberation Movement took the most seats with John Osbourne as chief minister. In 2004 Deborah Barnes Jones was appointed governor.

Nauru

An island that has been almost entirely excavated

People: 13,000. Nauruan 58%, other Pacific
Islanders 26%, other 16%. Most speak
Nauruan or English. Life expectancy: 62 years
Government: Republic, with a president
elected by the parliament
Economy: GDP per capita: $PPP 5,000.
Main export: phosphates

Nauru is a 21-square-kilometre coral island, the centre of which had a plateau of phosphate rock allegedly formed by guano (bird droppings), leaving just a narrow coastal strip of fertile land.

Nauruans grew wealthy with the income from phosphate exports. These have in the past funded generous social services but also encouraged an unhealthy, sedentary lifestyle. Most of the mining was done by immigrants from China and from neighbouring islands.

Now that the deposits are virtually exhausted, Nauru has found itself in desperate straits. By 2002 the country was declared virtually bankrupt and was having to be propped up by Australian aid, partly in exchange for which Australia has been allowed to establish a detention centre on the island to process asylum seekers. Attempts to develop its offshore banking centre were thwarted in 2003 by a money-laundering scandal.

Independent since 1968, Nauru has a 20-member parliament that elects one member as president. The political culture is very unstable. Since 2004, following the 16th change of government in ten years, the president has been Ludwig Scotty.

Niue

The largest uplifted coral island, linked to New Zealand

People: 2,166. Polynesian with some Europeans and others.
Government: Self-governing, in free association with New Zealand.
Capital: Alofi
Economy: GDP per capita: $PPP 3,600.
Main exports: root crops, coconuts

Though geographically part of the Cook Islands in the South Pacific, Niue is politically separate. This is the world's largest uplifted coral island.

Most people are Polynesian and speak their own Niue language. They live predominantly from agriculture, working on family plantations, growing subsistence crops such as yams and taro, as well as cash crops that include coconuts and passion fruit. The island also has a sawmill as well as some fruit processing factories.

Lacking local opportunities many people have migrated to New Zealand, where around 14,000 Niueans now live. Niue's population has halved over the past 30 years.

The main exports are root crops, coconuts, honey, and handicrafts. The island is heavily dependent on remittances and on aid from New Zealand, which at times has made up around two-thirds of GDP.

Since 1974, Niue has been a self-governing territory in free association with New Zealand, which appoints a high commissioner. The island has its own nine-member legislative assembly. Following the 2002 elections, Young Vivian of the Niue People's Party became prime minister.

Norfolk Island

Australian island settled by convicts and mutineers

People: 1,841. European and Polynesian. Around 30% were born on the Australian mainland and 23% in New Zealand
Government: External territory of Australia.
Capital: Kingston
Economy: Main exports: agricultural products, postage stamps

Norfolk Island is a volcanic island which lies north-west of New Zealand in the South Pacific. Though most of the land has now been cleared, the island still has many of its distinctive pine trees, and most of the coastline consists of cliffs. Originally a penal colony, Norfolk Island was later settled by descendants of mutineers from *The Bounty* who migrated from Pitcairn Island. Around 80% are citizens of Australia; most of the rest are citizens of New Zealand.

There is some agriculture, but the main source of income nowadays is tourism from Australia and New Zealand, with around 40,000 visitors per year.

Relations with Australia have at times been strained and there have been several efforts to clarify the island's status. It is not a dependent territory but an 'integral part of the Commonwealth of Australia'.

The Australian governor-general appoints an administrator, but the island has its own nine-member legislative assembly and raises its own taxes. In 1991, the islanders voted against any change in their constitutional status. Only since 1992 have they been able to vote in Australian federal elections.

Northern Mariana Islands

Offers duty-free entry to the USA for garment makers

People: 82,000. Chamorro, Carolinian, and Asian. Life expectancy: 76 years
Government: Self-governing commonwealth in political union with the USA. Capital: Saipan
Economy: GDP per capita: $PPP 12,500. Main export: garments

The country consists of 22 volcanic or coral islands in the Pacific. The largest indigenous group are the Chamorro. They and other islanders are US citizens. But more than half the population are immigrant workers, particularly from the Philippines and more recently from China.

Few people work in agriculture. Many more now work in the tourist industry, which caters to around 300,000 visitors per year. But many of these are these from Japan and Korea and the numbers dropped sharply in 2005 when Japan Airlines pulled out.

The main employer has been a foreign-owned garments industry. But this is being phased out now that other countries are being granted equivalent access to US markets.

As a commonwealth of the USA, the islanders get substantial aid. They will probably need more. The closure of garment factories and a decline in tourism has cut revenues. In 2006 the islands were asking the US for a $140-million bailout.

The islanders elect their own governor, who since 2005 has been the Republican, Benigno R. Fitial. There is also a legislature, though this often just rubber-stamps the governor's proposals.

Palau

Probably the world's most intensively governed country

People: 20,000. Palauans speaking Palauan, English, or other languages.
Life expectancy: 70 years
Government: Republic in free association with the US. Capital: Koror
Economy: GDP per capita: $PPP 5,800. Main exports: shellfish, tuna

Palau consists of more than 300 volcanic or coral islands in the Pacific, most of which are enclosed by a barrier reef. Palauans are Micronesians. More than half live on the island of Koror.

Many Palauans are still involved in subsistence agriculture and fishing, but the main sources of cash income are government employment and tourism. Around one-third of the workforce are employed by the government, which benefits from substantial US aid. There is also income from fishing licences sold to foreign fleets. Around 50,000 tourists visit each year.

Palau's emergence from the status of UN Trust Territory was a long and tortuous process—during which one president was assassinated—and was completed only when the country became independent in 1994 and joined the UN. It has a compact of free association with the USA.

The national president since 2000 has been Tommy Remengesau, but there is also a two-tier legislature, and a 16-member Council of Chiefs. In addition there is an elected government and legislature in each of the country's 16 component states, one of which has a population of 100.

St Helena

A remote but strategic island in the South Atlantic

People: *7,502, of mixed European, African, and Asian descent. Language: English* **Government:** *Dependency of the UK. Capital: Jamestown. Grouped with Ascension Island and Tristan da Cunha* **Economy:** *GDP per capita: $PPP 2,500. Main export: tuna*

St Helena is a mountainous island of volcanic origin, almost 2,000 kilometres from the coast of Angola, with a mild subtropical climate. Its remote but strategic location has in the past made this an important port of call for sailors and notably for aircrews during the Second World War. The most famous visitor was Napoleon Bonaparte who was exiled here in 1815.

The island has few resources. Many people make a living from fishing, primarily for tuna which is frozen for export. Agriculture is limited, though some farmers can grow maize, potatoes, and other crops and raise some livestock. With few opportunities, many people emigrate. The islanders have requested automatic British citizenship, but this has been refused. One promising development, however, has been the decision to build an airport by 2010.

St Helena, as a British dependency, has an appointed governor, but it also has a legislative council which includes 12 elected members. There have been pressures for greater autonomy. St Helena itself has other islands as dependencies, including Ascension Island, which has a US airfield, and Tristan da Cunha.

St Kitts and Nevis

Two Caribbean islands that have stayed together—just

People: *39,000, mostly black and English-speaking. Life expectancy: 72 years* **Government:** *Federation under a constitutional monarchy. Capital: Basseterre* **Economy:** *GDP per capita: $PPP 8,800. Main exports: machinery, food, electronics, tobacco*

St Kitts (short for St Christopher) and Nevis is a federation of two volcanic islands in the eastern Caribbean. Three-quarters of the population live on St Kitts. Standards of human development are high.

On St Kitts, many people used to work on sugar plantations but most of these have now been closed. As an alternative the government has encouraged small-scale manufacturing and St Kitts has now two electronics companies.

A more promising source of employment is tourism. The islands attract around 120,000 people each year, mostly fairly up-market, who occasionally stray from the beaches to visit historic sites. Nevis in particular has also become a centre for financial and business services with more than 15,000 registered companies. However the islands have also attracted drug traffickers and organized crime.

The 1997, 2000, and 2004 elections to the 11-person National Assembly were won by the St Kitts-Nevis Labour Party with Dr Denzil Douglas as prime minister. Nevis is richer, and itching to separate from St Kitts. A referendum in 1998 failed to get the requisite two-thirds majority.

St Pierre and Miquelon

French fishing islands in the North Atlantic

People: 7,000. French, of Basque and Breton origin. Life expectancy: 79 years
Government: Territorial collectivity of France, but self-government through a local assembly. Capital: Saint-Pierre
Economy: GDP per capita: $PPP 7,000. Main exports: fish and fish products

St Pierre and Miquelon comprises two main islands off the southern coast of Newfoundland, Canada. The islands are fairly bleak and their chief importance is as a French presence in the North Atlantic, offering an opportunity to exploit the rich fishing grounds.

Most of the people are descendants of fishing communities who settled here from France. Their main activity is still fishing for cod. At times this has led to disputes with Canada, though these have now been settled with agreements on quotas. Even so, with the fishing industry in long-term decline, there are fewer opportunities and many of the islanders have been emigrating to Canada. The territory is heavily subsidized by France.

As they live in a territorial collectivity of France, the islanders are entitled to send a deputy to the French National Assembly. Locally, they have their own prefect, who since 2002 has been Claude Valleix.

They also elect their own 19-member general council. The political parties include some local parties and others that correspond to those in metropolitan France. After the 2000 elections the most seats went to Ensemble pour construire.

San Marino

One of the world's smallest and oldest republics

People: 29,000. Sammarinese and Italian, speaking Italian. Life expectancy: 82 years
Government: Republic. The joint heads of state are the elected captains-regent. Capital: San Marino
Economy: GDP per capita: $PPP 34,600. Main exports: stone, ceramics, textiles

Enclosed within Italy, San Marino lies near the Adriatic coast on the slopes of Mount Titano. Although an independent republic, San Marino is necessarily linked in many ways to Italy, and a high proportion of residents are Italian.

Agriculture is still an important source of income, based on cereals and livestock. But the country also has a range of industries. The main one used to be stone quarrying. Legend has it that the state of San Marino was founded by a 4th-century stone-cutter.

Now that the quarries are mostly exhausted, manufacturing industry is devoted to a range of other products including ceramics, textiles, and electronics. San Marino has growing banking and tourist industries, with three million visitors per year, and also does a good trade in postage stamps which provide around one-tenth of government revenue.

San Marino has an elected 60-member 'great and general council'. The 2006 elections resulted in a coalition between the Christian Democrat and Socialist parties. The council chooses two of its members to serve for six months as 'captains-regent' who are joint heads of state.

Seychelles

Tourist island where Heinz is the largest single employer

People: *81,000. Seychellois, a blend of Asian, African, and European.*
Life expectancy: 72 years.
Languages: English, French, and Creole
Government: *Republic. Capital: Victoria*
Economy: *GDP per capita: $PPP 7,800.*
Main export: canned tuna

The Seychelles consists of over 100 islands in the Indian ocean, of which two-thirds are uninhabited. Most Seychellois, who for Africa have a high standard of living, live on the island of Mahé.

The mainstay of the economy is tourism, which employs around one-third of the labour force. Some 120,000 visitors arrived in 2004, mostly from Europe. The Seychelles has little agricultural land but has extensive fishing grounds and the tuna-fish cannery, majority-owned by Heinz, is the largest employer and biggest export earner—though 40% of the workforce has to be imported from China and elsewhere.

Following a coup in 1977 politics was dominated by Albert René, who with his People's Progressive Front (SPPF) ran the country as a one-party socialist state until 1993 when he reverted to multiparty democracy and switched to free-market policies.

René won a series of presidential elections and the SPPF dominated the National Assembly. Rene stood down in 2004, making way for James Michel who won the 2006 presidential election. The main opposition party is the Seychelles National Party, led by Wavel Ramkalawan.

Tokelau

Three scattered atolls in the South Pacific

People: *1,400. Most are Polynesian, speaking Tokelauan and English.*
Government: *Overseas dependency of New Zealand. No capital; there are centres for administration on each island*
Economy: *GDP per capita: $PPP 1,000.*
Main exports: stamps, handicrafts

Tokelau consists of three widely separated atolls in the South Pacific—Atafu, Nukunonu, and Fakaofo. Each has a series of islets around a lagoon, which would be at risk from any rise in the sea level.

The people are of Polynesian origin. Most are Christian. They have some subsistence farming though the soil is thin and infertile. The main crops are coconuts, copra, breadfruit, papayas, and bananas. There is also some livestock-raising and fishing, and a few people work in small enterprises producing handicrafts for export.

Tokelau also gains income from issuing licences for fishing in its waters and selling postage stamps, as well as from remittances from overseas workers. But the budget has to be balanced with substantial aid from New Zealand and elsewhere.

Tokelau is a dependent territory of New Zealand and has an appointed administrator. In addition there is an assembly, the General Fono, for which elections are held every three years. However, there are now moves to change Tokelau's status to make it a self-governing territory in free association with New Zealand and a new constitution is being drafted.

Turks and Caicos

Caribbean islands turning towards Canada

People: 21,000. *Black and European, speaking English, with unauthorized Haitian immigrants. Life expectancy: 75 years*
Government: *Overseas territory of the UK.*
Capital: Grand Turk
Economy: *GDP per capita: $PPP 11,500. Main export: seafood*

The Turks and Caicos islands are two groups of about 40 islands south of the Bahamas. Flat and sandy, the islands are surrounded by coral reefs. Only eight are inhabited.

Two-thirds of the residents are black, the 'belongers'. The rest are expatriates or unauthorized Haitian immigrants. Around 20% work in agriculture, growing vegetables and citrus fruits. Others fish for lobster and other seafood that are the main source of export earnings. Tourism is also an important source of income, mostly on the island of Providenciales which has a Club Med resort. Finally there is a financial services industry that covers more than 14,000 registered companies.

As an overseas territory of the UK, the islands have their own governor, as well as a local executive council and a 13-member legislative council. A currently mooted change of allegiance, however, could see the country cut ties with the UK and enter into 'free association' with Canada.

There are two political parties, the People's Democratic Movement (PDM) and the Progressive National Party (PNP). The PNP won the elections in 2003, with Michael Misick as chief minister.

Tuvalu

Owner of a valuable internet domain name—tv

People: 11,000, mostly Polynesian.
Life expectancy: 66 years
Government: *Constitutional monarchy.*
Capital: Funafuti
Economy: *GDP per capita: $PPP 1,100. Main export: copra. Relies heavily on remittances from emigrant workers*

Though its name means 'eight together', Tuvalu actually comprises nine scattered Pacific islands and atolls. Many people make their living from fishing or from subsistence agriculture, but around half those in employment work for the government.

The government gets 40% of its revenue from fishing licences from foreign vessels. But it has also made considerable sums from sales of postage stamps and coins, and more recently from selling access to its international telephone codes, initially for phone sex. Its internet domain name, 'tv', has been leased for $5 million per year.

Tuvalu was formerly linked with Kiribati as the 'Ellice' in the Gilbert and Ellice Islands colony. Its head of state is the British monarch who is represented by a local governor-general.

The country has a 12-member directly-elected parliament. Since 2004 the prime minister has been Maatia Toafa.

In 2000, Tuvalu joined the United Nations. It hopes to use this status to promote international action on climate change since global warming through a rise in sea levels threatens to submerge parts of the country.

Vatican

A sovereign state in Rome, the headquarters of the Holy See

People: 783. Italian, Swiss, and many others, speaking Italian and Latin.
Government: Monarchical-sacerdotal.
Head of state: Pope Benedict XVI
Economy: GDP per capita: data not available. Currency: Euro.
Main exports: stamps, religious mementos

The Vatican is a 44-hectare enclave on the banks of the River Tiber in Rome. This is a territorial state which is the headquarters of the Holy See, the central government of the Roman Catholic Church.

The head of state is the Pope, but most of the temporal duties, including international relations, are the responsibility of the secretary of state, currently Cardinal Tarcisio Bertone, The Vatican City is administered by a governor.

Those who live and work within the Vatican become citizens. Some 3,000 other workers commute in daily from Rome. Almost everything needed within the Vatican has to be imported. But the Vatican has its own diplomatic service, currency, radio station, and daily newspaper. The Swiss Guard are responsible for internal security.

The Roman Catholic Church derives its income from members' contributions, an extensive investment portfolio, and properties all over the world. These are administered by the Institute for Religious Works. In 2004, the Vatican's expenditure was $246 million, of which $52 million was contributed by Catholic dioceses around the world.

Wallis and Futuna

Pacific islands heavily dependent on French aid

People: 16,025. Polynesian, speaking Wallisian and French.
Government: Overseas collectivity of France. Capital: Mata-Utu
Economy: GDP per capita: $PPP 3,800.
Main exports: taro and other crops. Relies heavily on French aid

In addition to the islands of Wallis and Futuna the country includes the island of Alofi. These are formerly volcanic islands in the South Pacific. The people are mostly Polynesian, with a few French. The Roman Catholic Church has a strong influence.

Most Wallisians make their living from subsistence agriculture, growing taro, yams, bananas, and other crops. The main cash crop is coconuts, grown for copra. They also fish for tuna. A few other people work for the government. The country relies for survival on French aid.

With limited work prospects, many people have emigrated. Some 20,000 are in New Caledonia—more than live at home—and their remittances are a major source of income.

The islanders send a deputy and a senator to the French National Assembly, but they also have their own territorial assembly. In addition to the equivalents of the parties in metropolitan France, there is also a local party, the Union locale populaire, though there is no pressure for independence. Wallis and Futuna also has a territorial council made up of three traditional kings and three other appointed members.

Indicator tables

Notes on indicator tables

These tables summarize economic and social data produced by the United Nations Development Programme (UNDP). Full tables are available in the annual UNDP *Human Development Report*. An entry of '.' indicates that data are unavailable.

Table 1. Income and poverty

GDP $PPP—Gross domestic product per capita at purchasing power parity (PPP). For an explanation of GDP and PPP see pages ix–x. While this gives some indication of the relative wealth of countries, it does not take into account the distribution of income: many Indians, for example, earn more than many Americans. Nor does it necessarily offer guidance to quality of life: people in Cuba, for example, may have a low income but good health services. Dissatisfaction with GDP as an indicator led to the creation of the UNDP human development index.

HDI ranking—This is the human development index ranking, based on a composite of indicators for income, health, and educational attainment. In 2005, the ranking included 177 countries. This list is not exhaustive because it excludes non-UN member countries, such as dependencies, and also those countries for which there were insufficient data. Although the ranking was produced in 2005 it is based on data for earlier years.

Poverty—The proportion of people below the nationally determined poverty line. In developing countries the poverty line is usually set as an absolute amount of income, while in industrialized countries it is usually set relative to average incomes, so poverty rates are not comparable between poor and rich countries. See the note on poverty on page x. The data are for the latest available years.

Distribution of income—The ratio of proportion of national income going to the highest and to the lowest 10% of income earners. Note that this refers to income; not to wealth or total assets. Contrasts in wealth can be much more marked, but more difficult to measure. The data are for the latest available years.

Gini-index—This is another measure of income distribution. It takes a value between 0 (absolute equality) and 100 (one person owns everything). The figure tends to be higher in developing countries, though there can be significant differences even between these countries: those in Latin America, for example, tend to be more unequal than those in Asia.

Military expenditure—This shows military expenditure as a proportion of GDP. This proportion tends to be higher in developing countries, though absolute expenditures are far higher in richer countries. World military expenditure in 2005 was $1,118 billion of which the US accounted for almost half.

Table 2. Health and population

Life expectancy at birth—How long a newborn child can expect to live, on average. World average life expectancy for 2003 was 67 years. Women have a biological advantage that should enable them to live six or seven years longer than men, so there are normally more women than men. However, where there is severe discrimination against women that affects their health the differences can be far less. In Nepal, for example, the discrimination is so great that, on average, men outlive women.

Infant mortality—Out of every thousand live births, the number of children who die before their first birthday. This figure is sensitive to social and economic conditions, so is one of the most basic indicators of development. In 2003, the average for the richer countries was 11 and for developing countries was 60.

Malnutrition—The proportion of children under five years old who weigh less than they should for their age. Another indicator of malnutrition is stunting, which refers to low height for age. Children can also suffer from deficiencies in important 'micronutrients' such as Vitamin A or iodine. Children are most vulnerable to malnutrition in the first two years of life. Opportunities for growth lost during that period cannot be recovered: stunted children grow up as stunted adults. The data refer to the latest available year, between 1995 and 2003.

Total fertility rate—The number of children that a woman in each country on average bears throughout her life. A figure of 2.1 is the 'replacement rate' that would maintain a steady population. Globally, the average figure is 2.6 and world population is growing by 1.1% per year. But for many countries the fertility rate is far lower so their populations are destined to fall if they do not replenish them through immigration. The fertility rate is very low in a number of former socialist countries: in Russia, for example, it is only 1.3. It is also low in some West European

countries: 1.3 in Italy, for example.

Safe water—The proportion of the population with reasonable access to safe drinking water, either treated surface water, or uncontaminated water from other sources such as springs or wells.

Sanitation—The proportion of the population with reasonable access to sanitary means of excreta and waste disposal. Data on water and sanitation need to be used with care since standards are not always internationally comparable.

Table 3. Education and telecommunications

Adult literacy rate—The proportion of the population aged 15 and above who can read a short, simple statement on their everyday life. This information is collected either from national censuses or household surveys. However some countries do not collect this information and instead substitute school attendance data.

Primary and secondary net enrolment—The proportion of children of the appropriate age who are actually enrolled in primary or secondary school. Where this figure is greater than 100% this is because of weaknesses in data collection.

Telephones—Fixed lines and cellular subscribers per thousand people. Globally, per thousand people there are now 184 fixed lines but 226 cellular subscribers. Most cellular phones in developing countries are prepaid.

Internet users—Number of users per thousand people. Globally the average is 120.

Table 1. Indicators for income and poverty

	GDP $PPP per capita 2003	HDI ranking out of 177 2003	Poverty % of population 1990–2003	GDP annual growth, % 1990–2003	Income distribution Ratio of richest 10% to poorest 10%	Gini-index 1990–2003	Military exp. % GDP 2003
Albania	4,584	72	::	5.1	6	28	5.9
Algeria	6,107	103	12	0.6	10	35	1.5
Angola	2,344	160	::	0.4	::	::	5.8
Antigua and Barbuda	10,294	60	::	1.6	::	::	::
Argentina	12,106	34	::	1.3	39	52	1.2
Armenia	3,671	83	::	2.8	12	38	::
Australia	29,632	3	::	2.6	13	35	2.1
Austria	30,094	17	::	1.8	8	30	1.0
Azerbaijan	3,617	101	::	-2.6	10	37	::
Bahamas	17,159	50	::	0.3	::	::	::
Bahrain	17,479	43	::	1.5	::	::	5.1
Bangladesh	1,770	139	50	3.1	7	32	1.0
Barbados	15,720	30	::	1.4	::	::	::
Belarus	6,052	67	::	0.9	7	30	::
Belgium	28,335	9	::	1.8	8	25	2.4
Belize	6,950	91	::	2.2	::	::	1.2

Benin	1,115	162	33	2.2	:	:	1.8
Bhutan	1,969	134	:	3.6	:	:	:
Bolivia	2,587	113	63	1.3	25	45	2.4
Bosnia and Herzegovina	5,967	68	64	11.9	5	26	:
Botswana	8,714	131	:	2.7	78	63	4.1
Brazil	7,790	63	:	1.2	68	59	2.5
Brunei	19,210	33	:	:	:	:	:
Bulgaria	7,731	55	:	0.6	10	32	3.5
Burkina Faso	1,174	175	45	1.7	26	48	3.0
Burma	:	129	36	5.7	:	:	3.4
Burundi	648	169	:	-3.5	19	33	3.4
Cambodia	2,078	130	:	4.0	12	40	3.1
Cameroon	2,118	148	40	0.2	16	45	1.5
Canada	30,677	5	:	2.3	10	33	2.0
Cape Verde	5,214	105	:	3.3	:	:	:
Central African Republic	1,089	171	:	-0.4	69	61	:
Chad	1,210	173	64	(.)	:	:	:
Chile	10,274	37	17	4.1	41	57	4.3
China	5,003	85	5	8.5	18	45	2.7
Colombia	6,702	69	:	0.4	58	58	2.2

Table 1. Indicators for income and poverty

	GDP $PPP per capita 2003	HDI ranking out of 177 2003	Poverty % of population 1990–2003	GDP annual growth, % 1990–2003	Income distribution Ratio of richest 10% to poorest 10%	Gini-index 1990–2003	Military exp. % GDP 2003
Comoros	1,714	132	::	-1.3	::	::	::
Congo	965	142	::	-1.4	::	::	::
Congo, Dem. Rep.	697	167	::	-6.3	::	::	::
Costa Rica	9,606	47	22	2.6	25	47	0.0
Côte d'Ivoire	1,476	163	37	-0.4	17	45	1.3
Croatia	11,080	45	::	2.1	7	29	::
Cuba	::	52	::	3.5	::	::	::
Cyprus	18,776	29	::	3.2	::	::	5.0
Czech Republic	16,357	31	::	1.5	5	25	::
Denmark	31,465	14	::	1.9	8	25	2.0
Djibouti	2,086	150	45	-3.3	::	::	6.3
Dominica	5,448	70	::	1.2	::	::	::
Dominican Republic	6,823	95	29	4.0	18	47	::
Ecuador	3,641	82	35	0.1	45	44	1.9
Egypt	3,950	119	17	2.5	8	34	3.9
El Salvador	4,781	104	48	2.1	47	53	2.7

Equatorial Guinea	19,780	121	::	16.8	::	::	::
Eritrea	849	161	53	1.0	::	::	::
Estonia	13,539	38	::	3.3	15	37	::
Ethiopia	711	170	44	2.0	7	30	8.5
Fiji	5,880	92	::	1.8	::	::	2.3
Finland	27,619	13	::	2.5	6	27	1.6
France	27,677	16	::	1.6	9	33	3.5
Gabon	6,397	123	::	-0.4	::	::	::
Gambia	1,859	155	64	-0.1	20	48	1.1
Georgia	2,588	100	::	-2.7	12	37	::
Germany	27,756	20	::	1.3	7	28	2.8
Ghana	2,238	138	40	1.8	14	41	0.4
Greece	19,954	24	::	2.1	10	35	4.7
Grenada	7,559	66	::	2.4	::	::	::
Guatemala	4,148	117	56	1.1	55	60	1.5
Guinea	2,097	156	40	1.6	12	40	::
Guinea-Bissau	711	172	49	-2.4	19	47	::
Guyana	4,230	107	::	3.6	::	::	0.9
Haiti	1,742	153	65	-2.8	::	::	::
Honduras	2,665	116	53	0.2	49	55	::

Table 1. Indicators for income and poverty

	GDP $PPP per capita 2003	HDI ranking out of 177 2003	Poverty % of population 1990–2003	GDP annual growth, % 1990–2003	Income distribution Ratio of richest 10% to poorest 10%	Gini-index 1990–2003	Military exp. % GDP 2003
Hungary	14,584	35	..	2.6	6	27	2.8
Iceland	31,243	2	..	2.1	..	..	0.0
India	2,892	127	29	4.0	7	33	2.7
Indonesia	3,361	110	27	2.0	8	34	1.8
Iran	6,995	99	..	2.1	17	43	2.9
Ireland	37,738	8	..	6.7	10	36	1.2
Israel	20,033	23	..	1.6	12	36	12.4
Italy	27,119	18	..	1.5	12	36	2.1
Jamaica	4,104	98	19	(.)	11	38	..
Japan	27,967	11	..	1.0	5	25	0.9
Jordan	4,320	90	12	0.9	9	36	9.9
Kazakhstan	6,671	80	..	0.4	8	32	..
Kenya	1,037	154	42	-0.6	14	43	2.9
Korea, South	17,971	28	..	4.6	8	32	3.7
Kuwait	18,047	44	..	-2.3	..	..	48.5
Kyrgyzstan	1,751	109	..	-2.4	9	35	..

Laos	1,759	133	39	3.7	10	37	..
Latvia	10,270	48	..	2.2	9	34	..
Lebanon	5,074	81	..	2.9	..	..	7.6
Lesotho	2,561	149	49	2.3	105	63	4.5
Libya	..	58	..	..	..	..	..
Lithuania	11,702	39	..	0.5	8	32	..
Luxembourg	62,298	4	..	3.6	..	..	0.9
Macedonia	6,794	59	..	-0.7	7	28	..
Madagascar	809	146	71	-0.9	19	48	1.2
Malawi	605	165	65	0.9	23	50	1.3
Malaysia	9,512	61	16	3.4	22	49	2.6
Maldives	..	96	..	4.7	..	..	..
Mali	994	174	64	2.4	23	51	2.1
Malta	17,633	32	..	3.3	..	..	0.9
Mauritania	1,766	152	46	1.6	12	39	3.8
Mauritius	11,287	65	11	4.0	..	..	0.3
Mexico	9,168	53	10	1.4	45	55	0.5
Moldova	1,510	115	..	-5.7	10	37	..
Mongolia	1,850	114	36	-2.5	18	30	5.7
Morocco	4,004	124	19	1.0	12	40	4.1

Table 1. Indicators for income and poverty

	GDP $PPP per capita 2003	HDI ranking out of 177 2003	Poverty % of population 1990–2003	GDP annual growth, % 1990–2003	Income distribution Ratio of richest 10% to poorest 10%	Gini-index 1990–2003	Military exp. % GDP 2003
Mozambique	1,117	168	69	4.6	13	40	5.9
Namibia	6,180	125	..	0.9	129	71	..
Nepal	1,420	136	42	2.2	9	37	0.9
Netherlands	29,371	12	..	2.1	9	31	2.5
New Zealand	22,582	19	..	2.1	13	36	1.9
Nicaragua	3,262	112	48	0.9	16	43	10.6
Niger	835	177	63	-0.6	46	51	..
Nigeria	1,050	158	34	(.)	25	51	0.9
Norway	37,670	1	..	2.9	6	26	2.9
Oman	13,584	71	..	0.9	..	..	16.5
Pakistan	2,097	135	33	1.1	8	33	5.8
Palestine	..	102	..	-6.0	..	..	..
Panama	6,854	56	37	2.4	62	56	1.3
Papua New Guinea	2,619	137	38	0.2	24	51	2.1
Paraguay	4,684	88	22	-0.6	73	58	1.0
Peru	5,260	79	49	2.1	50	50	0.1

Philippines	4,321	84	37	1.2	17	46	1.4
Poland	11,379	36	:	4.2	9	34	2.7
Portugal	18,126	27	:	2.2	15	39	2.7
Qatar	19,844	40	:	:	:	:	:
Romania	7,277	64	:	0.6	8	30	4.6
Russia	9,230	62	17	-1.5	7	31	12.3
Rwanda	1,268	159	51	0.7	6	29	3.7
St Lucia	5,709	76	:	0.3	:	:	:
St Kitts and Nevis	12,404	49	:	3.1	:	:	:
St Vincent & Grenadines	6,123	87	:	1.8	:	:	:
Samoa	5,854	74	:	2.4	:	:	:
São Tomé and Príncipe	1,231	126	:	-0.2	:	:	:
Saudi Arabia	13,226	77	:	-0.6	:	:	12.8
Senegal	1,648	157	33	1.3	13	41	2.0
Seychelles	10,232	51	:	2.2	:	:	4.0
Sierra Leone	548	176	68	-5.3	87	63	1.4
Singapore	24,481	25	:	3.5	18	43	4.9
Slovakia	13,494	42	:	2.4	7	26	:
Slovenia	19,150	26	:	3.1	6	28	:
Solomon Islands	1,753	128	:	-2.5	:	:	:

Table 1. Indicators for income and poverty

	GDP $PPP per capita	HDI ranking out of 177	Poverty % of population	GDP annual growth, %	Income distribution Ratio of richest 10% to poorest 10%	Gini-index	Military exp. % GDP
	2003	2003	1990–2003	1990–2003		1990–2003	2003
South Africa	10,346	120	..	0.1	33	58	3.8
Spain	22,391	21	..	2.4	9	33	1.8
Sri Lanka	3,778	93	25	3.3	8	33	2.1
Sudan	1,910	141	..	3.3	..	..	3.6
Suriname	..	86	..	0.9	..	..	..
Swaziland	4,726	147	40	0.2	50	61	2.1
Sweden	26,750	6	..	2.0	6	25	2.6
Switzerland	30,552	7	..	0.5	10	33	1.8
Syria	3,576	106	35	1.4	..	..	6.9
Tajikistan	1,106	122	..	-6.5	8	33	..
Tanzania	621	164	36	1.0	11	38	..
Thailand	7,595	73	13	2.8	13	43	2.6
Timor-Leste	..	140	..	..	..	..	..
Togo	1,696	143	32	0.4	..	..	3.1
Tonga	6,992	54	..	2.0	..	..	..
Trinidad and Tobago	10,766	57	21	3.2	14	40	..

Tunisia	7,161	89	8	3.1	13	40	2.0
Turkey	6,772	94	..	1.3	13	40	3.5
Turkmenistan	5,938	97	..	-1.3	12	41	..
Uganda	1,457	144	55	3.9	15	43	3.0
Ukraine	5,491	78	..	-4.7	6	29	..
United Arab Emirates	22,420	41	..	-2.1	..	..	6.2
United Kingdom	27,147	15	..	2.5	14	36	4.0
United States	37,562	10	..	2.1	16	41	5.3
Uruguay	8,280	46	..	0.9	19	45	2.5
Uzbekistan	1,744	111	..	-0.5	6	27	..
Vanuatu	2,944	118	..	-0.3	..	..	..
Venezuela	4,919	75	31	-1.5	63	49	..
Vietnam	2,490	108	51	5.9	9	37	7.9
Yemen	889	151	42	2.4	9	33	7.9
Zambia	877	166	73	-0.9	42	53	3.7
Zimbabwe	2,443	145	35	-0.8	22	57	4.5

Table 2. Indicators for health and population

	Life expectancy years, 2003		Infant mortality per '000 live births	Malnutrition % children under 5	Fertility rate births per woman	Safe water % access	Sanitation % access
	Female	Male	2003	2003	2000–05	2002	2002
Albania	77	71	18	14	2.3	97	89
Algeria	72	70	35	6	2.5	87	92
Angola	42	39	154	31	6.8	50	30
Antigua and Barbuda	..	..	11	10	..	91	95
Argentina	78	71	17	5	2.4	..	..
Armenia	75	68	30	3	1.3	92	84
Australia	83	78	6	..	1.7	100	100
Austria	82	76	4	..	1.4	100	100
Azerbaijan	71	63	75	7	1.9	77	55
Bahamas	73	67	11	..	2.3	97	100
Bahrain	76	73	12	9	2.5	..	..
Bangladesh	64	62	46	48	3.2	75	48
Barbados	79	71	11	6	1.5	100	99
Belarus	74	62	13	..	1.2	100	..
Belgium	82	76	4	..	1.7	..	..
Belize	75	70	33	6	3.2	91	47

Benin	55	53	91	23	5.9	68	32
Bhutan	64	62	70	19	4.4	62	70
Bolivia	66	62	53	8	4.0	85	45
Bosnia & Herzegovina	77	71	14	4	1.3	98	93
Botswana	37	36	82	13	3.2	95	41
Brazil	75	67	33	6	2.3	89	75
Brunei	79	74	5	..	2.5	..	..
Bulgaria	76	69	14	..	1.2	100	100
Burkina Faso	48	47	107	34	6.7	51	12
Burma	43	41	109	24	5.5	42	27
Burundi	45	43	114	45	6.8	79	36
Cambodia	60	52	97	45	4.1	34	16
Cameroon	47	45	95	21	4.6	63	48
Canada	82	77	5	..	1.5	100	100
Cape Verde	73	67	26	14	3.8	80	42
Central African Republic	40	38	115	24	5.0	75	27
Chad	45	43	117	28	6.7	34	8
Chile	81	75	8	1	2.0	95	92
China	74	70	30	10	1.7	77	44
Colombia	75	69	18	7	2.6	92	86

Table 2. Indicators for health and population

	Life expectancy years, 2003		Infant mortality per '000 live births	Malnutrition % children under 5	Fertility rate births per woman	Safe water % access	Sanitation % access
	Female	Male	2003	2003	2000–05	2002	2002
Comoros	65	61	54	25	4.9	94	23
Congo	53	51	81	14	6.3	46	9
Congo, Dem. Rep.	44	42	129	31	6.7	46	29
Costa Rica	81	76	8	5	2.3	97	92
Côte d'Ivoire	47	45	117	21	5.1	84	40
Croatia	78	71	6	1	1.3	..	..
Cuba	79	76	6	4	1.6	91	98
Cyprus	81	76	4	..	1.6	100	100
Czech Republic	79	72	4	1	1.2	..	..
Denmark	79	75	3	..	1.8	100	..
Djibouti	54	52	97	18	5.1	80	50
Dominica	..	..	12	5	..	97	83
Dominican Republic	71	64	29	5	2.7	93	57
Ecuador	77	71	24	12	2.8	86	72
Egypt	72	68	33	9	3.3	98	68
El Salvador	74	68	32	10	2.9	82	63

Equatorial Guinea	44	43	97	19	5.9	44	53
Eritrea	56	52	45	40	5.5	57	9
Estonia	77	66	8	..	1.4	..	..
Ethiopia	49	47	112	47	5.9	22	6
Fiji	70	66	16	8	2.9	..	98
Finland	82	75	4	..	1.7	100	100
France	83	76	4	..	1.9	..	..
Gabon	55	54	60	12	4.0	87	36
Gambia	57	54	90	17	4.7	82	53
Georgia	74	67	41	3	1.5	76	83
Germany	82	76	4	..	1.3	100	..
Ghana	57	56	59	25	4.4	79	58
Greece	81	76	4	..	1.3	..	..
Grenada	..	..	18	..	..	95	97
Guatemala	71	64	35	23	4.6	95	61
Guinea	54	53	104	23	5.9	51	13
Guinea-Bissau	46	43	126	25	7.1	59	34
Guyana	66	60	52	14	2.3	83	70
Haiti	52	51	76	17	4.0	71	34
Honduras	70	66	32	17	3.7	90	68

Table 2. Indicators for health and population

| | Life expectancy years, 2003 | | Infant mortality per '000 live births | Malnutrition % children under 5 | Fertility rate births per woman | Safe water % access | Sanitation % access |
	Female	Male	2003	2003	2000–05	2002	2002
Hungary	77	69	7	2	1.3	99	95
Iceland	83	79	3	..	2.0	100	..
India	65	62	63	47	3.1	86	30
Indonesia	69	65	31	26	2.4	78	52
Iran	72	69	33	11	2.1	93	84
Ireland	80	75	6	..	1.9	..	..
Israel	82	78	5	..	2.9	100	..
Italy	83	77	4	..	1.3	..	..
Jamaica	73	69	17	4	2.4	93	80
Japan	85	78	3	..	1.3	100	100
Jordan	73	70	23	4	3.5	91	93
Kazakhstan	69	58	63	4	2.0	86	72
Kenya	46	48	79	20	5.0	62	48
Korea, South	81	73	5	..	1.2	92	..
Kuwait	80	75	8	10	2.4	..	..
Kyrgyzstan	71	63	59	11	2.7	76	60

Laos	56	53	82	40	4.8	43	24
Latvia	77	66	10	..	1.3	..	..
Lebanon	74	70	27	3	2.3	100	98
Lesotho	38	35	63	18	3.6	76	37
Libya	76	72	13	5	3.0	72	97
Lithuania	78	67	8	..	1.3	..	..
Luxembourg	82	75	5	..	1.7	100	..
Macedonia	76	71	10	6	1.5	..	..
Madagascar	57	54	78	33	5.4	45	33
Malawi	40	40	112	22	6.1	67	46
Malaysia	76	71	7	12	2.9	95	..
Maldives	66	67	55	30	4.3	84	58
Mali	49	47	122	33	6.9	48	45
Malta	81	76	5	..	1.5	100	..
Mauritania	54	51	120	32	5.8	56	42
Mauritius	76	69	16	15	2.0	100	99
Mexico	78	73	23	8	2.4	91	77
Moldova	71	64	26	3	1.2	92	68
Mongolia	66	62	56	13	2.4	62	59
Morocco	72	68	36	9	2.8	80	61

Table 2. Indicators for health and population

	Life expectancy years, 2003		Infant mortality per'000 live births	Malnutrition % children under 5	Fertility rate births per woman	Safe water % access	Sanitation % access
	Female	Male	2003	2003	2000–05	2002	2002
Mozambique	63	58	76	35	2.5	80	73
Namibia	49	48	48	24	4.0	80	30
Nepal	62	61	61	48	3.7	84	27
Netherlands	81	76	5	..	1.7	100	100
New Zealand	81	77	5	..	2.0	..	..
Nicaragua	72	67	30	10	3.3	81	66
Niger	44	44	154	40	7.9	46	12
Nigeria	44	43	98	29	5.8	60	38
Norway	74	71	22	4	5.6	94	76
Oman	76	73	10	24	3.8	79	89
Pakistan	82	77	3	..	1.8	100	..
Palestine	63	63	81	38	4.3	90	54
Panama	77	72	18	7	2.7	91	72
Papua New Guinea	56	55	69	35	4.1	39	45
Paraguay	73	69	25	5	3.9	83	78
Peru	73	68	26	7	2.9	81	62

Philippines	73	68	27	31	3.2	85	73
Poland	78	70	6	:	1.3	:	:
Portugal	81	74	4	:	1.5	:	:
Qatar	76	71	11	6	3.0	100	100
Romania	75	68	18	6	1.3	57	51
Russia	72	59	16	3	1.3	96	87
Rwanda	74	68	23	:	2.3	:	:
St Kitts and Nevis	:	:	19	:	:	99	96
St Lucia	74	71	16	14	2.2	98	89
St Vincent & Grenadines	77	72	13	29	2.0	78	91
Samoa	74	67	19	:	4.4	88	100
São Tomé and Príncipe	64	62	75	13	4.1	79	24
Saudi Arabia	74	70	22	14	4.1	:	:
Senegal	57	55	78	23	5.0	72	52
Seychelles	:	:	11	6	:	87	:
Sierra Leone	42	39	166	27	6.5	57	39
Singapore	81	77	3	14	1.4	:	:
Slovakia	78	70	7	:	1.2	100	100
Slovenia	80	73	4	:	1.2	:	:
Solomon Islands	63	62	19	21	4.3	70	31

Table 2. Indicators for health and population

	Life expectancy years, 2003		Infant mortality per '000 live births	Malnutrition % children under 5	Fertility rate births per woman	Safe water % access	Sanitation % access
	Female	Male	2003	2003	2000–05	2002	2002
South Africa	50	47	53	12	2.8	87	67
Spain	83	76	4	..	1.3	..	..
Sri Lanka	46	42	118	27	5.7	73	41
Sudan	58	55	63	17	4.4	69	34
Suriname	73	66	30	13	2.6	92	93
Swaziland	33	32	105	10	4.0	52	52
Sweden	82	78	3	..	1.6	100	100
Switzerland	83	78	4	..	1.4	100	100
Syriac	75	72	16	7	3.5	79	77
Tajikistan	66	61	92	..	3.8	58	53
Tanzania	46	46	104	29	5.0	73	46
Thailand	74	66	23	19	1.9	85	99
Timor-Leste	57	55	87	43	7.8	52	33
Togo	56	52	78	25	5.4	51	34
Tonga	74	71	15	..	3.5	100	97
Trinidad and Tobago	73	67	17	7	1.6	91	100

Tunisia	75	71	19	4	2.0	82	80
Turkey	71	67	33	8	2.5	93	83
Turkmenistan	67	58	79	12	2.8	71	62
Uganda	48	47	81	23	7.1	56	41
Ukraine	73	60	15	3	1.1	98	99
United Arab Emirates	81	76	7	14	2.5	..	100
United Kingdom	81	76	5	..	1.7	..	..
United States	80	75	7	1	2.0	100	100
Uruguay	79	72	12	5	2.3	98	94
Uzbekistan	70	63	57	8	2.7	89	57
Vanuatu	71	67	31	20	4.2	60	..
Venezuela	76	70	18	4	2.7	83	68
Vietnam	73	69	19	33	2.3	73	41
Yemen	62	59	82	46	6.2	69	30
Zambia	37	38	12	28	5.7	55	45
Zimbabwe	37	37	78	13	3.6	83	57

Table 3. Indicators for education and telecommunications

	Adult literacy rate 2002, %		Primary net enrolment %	Secondary net enrolment %	Mainline phones per '000	Mobile phones users per '000	Internet users per '000
	Female	Male	2001/02	2001/02	2002	2002	2002
Albania	98	99	95	77	83	358	10
Algeria	60	80	95	67	69	45	..
Angola	54	82	61	..	7	..	..
Antigua and Barbuda	..	..	..	..	..	..	..
Argentina	97	97	..	81	..	..	..
Armenia	99	100	94	83	148	30	37
Australia	..	..	97	88	542	719	567
Austria	..	..	90	89	481	879	462
Azerbaijan	98	100	80	76	114	128	..
Bahamas	..	95	86	76	415	367	265
Bahrain	83	93	90	87	268	638	216
Bangladesh	31	50	84	45	5	10	2
Barbados	100	100	100	90	497	519	371
Belarus	99	100	94	85	311	113	141
Belgium	..	..	100	97	489	793	386
Belize	77	77	99	69	113	205	..

Benin	23	46	58	20	9	34	10
Bhutan	..	..	..	..	34	11	20
Bolivia	80	93	95	71	72	152	..
Bosnia & Herzegovina	91	98	..	..	245	274	..
Botswana	82	76	81	54	75	297	..
Brazil	89	88	97	75	223	264	..
Brunei	90	95	..	..	..	..	..
Bulgaria	98	99	90	88	380	466	206
Burkina Faso	8	19	36	9	5	19	4
Burma	86	94	84	35	7	1	1
Burundi	52	67	57	9	3	9	2
Cambodia	64	85	93	24	3	35	2
Cameroon	60	77	..	..	..	66	..
Canada	..	..	100	98	651	419	..
Cape Verde	68	85	99	58	156	116	44
Central African Republic	34	65	..	..	..	10	1
Chad	13	41	63	10	..	8	..
Chile	96	96	85	81	221	511	272
China	87	95	..	..	209	215	63
Colombia	95	94	87	55	179	141	53

Table 3. Indicators for education and telecommunications

	Adult literacy rate 2002, %		Primary net enrolment %	Secondary net enrolment %	Mainline phones per '000	Mobile phones users per '000	Internet users per '000
	Female	Male	2001/02	2001/02	2002	2002	2002
Comoros	49	64	55	..	17	3	6
Congo	77	89	54	..	2	94	4
Congo, Dem. Rep.	52	80	..	..	..	19	..
Costa Rica	96	96	90	53	278	181	288
Côte d'Ivoire	38	60	61	21	14	77	14
Croatia	97	99	89	87	..	584	232
Cuba	97	97	94	86	64	3	9
Cyprus	95	99	96	93	572	744	337
Czech Republic	..	..	87	91	360	965	308
Denmark	..	..	100	96	669	883	541
Djibouti	..	..	36	21	15	34	10
Dominica	..	..	81	92	..	..	..
Dominican Republic	87	88	96	36	115	272	102
Ecuador	90	92	100	50	122	189	46
Egypt	44	67	91	81	127	84	44
El Salvador	77	82	90	49	113	173	83

Equatorial Guinea	76	92	85	26	18	76	..
Eritrea	..	68	45	22	9	0	7
Estonia	100	100	95	88	341	777	444
Ethiopia	34	49	51	18	6	1	1
Fiji	91	95	100	76	124	133	67
Finland	..	..	100	95	492	910	534
France	..	..	99	94	566	696	366
Gabon	..	..	78	..	29	224	26
Gambia	..	45	79	33	..	..	..
Georgia	..	..	89	61	134	145	24
Germany	..	..	83	88	657	785	473
Ghana	46	63	59	36	13	36	..
Greece	88	94	99	86	454	902	150
Grenada	..	..	84	104	290	376	169
Guatemala	63	75	87	30	77	165	..
Guinea	..	..	66	21	3	14	5
Guinea-Bissau	..	55	45	9	8	1	15
Guyana	..	99	99	76	..	..	..
Haiti	50	54	..	..	17	38	18
Honduras	80	80	87	..	49	55	40

Table 3. Indicators for education and telecommunications

	Adult literacy rate 2002, %		Primary net enrolment %	Secondary net enrolment %	Mainline phones per '000	Mobile phones users per '000	Internet users per '000
	Female	Male	2001/02	2001/02	2002	2002	2002
Hungary	99	99	91	94	349	769	232
Iceland	..	..	100	86	660	966	675
India	48	73	87	..	46	25	17
Indonesia	83	93	92	54	39	87	38
Iran	70	84	86	..	220	51	72
Ireland	..	..	96	83	491	880	317
Israel	96	98	99	89	458	961	..
Italy	..	..	100	91	484	1,018	337
Jamaica	91	84	95	75	..	680	..
Japan	..	..	100	101	472	679	483
Jordan	85	95	92	80	114	242	81
Kazakhstan	99	100	92	87	141	..	..
Kenya	70	78	67	25	10	50	..
Korea, South	..	..	100	88	538	701	610
Kuwait	81	85	83	77	196	572	228
Kyrgyzstan	98	99	89	..	76	27	38

Laos	61	77	85	35	12	20	3
Latvia	100	100	86	88	285	526	404
Lebanon	..	92	91	..	200	234	143
Lesotho	90	74	86	23	16	47	14
Libya	71	92	..	..	136	23	29
Lithuania	100	100	91	94	239	630	202
Luxembourg	..	..	90	80	797	1,194	377
Macedonia	94	98	91	81	252	372	60
Madagascar	65	76	79	12	4	17	4
Malawi	54	75	..	29	8	13	3
Malaysia	85	92	93	70	182	442	344
Maldives	97	97	92	51	..	..	..
Mali	12	27	45	..	..	23	..
Malta	89	86	96	87	521	725	..
Mauritania	43	60	68	16	14	127	4
Mauritius	81	88	97	74	285	267	123
Mexico	89	92	99	63	160	295	120
Moldova	95	98	79	69	219	132	80
Mongolia	98	98	79	77	56	130	58
Morocco	38	63	90	36	40	244	33

Table 3. Indicators for education and telecommunications

	Adult literacy rate 2002, %		Primary net enrolment %	Secondary net enrolment %	Mainline phones per '000	Mobile phones users per '000	Internet users per '000
	Female	Male	2001/02	2001/02	2002	2002	2002
Mozambique	31	62	55	12	..	23	..
Namibia	84	87	78	44	66	116	34
Nepal	35	63	71	..	16	2	..
Netherlands	..	..	99	89	614	768	522
New Zealand	..	..	100	93	448	648	526
Nicaragua	77	77	86	39	37	85	..
Niger	9	20	38	6	..	6	..
Nigeria	59	74	67	29	7	26	6
Norway	..	..	100	96	713	909	346
Oman	65	82	72	69	88	228	..
Pakistan	35	62	59	..	27	18	..
Palestine	87	96	91	84	87	133	40
Panama	91	93	100	63	122	268	62
Papua New Guinea	51	63	73	24	..	..	..
Paraguay	90	93	89	51	46	299	20
Peru	82	94	100	69	67	106	104

Philippines	93	93	94	59	41	270	..
Poland	..	100	98	83	307	451	232
Portugal	..	..	100	85	411	898	..
Qatar	..	0	95	82	261	533	199
Romania	96	98	89	81	199	324	184
Russia	99	100	90	..	253	249	..
Rwanda	59	71	87	..	..	16	..
St Kitts and Nevis	..	..	100	95	..	..	..
St Lucia	91	90	99	76	..	..	..
St Vincent & Grenadines	..	..	90	58	273	529	..
Samoa	98	99	98	62	73	58	..
São Tomé and Príncipe	..	..	97	29	46	32	99
Saudi Arabia	69	87	54	53	155	321	67
Senegal	29	51	58	..	22	56	22
Seychelles	92	91	100	100	256	595	..
Sierra Leone	21	40	..	..	..	..	..
Singapore	89	97	..	..	450	852	509
Slovakia	100	100	86	88	241	684	256
Slovenia	100	100	93	93	407	871	401
Solomon Islands	..	..	..	..	13	3	5

Table 3. Indicators for education and telecommunications

	Adult literacy rate 2002, %		Primary net enrolment %	Secondary net enrolment %	Mainline phones per '000	Mobile phones users per '000	Internet users per '000
	Female	Male	2001/02	2001/02	2002	2002	2002
South Africa	81	84	89	66	..	364	..
Spain	..	..	100	96	429	916	239
Sri Lanka	89	92	..	..	49	73	13
Sudan	50	69	46	..	27	20	9
Suriname	84	92	97	64	152	320	44
Swaziland	78	80	75	32	44	84	26
Sweden	..	..	100	100	..	980	..
Switzerland	..	..	99	87	727	843	398
Syria	74	91	98	43	..	68	35
Tajikistan	99	100	94	83	37	7	1
Tanzania	62	78	82	..	4	25	7
Thailand	91	95	85	..	105	394	111
Timor-Leste	..	..	..	20	..	..	..
Togo	38	69	91	27	12	44	42
Tonga	99	99	100	72	..	..	..
Trinidad and Tobago	98	99	91	72	..	399	..

Tunisia	65	83	97	65	118	197	64
Turkey	81	96	86	::	268	394	85
Turkmenistan	98	99	::	::	77	::	::
Uganda	59	79	::	17	2	30	5
Ukraine	99	100	84	85	233	136	::
United Arab Emirates	81	76	83	71	281	736	275
United Kingdom	::	::	100	95	::	912	::
United States	::	::	92	88	624	546	556
Uruguay	98	97	90	73	::	::	::
Uzbekistan	99	100	::	::	67	13	19
Vanuatu	::	0	94	28	31	38	36
Venezuela	93	93	91	59	111	273	60
Vietnam	87	94	94	65	54	34	43
Yemen	29	70	72	35	::	35	::
Zambia	60	76	68	23	8	22	6
Zimbabwe	86	94	79	34	::	::	::

Oxford Paperback Reference

A Dictionary of Psychology
Andrew M. Colman

Over 10,500 authoritative entries make up the most wide-ranging
dictionary of psychology available.

'impressive ... certainly to be recommended'
Times Higher Educational Supplement

'Comprehensive, sound, readable, and up-to-date, this is probably the
best single-volume dictionary of its kind.'
Library Journal

A Dictionary of Economics
John Black

Fully up-to-date and jargon-free coverage of economics. Over 2,500
terms on all aspects of economic theory and practice.

A Dictionary of Law

An ideal source of legal terminology for systems based on English law.
Over 4,000 clear and concise entries.

'The entries are clearly drafted and succinctly written ... Precision for the
professional is combined with a layman's enlightenment.'
Times Literary Supplement

Oxford Paperback Reference

The Kings of Queens of Britain
John Cannon and Anne Hargreaves

A detailed, fully-illustrated history ranging from mythical and pre-conquest rulers to the present House of Windsor, featuring regional maps and genealogies.

A Dictionary of Dates
Cyril Leslie Beeching

Births and deaths of the famous, significant and unusual dates in history – this is an entertaining guide to each day of the year.

'a dipper's blissful paradise ... Every single day of the year, plus an index of birthdays and chronologies of scientific developments and world events.'

Observer

A Dictionary of British History
Edited by John Cannon

An invaluable source of information covering the history of Britain over the past two millennia. Over 3,600 entries written by more than 100 specialist contributors.

Review of the parent volume
'the range is impressive ... truly (almost) all of human life is here'
Kenneth Morgan, *Observer*

More History titles from OUP

The Oxford History of the French Revolution
William Doyle

'A work of breath-taking range ... It is the fullest history to appear of
the Revolutionary era, of the events preceding it and of its impact
on a wider world. Masterfully written.'

Observer

The Twentieth Century World
William R. Keylor

The complete guide to world history during the last century.

Tudor England
John Guy

'Lucid, scholarly, remarkably accomplished ... an excellent
overview.'

The Sunday Times

VISIT THE HIGHER EDUCATION HISTORY WEB SITE AT
www.oup.com/uk/best.textbooks/history

More Social Science titles from OUP

The Globalization of World Politics
John Baylis and Steve Smith

The essential introduction for all students of international relations.

'The best introduction to the subject by far. A classic of its kind.'
Dr David Baker, University of Warwick

Macroeconomics
A European Text
Michael Burda and Charles Wyplosz

'Burda and Wyplosz's best-selling text stands out for the breadth of its coverage, the clarity of its exposition, and the topicality of its examples. Students seeking a comprehensive guide to modern macroeconomics need look no further.'
Charles Bean, Chief Economist, Bank of England

Economics
Richard Lipsey and Alec Chrystal

The classic introduction to economics, revised every few years to include the latest topical issues and examples.

VISIT THE COMPANION WEB SITES FOR THESE CLASSIC TEXTBOOKS AT:

www.oup.com/uk/booksites

Oxford Companions

'Opening such books is like sitting down with a knowledgeable friend. Not a bore or a know-all, but a genuinely well-informed chum ... So far so splendid.'

Sunday Times [of *The Oxford Companion to Shakespeare*]

For well over 60 years Oxford University Press has been publishing Companions that are of lasting value and interest, each one not only a comprehensive source of reference, but also a stimulating guide, mentor, and friend. There are between 40 and 60 Oxford Companions available at any one time, ranging from music, art, and literature to history, warfare, religion, and wine.

Titles include:

The Oxford Companion to English Literature
Edited by Margaret Drabble
'No guide could come more classic.'

Malcolm Bradbury, *The Times*

The Oxford Companion to Music
Edited by Alison Latham
'probably the best one-volume music reference book going'

Times Educational Supplement

The Oxford Companion to Western Art
Edited by Hugh Brigstocke
'more than meets the high standard set by the growing number of Oxford Companions'

Contemporary Review

The Oxford Companion to Food
Alan Davidson
'the best food reference work ever to appear in the English language'

New Statesman

The Oxford Companion to Wine
Edited by Jancis Robinson
'the greatest wine book ever published'

Washington Post

Oxford Paperback Reference

The Concise Oxford Dictionary of English Etymology
T. F. Hoad

A wealth of information about our language and its history, this reference source provides over 17,000 entries on word origins.

'A model of its kind'

Daily Telegraph

A Dictionary of Euphemisms
R. W. Holder

This hugely entertaining collection draws together euphemisms from all aspects of life: work, sexuality, age, money, and politics.

Review of the previous edition
'This ingenious collection is not only very funny but extremely instructive too'

Iris Murdoch

The Oxford Dictionary of Slang
John Ayto

Containing over 10,000 words and phrases, this is the ideal reference for those interested in the more quirky and unofficial words used in the English language.

'hours of happy browsing for language lovers'

Observer

Oxford Paperback Reference

The Concise Oxford Companion to English Literature
Margaret Drabble and Jenny Stringer

Based on the best-selling *Oxford Companion to English Literature*, this is an indispensable guide to all aspects of English literature.

Review of the parent volume
'a magisterial and monumental achievement'

Literary Review

The Concise Oxford Companion to Irish Literature
Robert Welch

From the ogam alphabet developed in the 4th century to Roddy Doyle, this is a comprehensive guide to writers, works, topics, folklore, and historical and cultural events.

Review of the parent volume
'Heroic volume ... It surpasses previous exercises of similar nature in the richness of its detail and the ecumenism of its approach.'

Times Literary Supplement

A Dictionary of Shakespeare
Stanley Wells

Compiled by one of the best-known international authorities on the playwright's works, this dictionary offers up-to-date information on all aspects of Shakespeare, both in his own time and in later ages.

Oxford Paperback Reference

The Concise Oxford Dictionary of Quotations
Edited by Elizabeth Knowles

Based on the highly acclaimed *Oxford Dictionary of Quotations*, this
paperback edition maintains its extensive coverage of literary and
historical quotations, and contains completely up-to-date material. A
fascinating read and an essential reference tool.

The Oxford Dictionary of Humorous Quotations
Edited by Ned Sherrin

From the sharply witty to the downright hilarious, this sparkling
collection will appeal to all senses of humour.

Quotations by Subject
Edited by Susan Ratcliffe

A collection of over 7,000 quotations, arranged thematically for easy
look-up. Covers an enormous range of nearly 600 themes from 'The
Internet' to 'Parliament'.

The Concise Oxford Dictionary of Phrase and Fable
Edited by Elizabeth Knowles

Provides a wealth of fascinating and informative detail for over 10,000
phrases and allusions used in English today. Find out about anything
from the 'Trojan horse' to 'ground zero'.

Oxford Paperback Reference

The Concise Oxford Dictionary of World Religions
Edited by John Bowker

Over 8,200 entries containing unrivalled coverage of all the major world religions, past and present.

'covers a vast range of topics ... is both comprehensive and reliable'
The Times

The Oxford Dictionary of Saints
David Farmer

From the famous to the obscure, over 1,400 saints are covered in this acclaimed dictionary.

'an essential reference work'
Daily Telegraph

The Concise Oxford Dictionary of the Christian Church
E. A. Livingstone

This indispensable guide contains over 5,000 entries and provides full coverage of theology, denominations, the church calendar, and the Bible.

'opens up the whole of Christian history, now with a wider vision than ever'

Robert Runcie, former Archbishop of Canterbury

OXFORD

Oxford Paperback Reference

A Dictionary of Chemistry

Over 4,200 entries covering all aspects of chemistry, including physical chemistry and biochemistry.

'It should be in every classroom and library ... the reader is drawn inevitably from one entry to the next merely to satisfy curiosity.'

School Science Review

A Dictionary of Physics

Ranging from crystal defects to the solar system, 3,500 clear and concise entries cover all commonly encountered terms and concepts of physics.

A Dictionary of Biology

The perfect guide for those studying biology – with over 4,700 entries on key terms from biology, biochemistry, medicine, and palaeontology.

'lives up to its expectations; the entries are concise, but explanatory'

Biologist

'ideally suited to students of biology, at either secondary or university level, or as a general reference source for anyone with an interest in the life sciences'

Journal of Anatomy

More Art Reference from Oxford

The Grove Dictionary of Art

The 34 volumes of *The Grove Dictionary of Art* provide unrivalled coverage of the visual arts from Asia, Africa, the Americas, Europe, and the Pacific, from prehistory to the present day.

'succeeds in performing the most difficult of balancing acts, satisfying specialists while ... remaining accessible to the general reader'

The Times

The Grove Dictionary of Art – Online
www.groveart.com

This immense cultural resource is now available online. Updated regularly, it includes recent developments in the art world as well as the latest art scholarship.

'a mammoth one-stop site for art-related information'

Antiques Magazine

The Oxford History of Western Art
Edited by Martin Kemp

From Classical Greece to postmodernism, *The Oxford History of Western Art* is an authoritative and stimulating overview of the development of visual culture in the West over the last 2,700 years.

'here is a work that will permanently alter the face of art history ... a hugely ambitious project successfully achieved'

The Times

The Oxford Dictionary of Art
Edited by Ian Chilvers

The Oxford Dictionary of Art is an authoritative guide to the art of the western world, ranging across painting, sculpture, drawing, and the applied arts.

'the best and most inclusive single-volume available'

Marina Vaizey, *Sunday Times*